THE PHYSICS AND CHEMISTRY OF NANOSOLIDS

THE PHYSICS AND CHEMISTRY OF NANOSOLIDS

Frank J. Owens and Charles P. Poole, Jr.

A JOHN WILEY & SONS, INC., PUBLICATION

Published by John Wiley & Sons, Inc., Hoboken, New Jersey
Published simultaneously in Canada

For general information on our other products and services or for technical support, please contact our Customer Care Department within the United States at (800) 762-2974, outside the United States at (317) 572-3993 or fax (317) 572-4002.

Wiley also publishes its books in variety of electronic formats. Some content that appears in print may not be available in electronic formats. For more information about Wiley products, visit our web site at www.wiley.com.

Library of Congress Cataloging-in-Publication Data:

Owens, Frank J.
  The physics and chemistry of nanosolids / by Frank J. Owens and Charles P. Poole
    p. cm.
  ISBN   978-0-470-06740-6 (cloth)
  1.   Nanostructured materials.   I.   Poole, Charles P.   II.   Title
    TA418.9.N35084 2007
    620′.5—dc22

                                                                  2007019886

Printed in the United States of America

▰▰▰ CONTENTS

▮▮▮ PREFACE

The prefix *nano* in the word *nanotechnology* means a billionth (1×10^{-9}). *Nanoscience* is a science that studies the properties of matter that has at least one dimension of the order of a billionth of a meter, called a *nanometer*. Its importance lies in the fact that when solids have dimensions of nanometers, the properties of these solids change. Properties such as strength, melting temperature, color, electrical conductivity, thermal conductivity, reactivity, and magnetic properties are affected, and the magnitude of change depends on the size of the solid in the nanometer regime. The changes also depend on the number of dimensions that are nanometers in size. This introduces the possibility of using size in the nanometer regime to design and engineer materials with new and possibly technologically interesting properties. The development of applications based on nanoscience research is referred to as *nanotechnology*. Because of this, nanotechnology and nanoscience have generated much interest in recent years (as of 2007) in the materials science, chemistry, physics, and engineering communities. As a result, chemistry, physics, materials science, and engineering departments at universities are developing courses in the subject. However, there is no textbook available to meet the needs of the development of the courses that describes the changes in properties and explains why they occur. This book, *The Physics and Chemistry of Nanosolids*, is intended as a senior undergraduate or graduate-level textbook on how and why reducing the size of solids to nanodimesions changes their properties. An undergraduate course in modern physics, physical chemistry, or materials science that has some introduction to quantum theory, is a prerequisite. The objectives of the book are to describe how properties depend on size in the nanometer regime, and explain, using relatively simple models of the physics and chemistry of the solid state, why these changes occur. This means that the student needs some understanding of the physics and chemistry of macroscopic solids and models developed to explain properties such as the theory of phonon and lattice vibrations and electronic band structure. Many of the chapters examine these models to see what they predict when one or more dimensions of a solid have a nanometer length. In some instances this will lead to unexplored territory where no experimental information exists. Chapter 1 provides an overview of the basic principles of solids. For those students who have had a course in the chemistry or physics of solids or materials science, this chapter can be skipped or used as a brief review. For those who have not, the chapter should be used as a guide to what subjects they may need to study in further detail in books such as Kittel's *Introduction to Solid State Physics* or Ashcroft and Mermin's *Solid State Physics*. Chapter 2 describes the various experimental methods used to measure the properties of the nanosolids. This chapter is important because the results of measurements using many of the methods are discussed throughout the remainder of the book. Chapter 3 is an overview of changes in properties of nanometer-size materials. Chapters 4–5 for the most part present data on how various properties of solids are affected by nanosizing and examine, in the context of the theories of macroscopic solids, why these changes occur. Because of the importance of carbon in biological materials and the potential applications of carbon nanostructures, a separate chapter is devoted to this subject. Each chapter is followed by a series of exercises designed to enhance the student's understanding of the reasons for changes in properties in the nanometer region.

FRANK J. OWENS
CHARLES P. POOLE JR.

CHAPTER 1

Physics of Bulk Solids

In this text we will be discussing the physics and chemistry of nanostructures. The materials used to form these structures generally have bulk properties that become modified when their sizes are reduced to the nanorange; the present chapter presents background material on bulk properties of this type. Much of what is discussed here can be found in a standard text on solid-state physics.[1-4]

1.1. STRUCTURE

1.1.1. Size Dependence of Properties

Many properties of solids depend on the size range over which they are measured. Microscopic details become averaged when investigating bulk materials. At the macro- or large-scale range ordinarily studied in traditional fields of physics such as mechanics, electricity and magnetism, and optics, the sizes of the objects under study range from millimeters to kilometers. The properties that we associate with these materials are averaged properties, such as the density and elastic moduli in mechanics, the resistivity and magnetization in electricity and magnetism, and the dielectric constant in optics. When measurements are made in the micrometer or nanometer range many properties of materials change, such as mechanical, ferroelectric, and ferromagnetic properties. The aim of the present book is to examine characteristics of solids at the next lower level of size, namely, the nanoscale level, perhaps from 1 to 100 nm. Below this there is the atomic scale near 0.1 nm, followed by the nuclear scale near a femtometer (10^{-15} m). In order to understand properties at the nanoscale it is necessary to know something about the corresponding properties at the macroscopic and mesoscopic scales, and the present chapter aims to provide some of this background.

Many important nanostructures are composed of the group IV elements Si or Ge, types III–V semiconducting compounds such as GaAs or types II–VI semiconducting materials such as CdS, so these semiconductor materials will be used to illustrate some of the bulk properties that become modified when their dimensions are reduced

The Physics and Chemistry of Nanosolids. By Frank J. Owens and Charles P. Poole, Jr.
Copyright © 2008 John Wiley & Sons, Inc.

to the nanometer range. These Roman numerals IV, III, V, and, so on refer to columns of the periodic table. Tabulations of various bulk properties of these semiconductors are found in Appendix B.

1.1.2. Crystal Structures

Most solids are crystalline with their atoms arranged on sites of a regular lattice structure. They possess "long-range order" because this regularity extends throughout the entire crystal. In contrast to this amorphous materials such as glass and wax lack long-range order, but they have "short-range order," which means that the local environment of each atom is similar to that of other equivalent atoms, but this regularity does not persist over appreciable distances. In other words, each atom of a particular type might have, for example, six nearest neighbors, and these neighboring atoms might be at positions that approximate an octahedral configuration. Liquids also have short-range order, but lack long-range order, and this short-range order is undergoing continual rearrangements as a result of Brownian motion. Gases lack both long-range and short-range order. Their constituent molecules undergo rapid translational motion in all directions, so the disorder is continually rearranging at a rapid rate.

Figure 1.1 shows the five regular arrangements of lattice points that can occur in two dimensions, namely, the square (a), primitive rectangular (b), centered rectangular (c), hexagonal (d), and oblique (e) kinds. These arrangements are called *Bravais lattices*. The general or oblique Bravais lattice has two unequal lattice constants $a \neq b$ and an arbitrary angle θ between them. For the perpendicular case when $\theta = 90°$, the lattice becomes rectangular, and if in addition $a = b$, the lattice is called *square*. For the special case $a = b$ and $\theta = 60°$, the lattice is hexagonal, formed from equilateral triangles. Each lattice has a unit cell, indicated in the figures, which can replicate throughout the plane and generate the lattice.

A crystal structure is formed by associating with a lattice a regular arrangement of atoms or molecules. Figure 1.2 presents a two-dimensional crystal structure based on a primitive rectangular lattice containing two diatomic molecules A–B in each unit cell. A single unit cell can generate the overall lattice.

In three dimensions there are three lattice constants, a, b, and c, and three angles: α between b and c, β between a and c, and γ between lattice constants a and b. There are 14 Bravais lattices, ranging from the lowest symmetry triclinic type in which all three lattice constants and all three angles differ from each other ($a \neq b \neq c$

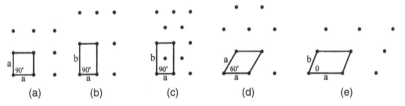

Figure 1.1. The five Bravais lattices that occur in two dimensions, with the unit cells indicated: (a) square; (b) primitive rectangular; (c) centered rectangular; (d) hexagonal; (e) oblique.

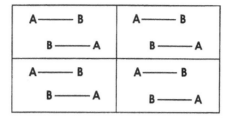

Figure 1.2. Sketch of a two-dimensional crystal structure based on a primitive rectangular lattice containing two diatomic molecules A–B in each unit cell.

and $\alpha \neq \beta \neq \gamma$), to the highest-symmetry cubic case in which all the lattice constants are equal and all the angles are 90° ($a = b = c$ and $\alpha = \beta = \gamma = 90°$). There are three Bravais lattices in the cubic system: a primitive or simple cubic (sc) lattice in which the atoms occupy the eight apices of the cubic unit cell, as shown in Fig. 1.3a; a body-centered cubic (bcc) lattice with lattice points occupied at the apices and in the center of the unit cell, as indicated in Fig. 1.3b; and a face-centered cubic (fcc) Bravais lattice with atoms at the apices and in the centers of the faces, as shown in Fig. 1.3c.

In two dimensions the most efficient way to pack identical circles (or spheres) is the equilateral triangle arrangement shown in Fig. 1.4a, corresponding to the hexagonal Bravais lattice of Fig. 1.1d. A second hexagonal layer of spheres can be positioned on top of the first to form the most efficient packing of two layers, illustrated in Fig. 1.4b. For efficient packing, the third layer can be placed either above the first layer with an atom at the location indicated by T or in the third possible arrangement with an atom above the position marked by X on the figure. In the first case a hexagonal lattice structure called *hexagonal close-packed* (hcp) is generated, and in the second case a fcc lattice results. The former is easy to identify in a unit cell, but the latter is not so easy to visualize while looking at a unit cell since the close-packed planes are oriented perpendicular to the [111] direction.

In three dimensions the unit cell of the fcc structure is the cube shown in Fig. 1.3c, which has a side (i.e., lattice constant) a and volume a^3. It has six face-centered atoms, each shared by two unit cells, and eight apical atoms, each shared by eight unit cells, corresponding to a total of four for this individual unit cell. Nearest

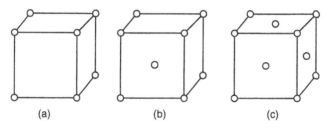

(a)   (b)   (c)

Figure 1.3. Unit cells of the three cubic Bravais lattices: (a) simple cubic (sc); (b) body-centered cubic (bcc); (c) and face-centered cubic (fcc).

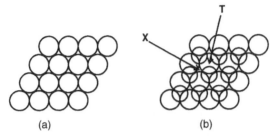

(a)   (b)

Figure 1.4. Close packing of spheres on a flat surface (a) for a monolayer and (b) with a second layer added. The circles of the second layer are drawn smaller for clarity. The location of an octahedral site is indicated by X, and the position of a tetrahedral site is designated by T in (b).

neighbors are a distance $a/(2)^{1/2}$ apart, so the atomic radius is defined as one-half of this value, namely, $r_A = a/2(2)^{1/2}$. Since there are four atoms per unit cell the volume density is $\rho_V = 4/a^3$. Each spherical atom has the individual volume $4\pi\, r_A^3/3$, so the packing fraction or percentage of the unit cell volume occupied by the atoms is $\pi/3(2)^{1/2} = 0.7406$. Figure 1.5 illustrates an alternate way to present the unit cell of the fcc lattice. There are one atom in the center, and there are 12 atoms at the edges. Each edge atom is shared by four unit cells so it counts as one-fourth of this particular cell, and the centrally located atom is not shared, corresponding to the expected total of four for this particular unit cell.

If only the 13 atoms shown in Fig. 1.5 are present, then the configuration constitutes the nanoparticle displayed in Fig. 1.6; this is discussed in the next section. This nanoparticle may be assumed to consist of three layers of two-dimensional close-packed planes perpendicular to its body diagonal or [111] axis, as indicated in Fig. 1.7.

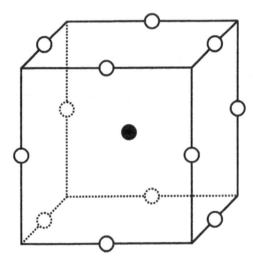

Figure 1.5. Face-centered cubic unit cell showing the 12 nearest-neighbor atoms that surround the atom (darkened circle) in the center.

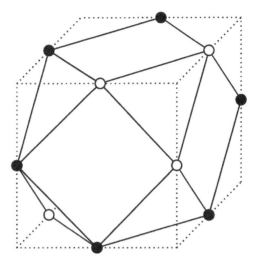

Figure 1.6. A 13-atom nanoparticle set in its fcc unit cell, showing the shape of the 14-sided polyhedron associated with the nanoparticle. The three open circles at the upper right correspond to the top layer, the six solid circles plus the atom (not pictured) in the center of the cube constitute the middle hexagonal layer, and the open circle at the lower left corner of the cube is one of the three atoms at the bottom layer of the cluster.

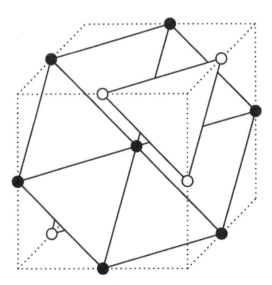

Figure 1.7. A 13-atom nanocluster set in its fcc centered cubic unit cell, delineating the planar configurations of the 13 included atoms. The 7-atom (darkened circles) plane in the center lies between the 3-atom (unshaded circles) plane at the upper right and the 3-atom partly hidden plane at the rear. Figure 1.6 provides a similar perspective.

There is a centrally located seven-atom close-packed plane (blackened circles), sandwiched between a pair of three-atom close-packed planes (unshaded circles). Two of the three atoms of the lower such plane are obscured from view on the figure. The distance between adjacent planes is one-third of the length $(3)^{1/2} a$ of the body diagonal, or $a/(3)^{1/2}$. The apex of the cube along this body diagonal is the distance $a/2(3)^{1/2}$ above the upper close-packed (three-atom) plane, and it is the position of an octahedral site equidistant between two close-packed planes. The atom configurations presented in Fig. 1.7 provide an easy way to visualize the relationships between fcc close-packed planes.

In the three-dimensional arrangements of close-packed spheres there are spaces or sites between the spheres where smaller atoms can reside. These sites exist between pairs of close-packed layers, so they are independent of the manner in which a third layer is added. In other words, they are the same for fcc and hcp lattices.

The point marked by X on Fig. 1.4b is called an *octahedral site* since it is equidistant from the three spheres O below it, and from the three spheres O above it, and these six nearest neighbors constitute the apices of an imaginary octahedron. An atom A at this site has the local coordination AO_6. The radius a_{oct} of this octahedral site is

$$a_{oct} = \frac{1}{4}(2 - (2)^{1/2})a = ((2)^{1/2} - 1)a_0 = 0.41411a_0 \qquad (1.1)$$

where a is the lattice constant and a_0 is the sphere radius. The number of octahedral sites is equal to the number of spheres. There are also smaller sites called *tetrahedral sites*, labeled T on the figure that are equally distant from four nearest-neighbor spheres, one below and three above, corresponding to AO_4 for the local coordination. These four spheres define an imaginary tetrahedron surrounding an atom A located at the site. This is a smaller site since its radius a_T is

$$a_T = \frac{1}{4}((3)^{1/2} - (2)^{1/2})a = [(3/2)^{1/2} - 1]a_0 = 0.2247a_0 \qquad (1.2)$$

There are twice as many tetrahedral sites as spheres in the structure. Many diatomic oxides and sulfides such as MgO, MgS, MnO, and MnS possess larger oxygen or sulfur anions in a perfect fcc arrangement with the smaller metal cations located at octahedral sites. This is an NaCl lattice, where we use the term *anion* for a negative ion (e.g., Cl^-), and cation for a positive ion (e.g., Na^+). The mineral spinel $MgAl_2O_4$ has a face-centered arrangement of divalent oxygens O^{2-} (radius 0.132 nm) with the Al^{3+} ions (radius 0.051 nm) occupying one-half of the octahedral sites and Mg^{2+} (radius 0.066 nm) located in one-eighth of the tetrahedral sites in a regular manner. In some spinels the oxygens deviate somewhat from a perfect fcc arrangement by moving slightly toward or away from the tetrahedral sites.

1.1.3. Face-Centered Cubic Nanoparticles

Most metals in the solid-state form close-packed lattices; thus Ag, Al, Au, Co, Cu, Pb, Pt, and Rh, as well as the rare gases Ne, Ar, Kr, and Xe, are fcc, and Mg, Nd, Os, Re, Ru, Y, and Zn are hcp. Some other metallic atoms crystallize in the more loosely packed bcc lattice, and a few such as Cr, Li, and Sr crystallize in all three structure types, depending on the temperature. An atom in each of the two close-packed lattices has 12 nearest neighbors. Figure 1.5 shows the 12 neighbors that surround an atom (darkened circle) located in the center of a cube for a fcc lattice. In Fig. 1.7 we show a redrawn version of the fcc unit cell of Fig. 1.6 that clarifies the layered structure of the atoms. The topmost layer is an equilateral triangle of atoms (open circles); the next layer below is a regular hexagon of atoms (blackened circles); and the bottom layer, another equilateral triangle of open circle atoms, is mostly hidden from view. These 13 atoms constitute the smallest theoretical nanoparticle for an fcc lattice. More precisely it should probably be called a *nanocluster*. Figure 1.6 shows the 14-sided polyhedron, called a *dekatessarahedron*, that is generated by connecting the nearest-neighbor atoms. This polyhedron has six square faces and eight equilateral triangle faces. Figure 1.8 shows this polyhedron viewed from the [111] or top direction with the coordinate axis perpendicular to each side indicted. The square sides are perpendicular to the x axis [100], the y axis [010], and the z axis [001], respectively, and the triangular faces are perpendicular to the indicated body diagonal directions. Figure 1.7 clarifies the layering scheme by connecting atoms only within two-dimensional close-packed layers. The three open circles in the triangular layer of Fig. 1.7 are the three atoms in the top layer of Fig. 1.8. Likewise, the seven darkened

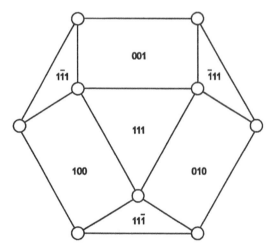

Figure 1.8. Perspective along the [111] direction of the faces or facets of the 13-atom fcc nanoparticle of Fig. 1.6. The three visible square faces are perpendicular to the x direction [100], the y direction [010], and the z direction [001], respectively. The four equilateral triangle faces are perpendicular to body diagonal directions, as indicated.

circles, including the atom in the center of the cube of Fig. 1.7, constitute the periphery of Fig. 1.8.

To construct the next-larger 55-atom fcc nanoparticle, a layer of 30 atoms is added around the three layers of Fig. 1.7, and six-atom triangular arrays are added to the top and bottom, as explained in Appendix C. The process is repeated to form the higher-level nanoparticles with the magic numbers of atoms listed in Table 1.1. A perspective sketch of the atoms on the surfaces of the 147-atom nanoparticle viewed along the [111] axis is provided in Fig. 1.9, with the crystallographic planes labeled by their indices. There are six square faces with square planar arrays of 16 atoms each, and eight equilateral triangle faces with planar hcp arrays of 10 atoms each. The atoms on all of these triangular and square faces have the nearest-neighbor separation $a/2^{1/2}$, where a is the lattice constant. Larger fcc nanoparticles with magic numbers 309, 561, ... of atoms have analogous arrays of atoms on their faces, with the numbers of surface atoms involved in each case listed in Table C.1. The numbers of atoms in the various planes of these nanoparticles are tabulated in Table C.2. The nanoparticles pictured in Fig. 3.18 (Fig. 4.15 of first edition), and in Fig. 2.13 (Fig. 3.13 of the first edition) clearly display both the planar square and the planar hexagonal arrays of atoms corresponding to those on the faces of Fig. 1.9.

TABLE 1.1. Diameter, Total Number of Atoms, Number on Surface, and Percentage on Surface of Rare Gas or Metallic Nanoparticles with FCC Close-Packed Structures and a Structural Magic Number of Atoms[a]

Shell	Diameter	Total Number	Number on Surface	Percentage on Surface
1	d	1	1	100
2	$3d$	13	12	92.3
3	$5d$	55	42	76.4
4	$7d$	147	92	62.6
5	$9d$	309	162	52.4
6	$11d$	561	252	44.9
7	$13d$	923	362	39.2
8	$13d$	1415	492	34.8
9	$17d$	2057	642	31.2
10	$19d$	2869	812	28.3
11	$21d$	3871	1002	25.9
12	$23d$	5083	1212	23.8
25	$49d$	4.90×10^4	5.76×10^3	11.7
50	$99d$	4.04×10^5	2.40×10^4	5.9
75	$149d$	1.38×10^6	5.48×10^4	4.0
100	$199d$	3.28×10^6	9.80×10^4	3.0

[a]The diameters d in nanometers for some representative fcc atoms are Ag 0.289, Al 0.286, Ar 0.372, Au 0.288, Cu 0.256, Fe 0.254, Kr 0.404, Pb 0.350, and Pd 0.275.

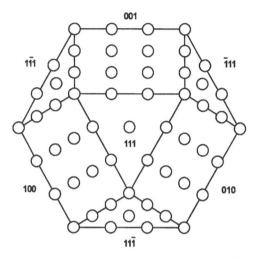

Figure 1.9. Arrays of atoms on triangular and square faces of the 147-atom fcc nanoparticle. The directions perpendicular to each face are indicated.

1.1.4. Large Face-Centered Cubic Nanoparticles

Larger fcc nanoparticles with the same polyhedral shape are obtained by adding more planes or layers, and the sequence of numbers in the resulting particles, $N = 1,13,55,147,309, \ldots$, listed in Table 1.1, are called *structural magic numbers*. For n layers the number of atoms N in this fcc nanoparticle is given by the formula

$$N = (1/3)[10n^3 - 15n^2 + 11n - 3] \tag{1.3}$$

and the number of atoms on the surface N_{surf} is

$$N_{surf} = 10n^2 - 20n + 12 \tag{1.4}$$

Equation (1.3) is valid for all n, but Eq. (1.4) is valid only for $n > 1$. For each value of n Table 1.1 lists the number of atoms on the surface, as well as the percentage of atoms on the surface. The table also lists the diameter of the each nanoparticle, which is given by the expression $(2n - 1)d$, where d is the distance between the centers of nearest-neighbor atoms and $d = a/(2),^{1/2}$, where a is the lattice constant.

Purely metallic fcc nanoparticles such as Au_{55} tend to be very reactive and have short lifetimes. They can be ligand-stabilized by adding atomic groups between their atoms and on their surfaces. The Au_{55} nanoparticle has been studied in the ligand-stabilized form $Au_{55}(PPh_3)_{12}Cl_6$, which has the diameter of ≈ 1.4 nm, where PPh_3 is an organic group. Further examples are the magic number nanoparticles $Pt_{309}(1,10\text{-phenantroline})_{36}O_{30}$, and $Pd_{561}(1,10\text{-phenantroline})_{36}O_{200}$.

The magic numbers that we have been discussing are called structural magic numbers because they arise from minimum-volume, maximum-density nanoparticles that approximate a spherical shape, and have close-packed structures characteristic of a bulk solid. These magic numbers take no account of the electronic structure of the constituent atoms in the nanoparticle. Sometimes the predominant factor in determining the minimum-energy structure of small nanoparticles is the interactions between the valence electrons of the constituent atoms with an averaged molecular potential, so that the electrons occupy orbital levels associated with this potential. Atomic cluster configurations in which these electrons fill closed shells are especially stable, and constitute electronic magic numbers. Their atomic structures differ from the fcc arrangement, as will be discussed in Section 3.2.

When mass spectra were recorded for sodium nanoparticles Na_N it was found that mass peaks corresponding to the first 15 electronic magic numbers $N = 3,9,20,36,61,$ ... were observed for cluster sizes up to $N = 1220$ atoms ($n = 15$), and fcc structural magic numbers starting with $N = 1415$ for $n = 8$ were observed for larger sizes.[5,6] The mass spectral data are plotted versus the cube root of the number of atoms $N^{1/3}$ on Fig. 1.10, and it is clear that the lines from both sets of magic numbers are approximately equally spaced, with the spacing between the structural magic numbers about 2.6 times that between the electronic ones. This result is evidence that small clusters tend to satisfy electronic criteria, and large structures tend to be structurally determined.

1.1.5. Tetrahedrally Bonded Semiconductor Structures

Types III–V and II–VI binary semiconducting compounds, such as GaAs and ZnS, respectively, crystallize with one atom situated on a fcc sublattice at positions 000, $\frac{1}{2}\frac{1}{2}0$, $\frac{1}{2}0\frac{1}{2}$, and $0\frac{1}{2}\frac{1}{2}$, and the other atom on a second fcc sublattice displaced from the first by the amount $\frac{1}{4}\frac{1}{4}\frac{1}{4}$ along the body diagonal, as shown in Fig. 1.11b. This is called the *zinc blende* or ZnS structure. It is clear from the figure that each Zn atom (white sphere) is centered in a tetrahedron of S atoms (black spheres), and likewise each S has four similarly situated Zn nearest neighbors. The small half-sized, dashed-line cube delineates one such tetrahedron. The same structure would result if the Zn and S atoms were interchanged. Figure 1.12 presents a more realistic sketch of the unit cell of ZnS with the small zinc atoms centered in the tetrahedral sites of

Sodium Nanoparticle Na_n Magic Numbers

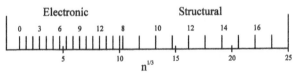

Figure 1.10. Dependence of the observed mass spectra lines from Na_N nanoparticles on the cube root $N^{1/3}$ of the number of atoms N in the cluster. The lines are labeled with the index n of their respective electronic and structural magic numbers. [Adapted from T. P. Martin, T. Bergmann, H. Gohlich, and T. Lange, *Chem. Phys. Lett.* **172**, 209 (1990).]

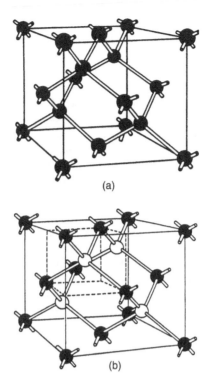

(a)

(b)

Figure 1.11. Unit cell of the diamond structure (a) containing only one type of atom, and corresponding unit cell of the zinc blende structure (b) containing two atom types. The rods represent the tetrahedral bonds between nearest-neighbor atoms. The small dashed-line cube in (b) delineates a tetrahedron. (From G. Burns, *Solid State Physics*, Academic Press, Boston, 1985, p. 148.)

the larger sulfur atoms. In contrast to this arrangement, the atoms in the semiconductor gallium arsenide have comparable radii $r_{Ga} = 0.122$ nm and $r_{As} = 0.124$ nm, and the lattice constant $a = 0.565$ nm is close to the value $4(r_{Ga} + r_{As})/(3)^{1/2} = 0.568$ nm, as expected, where data were used from Tables B.1 and B.2 (of Appendix B).

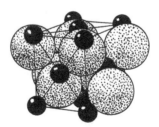

Figure 1.12. Packing of larger sulfur atoms and smaller zinc atoms in the zinc blende (ZnS) structure. Each atom is centered in a tetrahedron of the other atom type. (From R. W. G. Wyckoff, *Crystal Structures*, Vol. 1, Wiley, New York, 1963, p. 109.)

The elements Si and Ge crystallize in this same structure, in which the Si (or Ge) atoms occupy all the sites on the two sublattices, so there are eight identical atoms in the unit cell. This atom arrangement, sketched in Fig. 1.11a, is called the "diamond structure." Both Si and Ge have a valence of 4, so from bonding considerations it is appropriate for each to be bound to four other atoms in the shape of a regular tetrahedron.

In Table B.1 we see the lattice constants a listed for various compounds with the zinc blende structure, and Table B.2 provides the crystal radii of their monatomic lattices in which the atoms are uncharged, as well as the ionic radii for ionic compounds in which the atoms are charged. We see from Table B.2 that the negative anions are considerably larger than the positive cations, in accordance with the sketch of the unit cell presented in Fig. 1.12, and this size differential is greater for the III–V compounds than for II–VI compounds. However, these size changes for the negative and positive ions tend to balance each other so that the III–V compounds have the same range of lattice constants as do the II–VI compounds, with Si and Ge also in this range. Table B.4 gives the molecular masses, and Table B.5 gives the densities of these semiconductors. The three tables, Tables B.1, B.4, and B.5, show a regular progression in the values as one goes from left to right in a particular row, and as one goes from top to bottom in a particular column. This occurs because of the systematic increase in the size of the atoms in each group with increasing atomic number, as indicted in Table B.2.

There are two simple models for representing these AC binary compound structures. For an ionic model the lattice $A^{n-}C^{n+}$ consists of a fcc arrangement of the large anions A^{n-} with the small cations C^{n+} located in the tetrahedral sites of the anion fcc lattice. If the anions touch each other, their radii has the value $r = a/2(2)^{1/2}$, where a is the lattice parameter, and the radius a_T of the tetrahedral site $r_{tetr} = 0.2247\,a$ is given by Eq. (1.2). This is the case for the very small Al^{3+} cation in the AlSb structure. In all other cases the cations in Table B.2 are too large to fit in the tetrahedral site, so they push the larger anions further apart, and the latter no longer touch each other, in accordance with Fig. 1.12. In a covalent model for the structure consisting of neutral atoms A and C the atom sizes are comparable, as the data in Table B.2 indicate, and the structure resembles that of Si or Ge. Comparing these two models, we note that the distance between atom A at lattice position 000 and its nearest neighbor C at position $\frac{1}{4}\frac{1}{4}\frac{1}{4}$ is equal to $\frac{1}{4}(3)^{1/2}a$, and in Table B.3 we compare this crystallographically evaluated distance with the sums of radii of ions A^{n-}, C^{n+} from the ionic model, and with the sums of radii of neutral atoms A and C of the covalent model using the data of Table B.2. We see from the results in Table B.3 that neither model fits the data in all cases, but the neutral atom covalent model is closer to agreement, especially for ZnS, GaAs, and CdS. For comparison purposes we also list corresponding data for several alkali halides NaCl, KBr, and RbI, and alkaline-earth chalcogenides CaS and SrSe, which also crystallize in the cubic rock salt or NaCl structure. We see that all of these compounds fit the ionic model very well. In these compounds each atom type forms a fcc lattice, with the atoms of one fcc lattice located at octahedral sites of the other lattice. The octahedral site has the radius $r_{oct} = 0.41411\,a_0$ given by Eq. (1.1), which is larger than the tetrahedral one, $r_{tetr} = 0.2247\,a$, of Eq. (1.2).

Since the alkali halide and alkaline-earth chalcogenide compounds fit the ionic model so well, it is significant that neither model fits the structures of the semiconductor compounds. The extent to which the semiconductor crystals exhibit ionic or covalent bonding is not clear from crystallographic data. If the wavefunction describing the bonding is written in the form

$$\Psi = a_{cov}\psi_{cov} + a_{ion}\psi_{ion} \quad (1.5)$$

where the coefficients of the covalent and ionic wavefunction components are normalized

$$a_{cov}{}^2 + a_{ion}{}^2 = 1 \quad (1.6)$$

then $a_{cov}{}^2$ is the fractional covalency and $a_{ion}{}^2$ is the fractional ionicity of the bond. A chapter in a recent book by Karl Boer[7] tabulates the effective charges e^* associated with various II–VI and III–V semiconducting compounds, and this effective charge is related to the fractional covalency by the expression

$$a_{cov}{}^2 = \frac{8 - N + e^*}{8} \quad (1.7)$$

where $N = 2$ for II–VI and $N = 3$ for III–V compounds. The fractional charges all lie in the range from 0.43 to 0.49 for the compounds under consideration. Using the e^* tabulations in the Boer book and Eq. (1.7), we obtain the fractional covalencies of $a_{cov}{}^2 \sim 0.81$ for all the II–VI compounds, and $a_{cov}{}^2 \sim 0.68$ for all the III–V compounds listed in Tables B.1, B.4, B.5, and so on. These values are consistent with the better fit of the covalent model to the crystallographic data for these compounds.

We conclude this section with some observations that will be of use in later chapters. Table B.1 shows that the typical compound GaAs has the lattice constant $a = 0.565$ nm, so the volume of its unit cell is 0.180 nm^3, corresponding to about 22 of each atom type per cubic nanometer. The distances between atomic layers in the [100], [110], and [111] directions are, respectively, $a/2 = 0.28$ nm, $a/(2)^{1/2} = 0.40$ nm, and $a/(3)^{1/2} = 0.33$ nm, for GaAs. The various III–V semiconducting compounds under discussion form mixed crystals over broad concentration ranges, as do the group of II–VI compounds. In a mixed crystal of the type In$_x$Ga$_{1-x}$As, it is ordinarily safe to assume that Vegard's law is valid, whereby the lattice constant a scales linearly with the concentration parameter x. As a result, we have the following relationships

$$a(x) = a(\text{GaAs}) + [a(\text{InAs}) - a(\text{GaAs})]x$$
$$= 0.565 + 0.039x \quad (1.8)$$

where $0 < x < 1$. In the corresponding expression for the mixed semiconductor Al$_x$Ga$_{1-x}$As the term $\leq 0.003x$ replaces the term $+0.039x$, so the fraction of lattice mismatch $2|a_{AlAs} - a_{GaAs}|/(a_{AlAs} + a_{GaAs}) = 0.0054 = 0.54\%$ for this system is quite minimal compared to that $2|a_{InAs} - a_{GaAs}|/(a_{InAs} + a_{GaAs}) = 0.066 = 6.6\%$ of the $In_xGa_{1-x}As$ system, as calculated from Eq. (1.8). Table B.1 of Appendix B gives the lattice constants a for various III–V and II–VI semiconductors with the zinc blende structure.

1.1.6. Lattice Vibrations

We have discussed atoms in a crystal as residing at particular lattice sites, but in reality they undergo continuous fluctuations in the neighborhood of their regular positions in the lattice. These fluctuations arise from the heat or thermal energy in the lattice, and become more pronounced at higher temperatures. Since the atoms are bound together by chemical bonds, the movement of one atom about its site causes the neighboring atoms to respond to this motion. The chemical bonds act like springs that stretch and compress repeatedly during the oscillatory motion. The result is that many atoms vibrate in unison, and this collective motion spreads throughout the crystal. Every type of lattice has its own characteristic modes or frequencies of vibration called *normal modes*, and the overall collective vibrational motion of the lattice is a combination or superposition of many, many normal modes. For a diatomic lattice like GaAs, there are low-frequency modes called *acoustic modes*, in which the heavy and light atoms tend to vibrate in phase or in unison with each other, and high-frequency modes called *optical modes*, in which they tend to vibrate out of phase.

A simple model for analyzing these vibratory modes is a linear chain of alternating atoms with a large mass M and a small mass m joined to each other by springs ($\sim$) as follows:

$$\sim m \sim M \sim m \sim M \sim m \sim M \sim m \sim M \sim$$

When one of the springs stretches or compresses by an amount Δx, a force is exerted on the adjacent masses with the magnitude $C\,\Delta x$, where C is the spring constant. As the various springs stretch and compress in step with each other, longitudinal modes of vibration take place in which the motion of each atom is along the string direction. Each such normal mode has a particular frequency ω and a wavevector $k = 2\pi/\lambda$, where λ is the wavelength, and the energy E associated with the mode is given by $E = \hbar\omega$. There are also transverse normal modes in which the atoms vibrate back and forth in directions perpendicular to the line of atoms. Figure 1.13 shows the dependence of ω on k for the low-frequency acoustic and the high-frequency optical longitudinal modes. We see that the acoustic branch continually increases in frequency ω with increasing wavenumber k, and the optical branch continuously decreases in frequency. The two branches have respective limiting frequencies given by $(2C/M)^{1/2}$ and $(2C/m)^{1/2}$, with an energy gap between them at the edge of the Brillouin zone $k_{max} = \pi/a$, where a is the distance between atoms m and M

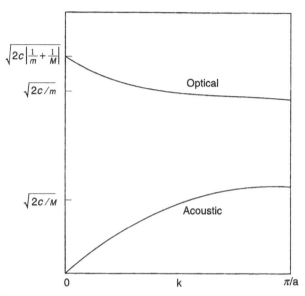

Figure 1.13. Dependence of the longitudinal normal-mode vibrational frequency ω on the wavenumber $k = 2\pi/\lambda$ for a linear diatomic chain of atoms with alternating masses $m < M$ having an equilibrium spacing a, and connected by bonds with spring constant C. [From C. P. Poole, Jr., *The Physics Handbook*, Wiley, New York, 1998, p. 53.)

at equilibrium. The Brillouin zone is a unit cell in wavenumber or reciprocal space, as will be explained in Section 1.3.2 later in this chapter. The optical branch vibrational frequencies are in the infrared region of the spectrum, generally with frequencies in the range from 10^{12} to 3×10^{14} Hz, and the acoustic branch frequencies are much lower. In three dimensions the situation is more complicated, and there are longitudinal acoustic (LA), transverse acoustic (TA), longitudinal optical (LO), and transverse optical (TO) modes.

The atoms in molecules also undergo vibratory motion, and a molecule containing N atoms has $3N - 6$ normal modes of vibration. Particular molecular groups such as hydroxyl —OH, amino —NH$_2$, and nitro —NO$_2$ have characteristic normal modes that can be used to detect their presence in molecules and solids.

The atomic vibrations that we have been discussing correspond to standing waves. This vibrational motion can also produce traveling waves in which localized regions of vibratory atomic motion travel through the lattice. Examples of such traveling waves are sound moving through the air, or seismic waves that start at the epicenter of an earthquake and travel thousands of miles to reach a seismograph detector that records the earthquake event many minutes later. Localized traveling waves of atomic vibrations in solids, called *phonons*, are quantized with the energy $h\omega = h\nu$, where $\nu = \omega/2\pi$ is the frequency of vibration of the wave. Phonons play an important role in the physics of the solid state.

1.2. SURFACES OF CRYSTALS

1.2.1. Surface Characteristics

When we consider the bulk properties of macroscopic crystals, the surface layers play a negligible role. If a simple cubic crystal in the shape of a cube contains 10^{24} atoms, then each face of the cube will contain 10^{16} atoms, for a total of 6×10^{16} on the surface. This is less than one atom in 10^7 on the surface, so the bulk is not appreciably affected by the presence of the surface. On the other hand, when we consider particles in the nanoscale range of dimensions, the percentage of the atoms on the surface can be large. We see from Table 1.1 that an aluminum nanoparticle with a diameter $49d = 14.0$ nm contains 4.9×10^4 atoms, 11.7% of which are on the surface. The largest aluminum nanoparticle listed in the table has a diameter of 56.9 nm and contains 3.28 million atoms, 3% of which are on the surface. As a result, the surface can influence the bulk properties. The same is true of thin films with nanometer thicknesses. For example, consider a fcc thin film consisting of 100 close-packed layers (the [111] axis is perpendicular to the surface). Such a film made of copper would be 21 nm thick. Since there is both a top surface and a bottom surface, it follows that 2% of the atoms are in a surface layer.

We have been discussing surface layers as if they are identical to the layers in the bulk, and this is seldom the case. Within the crystal the bonding orbitals of all the atoms are satisfied, and each type of atom is in the same environment as its counterparts. In contrast to this, the surface atoms have dangling bonds that are not satisfied, and this deficiency of chemical bonding can be characterized quantitatively by a surface energy per unit area. This surface energy results from the work that must be done to cleave the crystal and break the dangling surface bonds while forming the surface. Sometimes a surface atom reconstruction takes place to lower this surface energy. For example, the [111] surface of a silicon (or germanium) crystal consists of a hexagonal array of Si atoms, each of which, shown unshaded, has a single vertical dangling bond indicated by a dot in Fig. 1.14. Each surface atom on the figure is bonded to three Si atoms shown shaded in the layer below the surface. The energy is lowered by the surface reconstruction illustrated in Fig. 1.15, in which surface atoms move together and bond to each other in pairs to accommodate and satisfy the broken dangling bonds. Another phenomenon that sometimes takes place is surface structure relaxation in which the outer layer of atoms moves slightly toward or a short distance away from the layer below. Contraction takes place with most metal surfaces. The silicon [111] surface layer illustrated in Figs. 1.14 and 1.15 contracts by about 25%, and the three interlayer spacings further below compensate for this by expanding between 1% and 5%.

Other phenomena can occur at the surface. Sometimes there are steps on the surface beyond which a new layer has been added. Various types of defects can occur, such as added atoms in an irregular manner, or missing atoms at lattice sites. Surface atoms can bond together in ways that differ from those in the bulk, as took place in the reconstruction pictured in Fig. 1.15. Occasionally surface atoms will undergo a two-dimensional phase transition.

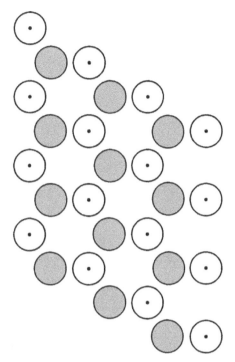

Figure 1.14. Sketch of the first two [111] outermost layers of the zinc blende structure showing unshaded sulfur atoms on the surface with vertical dangling bonds denoted by dots, and shaded zinc atoms one layer below the surface.

1.2.2. Surface Energy

Work must be done to cut a crystal into two pieces that expose a pair of new crystal faces or facets. The energy required to bring about this cleavage, called the *surface energy*, is proportional to the density of the various types of chemical bonds that are cut and transformed into dangling bonds. In the case of fcc crystals, each large atom has 12 large atom nearest neighbors to which it can be bonded. Each medium-sized atom in an octahedral site is bonded to six surrounding large atoms, and each small atom in a tetrahedral site is bonded to four large atoms. Which bonds break depends on the particular crystallographic plane where the cleavage occurs.

If the surface contains a density ρ_i of atoms of type i, each with n_i dangling bonds of energy $2\varepsilon_i$, then the surface energy density γ or energy per unit area is given by the following sum over the atom types

$$\gamma = \Sigma\rho_i n_i \varepsilon_i \tag{1.9}$$

where the factor of 2 takes into account the fact that the energy associated with a dangling bond is half of the corresponding chemical bond energy. If all of the atoms are

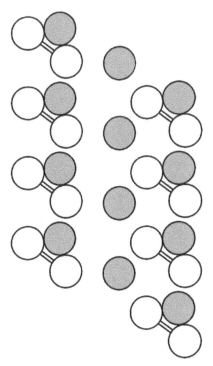

Figure 1.15. Sketch of the surface configuration of Fig. 1.14 after it has undergone a recon-structuring process during which surface sulfur atoms bond together in pairs.

identical as in the fcc and hcp lattices discussed above, this expression reduces to

$$\gamma = \rho n \varepsilon \tag{1.10}$$

In the next few sections we will evaluate this surface energy density for several special cases.

1.2.3. Face-Centered Cubic Surface Layers

We have been discussing how surface layers become modified relative to bulk layers in a crystal. Since many nanoparticles and nanofilms have a fcc structure, it will be instructive to consider the arrangement of the atoms on the surfaces of fcc cubic solids that are cleaved along particular crystallographic planes. Consider a solid with large atoms close-packed so that they touch all 12 nearest neighbors, and with the octahedral sites filled with medium-sized atoms in contact with their 6 nearest neighbors, and the tetrahedral sites occupied by small atoms in contact with their 4

nearest neighbors. The radii of these three types of atoms have the following values
from Eqs. (1.1) and (1.2)

$$r_L = 0.7072a \tag{1.11a}$$

$$r_{Oh} = 0.4142a \tag{1.11b}$$

$$r_T = 0.2247a \tag{1.11c}$$

where a is the lattice constant. Figure 1.16 shows an example of the octahedral sites
occupied by smaller atoms, as happens in the NaCl structure. The zinc blende structure of Fig. 1.11 has half of the tetrahedral sites and none of the octahedral sites occupied. It is clear from these figures that some of the octahedral sites lie on the surface of
the unit cell, while all of the tetrahedral ones are inside it. If the crystal is cleaved to
expose the [100] face, it is evident from Fig. 1.16 that the surface will contain 50%
larger and 50% smaller atoms, the latter occupying the octahedral sites of the former,
as shown in Fig. 1.17 (i.e., 4 small half-atoms + 1 large atom + 4 large quarter-
atoms). The [110] face displayed in Fig. 1.18 contains all three types of atoms.
This can be seen by comparison with Figs. 1.11b and 1.16. The upper two tetrahedrally coordinated atoms on Fig. 1.18 are shaded to indicate that they are absent
in the zinc blende structure. The [111] plane displayed in Fig. 1.4a contains only
the main large atoms since none of the tetrahedral or octahedral sites are in this
plane, as can be deduced from the positions of these sites specified in Fig. 1.4b.

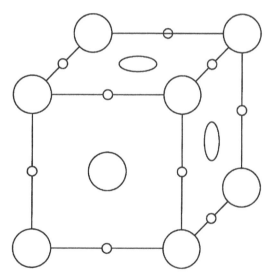

Figure 1.16. Unit cell consisting of fcc atoms (large circles) with the octahedral holes
occupied by smaller atoms (small circles).

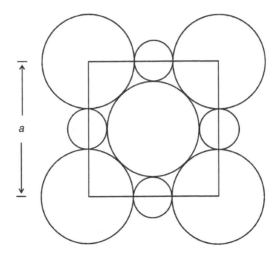

Figure 1.17. The [100] surface plane of the structure depicted in Fig. 1.16.

There are several special cases that can be clarified from the analysis above. If all of the atoms are of the same size and only octahedral sites are occupied, then the lattice reduces to a simple cubic one with the lattice constant $a/2$, and a unit cell one-eighth of the size. Since nanoparticles are rarely simple cubic, we will not discuss this case further. An important special case is when none of these extra sites are occupied so only the large atoms are present. This is the situation that

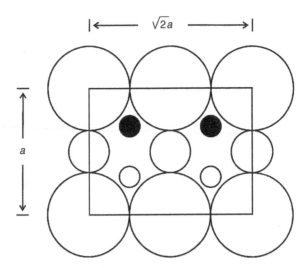

Figure 1.18. The [110] surface plane of a fcc structure formed by large atoms with the octahedral sites occupied by medium-sized atoms and the tetrahedral sites occupied by small atoms. The upper two small atoms shown shaded as well as the three medium-sized atoms are absent in the zinc blende structure.

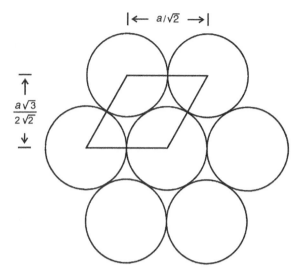

Figure 1.19. Surface unit cell of the close-packed lattice sketched in Fig. 1.4a.

corresponds to the fcc nanoparticles discussed in Section 1.1.3. The [100] face shown in Fig. 1.17 with the surface unit cell area a^2 contains two atoms, one entire atom in the center and four-quarters of an atom in the corners, so the atom density on the surface is $2/a^2$ (the smaller atoms in the figure are now absent). The center atom has four nearest neighbors inside the crystal face below it, four neighbors alongside it in the face, and four missing nearest neighbors above so four of its chemical bonds have been cut, and are now dangling bonds. This means that the number of dangling bonds per unit area is $8/a^2$, and the surface energy density from Eq. (1.10) is $\gamma = 8\varepsilon/a^2$. By similar reasoning, the [110] face unit cell shown in Fig. 1.18 with the surface unit cell area $(2)^{1/2}a^2$ contains two atoms, so the atom density on the surface is $(2)^{1/2}/a^2$. Each surface atom is bonded to five atoms below it and to two surface atoms, so it has five dangling bonds, for a dangling bond density of $5(2)^{1/2}/a^2$, and a surface energy density γ of $5(2)^{1/2}\varepsilon/a^2$. The [111] surface sketched in Fig. 1.19 is a little more complex to analyze. The rhombic unit cell outlined in the figure contains one atom and has the area $(3)^{1/2}a^2/4$. Each atom is bonded to three atoms below and six in the surface, so there are three dangling bonds, for a dangling bond density of $4(3)^{1/2}/a^2$, and a surface energy density γ of $4(3)^{1/2}\varepsilon/a^2$. These results are gathered together in Table 1.2.

1.2.4. Surfaces of Zinc Blende and Diamond Structures

The remaining structure that is of interest is the diamond structure depicted in Fig. 1.11. This consists of a set of fcc atoms together with a second set occupying half of the tetrahedral holes of the first set, as was discussed above. We will discuss surfaces of ZnS crystals in which the small zinc atoms reside in the tetrahedral

TABLE 1.2. Atom Density, Number of Bonds Broken per Atom, and Surface Energy Density for Various Facets or Surfaces of FCC centered cubic and diamond structures[a]

Structure and Facet	Atom Density	Bonds per Atom	Surface-energy Density
Fcc [100]	$2/a^2$	4	$8\varepsilon/a^2$
Fcc [110]	$(2)^{1/2}/a^2$	5	$5(2)^{1/2}\,\varepsilon/a^2$
Fcc [111]	$4/a^2$	3	$4(3)^{1/2}\,\varepsilon/a^2$
Diamond [100]	$2/a^2$	2	$4\varepsilon/a^2$
Diamond [110]	$2(2)^{1/2}/a^2$	1	$2(2)^{1/2}\,\varepsilon/a^2$
Diamond [111]	$4/a^2$	1	$4\varepsilon/(3)^{1/2}a^2$

[a]*Key*: $a=$ lattice constant; $\varepsilon =$ surface energy per dangling chemical bond.

sites of the dominant fcc sulfur lattice. The atom locations in various faces can be deduced from Figs. 1.11b and 1.12.

Figure 1.17, with the smaller octahedral site atoms omitted, depicts the [001] surface, which corresponds to the top face of the cubic unit cell in Figs. 1.11b and 1.12. The atom density of this face is $2/a^2$, the same as in the simple fcc case discussed above. It is clear from Fig. 1.11b that each surface atom has two dangling bonds, so the dangling bond density is $4/a^2$. Figure 1.18 sketches the [110] surface, which contains four large edge and apex S atoms of the surface unit cell, plus two smaller interior Zn atoms depicted by small open circles in the lower part of Fig. 1.18. In this figure the medium-sized octahedral site atoms and the shaded small tetrahedral site atoms are absent, so there are four atoms per surface unit cell of area $(2)^{1/2}a^2$. It is clear from Fig. 1.11b that each atom has two bonds within the surface and one below it, so there is one dangling bond per atom, for a dangling bond density of $2(2)^{1/2}/a^2$. For simplicity we presume a surface energy density $2(2)^{1/2}a^2$, assuming that the dangling bond energy ε_i of the zinc and sulfur atoms has the same value ε, which is really not the case. Figure 1.14 depicts the atoms associated with the [111] plane. The unshaded circles represent the atoms on the surface, and the shaded circles correspond to the first layer below the surface. Either type could be sulfur, and the other one would be zinc. The atom density is $4/(3)^{1/2}a^2$ as in the fcc cubic case discussed above. The dots centered in the open circles denote dangling bonds, one such bond per atom, directed upward perpendicular to the surface. Hence the dangling bond density is $4/(3)^{1/2}a^2$.

We showed above how the energy of the atoms close to the [111] face of a diamond or zinc blende crystal can decrease by a reconstruction in which the atoms rearrange to a lattice with a bimolecular unit cell formed from the pairs of atoms depicted in Fig. 1.15. These surface atoms bond in pairs so that there are no dangling bonds remain and the energy is lowered. In some cases the spacing between layers slightly varies between the surface and the bulk, with the layers near the surface undergoing an expansion outward or a contraction inward. In Fig. 1.14 the atoms with upward-directed bonds formed the surface. If the layer below had formed the surface, then each surface atom would have had three dangling bonds, a higher-energy situation, which is therefore unlikely to occur. Surfaces tend to form which minimize the density of dangling bonds.

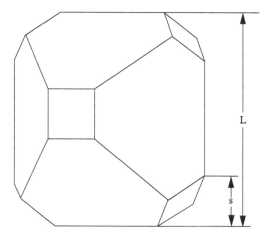

Figure 1.20. Truncated octahedron. For a fcc crystal the atoms on the square faces have a square lattice configuration, and the atoms on the hexagon shaped faces have a hexagonal close packed coniiguration. (From G. Burns, *Solid State Physics*, Academic Press, Orlando, FL, 1985, p. 676.)

Sometimes silicon forms a crystal with the external shape of an octahedron which has eight [111] faces that minimize the surface energy. On other occasions the octahedron is truncated in the manner illustrated in Fig. 1.20 since a truncated configuration has a lower surface energy for the same volume of crystal.

A favorable technique for studying surface structure is low-energy electron diffraction (LEED). Another technique is reflection high-energy electron diffraction (RHEED) carried out at grazing incidence. These techniques are discussed in Chapter 2.

1.2.5. Adsorption of Gases

Since surfaces generally have dangling bonds, they can be chemically very reactive. Owing to this reactivity, they readily adsorb gases in contact with them. Adsorbed molecules are held in place on the surface by a binding energy ΔH_{ads} called the *heat of adsorption*. When chemical bonding takes place between surface atoms and adsorbed atoms or molecules, the phenomenon is called *chemisorption*. This bonding can be of either a covalent or an ionic type. If a much weaker van der Waals type of bonding occurs the phenomenon is referred to as *physisorption*. It is customary to call the process *physisorption* when the heat of adsorption is less than $\sim 10\,\mathrm{kcal/mol}$, and to label it *chemisorption* when the heat of adsorption exceeds $10\,\mathrm{kcal/mol}$. For example, H_2 and O_2 molecules chemisorb on molybdenum with ΔH_{ads} values of 40 kcal/mol and 172 kcal/mol, respectively.

The surface of a solid can be considered as the outermost layer of atoms plus the region between 0.5 and 1.5 nm above and below it. This includes the bonding electrons on the inside and the dangling bonds on the outside. We are talking about an

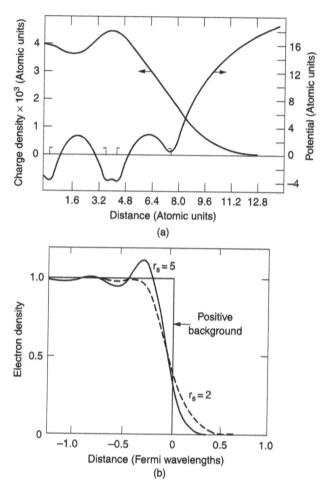

Figure 1.21. (a) Calculated potential energy and charge density as a function of z for metallic sodium. The Na^+ ion cores of the last two atomic layers are indicated by brackets. [From J. A. Applebaum and D. R. Hamann, *Phys. Rev.* **B6**, 2166 (1972)]. (b) Calculated electron density near the surface of metallic Na, where $r_s = 2$ Å is close to the Na value $r_s = 2.08$ Å. [From N. D. Lang and W. Kohn, *Phys. Rev.* **B1**, 4555 (1970). See also G. Burns (cited in Fig. 1.20 legend), p. 703.]

atomically clean surface, one exposed to an ultrahigh vacuum, and which contains no adsorbed atoms or molecules. An ultrahigh vacuum is one in which the pressure is less than 10^{-8} Pa ($<7.5 \times 10^{-11}$ torr). When a molecule from the surrounding gas strikes the surface, there is a probability S called the *sticking coefficient* that it will stick to the surface, that is, that it will be adsorbed. In a standard vacuum of 10^{-6} torr it will take surrounding gas molecules with a sticking coefficient $S = 1$ about one second to build up a monolayer, whereas at 10^{-10} torr the monolayer buildup time is much longer than an hour.

1.2.6. Electronic Structure of a Surface

The formation of a surface by cleaving a crystal in an ultrahigh vacuum can lead to a rearrangement of the electrons on the atoms that constitute the surface layer. We will discuss the case of a metallic surface. A double layer of charge forms because the center of gravity of the conduction electron charge differs from the center of gravity of the atomic nuclei. The charge density and potential energy at a [001] surface of sodium, which is fcc, vary with the distance perpendicular to the surface in the manner shown in Fig. 1.21a. The calculated electron density varies with this distance in the manner shown in Fig. 1.21b. The calculations were carried out for two values $r_s = 2$ Å and $r_s = 5$ Å of the equivalent electron radius (radius of average spherical volume occupied by a conduction electron), where r_s equals 2.08 Å and 1.41 Å for Na and Cu, respectively.

An important surface-related property of a metal is its workfunction φ, which is the energy required to remove an electron from the surface and place it an infinite distance away. The workfunction for polycrystalline sodium is 2.75 eV. When sodium is chemisorbed on the surface of another metal the workfunction varies with the surface coverage in the manner shown on Fig. 1.22. The figure shows calculations carried for sodium and for cesium chemisorbed on an aluminum substrate. As the coverage increases, the workfunction initially drops from its initial value for a bare aluminum surface, reaches a minimum, and then rises to the value of φ corresponding to that of the covering metal Na or Cs. The observed effect occurs because the adsorbed alkali atom Na loses an electron to the aluminum and becomes a positive ion or cation Na^+. The result is the creation of a dipole double layer with a polarity opposite that of the initial bare aluminum surface, so the magnitude of the net dipole moment density decreases, and φ also decreases.

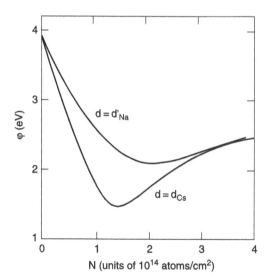

Figure 1.22. Calculated workfunction versus Na and Cs surface coverage for $r_s = 2$ Å, which is close to the value for aluminum. [From N. D. Lang, *Phys. Rev.* **B4**, 4234 (1971).]

1.2.7. Surface Quantum Well

A *quantum well* is a three-dimensional structure in which two dimensions are large, perhaps a centimeter or a millimeter, and the third is in the nanometer range. The surface of a solid, with or without the presence of a layer of adsorbed gas molecules, may be regarded as a quantum well. It has a regular structure, its lateral dimensions are macroscopic, and its thickness is several nanometers. It has characteristic charge and potential energy distributions, as discussed above. Quantum wells are discussed in Chapter 9.

1.3. ENERGY BANDS

1.3.1. Insulators, Semiconductors, and Conductors

When a solid is formed, the energy levels of the atoms broaden and form bands with forbidden gaps between them. The electrons can have energy values that exist within one of the bands, but cannot have energies corresponding to values in the gaps between the bands. The lower-energy bands due to the inner atomic levels are narrower and are all full of electrons, so they do not contribute to the electronic properties of a material. They are not shown in the figures. The outer or valence electrons that bind the crystal together occupy a *valence band*, which, for an insulating material, is full of electrons that cannot move since they are fixed in position in chemical bonds. There are no delocalized electrons to carry current, so the material is an insulator. The conduction band is far above the valence band in energy, as shown in Fig. 1.23a, so it is not thermally accessible, and remains essentially empty. In other words, the heat content of the insulating material at room temperature $T = 300$ K is not sufficient to raise an appreciable number of electrons from the valence band to the conduction band, so the number in the conduction band is negligible. Another way to express this is to say that the value of the gap energy E_g far exceeds the value $k_B T$ of the thermal energy, where k_B is Boltzmann's constant.

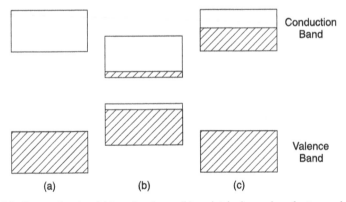

Figure 1.23. Energy bands of (a) an insulator, (b) an intrinsic semiconductor, and (c) a conductor. The crosshatching indicates the presence of electrons in the bands.

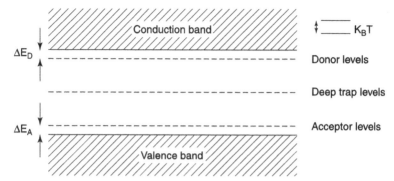

Figure 1.24. Sketch of the forbidden energy gap showing acceptor levels the typical distance Δ_A above the top of the valence band, donor levels the typical distance Δ_D below the bottom of the conduction band, and deep-trap levels nearer to the center of the gap. The value of the thermal energy k_BT is indicated on the right.

For a semiconductor, the gap between the valence and conduction bands is much less, as shown in Fig. 1.23b, so E_g is closer to the thermal energy k_BT, and the heat content of the material at room temperature can bring about the thermal excitation of some electrons from the valence band to the conduction band where they carry current. The density of electrons reaching the conduction band by this thermal excitation process is relatively low, but by no means negligible, so the electrical conductivity is small; hence the name *semiconducting*. A material of this type is called an *intrinsic semiconductor*. A semiconductor can be doped with donor atoms that donate electrons to the conduction band where they can carry current. The material can also be doped with acceptor atoms that obtain electrons from the valence band and leave behind positive charges called *holes* which can also carry current. The energy levels of these donors and acceptors lie in the energy gap, as shown in Fig. 1.24. The former produces N type, that is, negative charge or electron conductivity, and the latter produces P type, that is, positive charge or hole conductivity, as will be clarified in Section 1.4.1. These two types of conductivity in semiconductors are temperature dependent, as is the intrinsic semiconductivity.

A *conductor* is a material with a full valence band, and a conduction band partly full with delocalized conduction electrons that are efficient in carrying electric current. The positively charged metal ions at the lattice sites have given up their electrons to the conduction band, and constitute a background of positive charge for the delocalized electrons. Figure 1.23c shows the energy bands for this case. In actual crystals the energy bands are much more complicated than is suggested by the sketches of Fig. 1.23, with the bands depending on the direction in the lattice, as we shall see below.

1.3.2. Reciprocal Space

In Section 1.1.2 we discussed the structures of different types of crystals in ordinary or coordinate space. These provided us with the positions of the atoms in the lattice. To treat the motion of conduction electrons, it is necessary to consider a different type of space, which is mathematically called a *dual space* relative to the coordinate space. This dual or reciprocal space arises in quantum mechanics, and a brief qualitative description of this space is presented here.

The basic relationship between the frequency $f = \omega/2\pi$, the wavelength λ, and the velocity v of a wave is $\lambda f = v$. It is convenient to define the wavevector $k = 2\pi/\lambda$ to give $f = (k/2\pi)v$. For a matter wave, or the wave associated with conduction electrons, the momentum $p = mv$ of an electron of mass m is given by $p = (h/2\pi)k$, where Planck's constant h is a universal constant of physics. Sometimes a reduced Planck's constant $\hbar = h/2\pi$ is used, where $p = \hbar k$. Thus, for this simple case the momentum is proportional to the wavevector k, and k is inversely proportional to the wavelength with the units of reciprocal length, or reciprocal meters. We can define a reciprocal space called *k-space* to describe the motion of electrons.

If a one-dimensional crystal has a lattice constant a and a length that we take to be $L = 10a$, then the atoms will be present along a line at positions $x = 0, a, 2a, 3a, \ldots,$ $10a = L$. The corresponding wavevector k will assume the values $k = 2\pi/L, 4\pi/L,$ $6\pi/L, \ldots, 20\pi/L = 2\pi/a$. We see that the smallest value of k is $2\pi/L$ and the largest value is $2\pi/a$. The unit cell in this one-dimensional coordinate space has length a, and the important characteristic cell in reciprocal space, called the *Brillouin zone*, has the value $2\pi/a$. The electron sites within the Brillouin zone are at the reciprocal lattice points $k = 2\pi n/L$, where for our example $n = 1, 2, 3, \ldots, 10$, and $k = 2\pi/a$ at the Brillouin zone boundary where $n = 10$.

For a rectangular direct lattice in two dimensions with coordinates x and y, and lattice constants a and b, the reciprocal space is also two-dimensional with the wavevectors k_x and k_y. By analogy with the direct lattice case, the Brillouin zone in this two-dimensional reciprocal space has length $2\pi/a$ and width $2\pi/b$, as sketched in Fig. 1.25. The extension to three dimensions is straightforward. It is important to keep in mind that k_x is proportional to the momentum p_x of the conduction electron in the x direction, and similarly for the relationship between k_y and p_y.

1.3.3. Energy Bands and Gaps of Semiconductors

The electrical, optical, and other properties of semiconductors depend strongly on how the energy of the delocalized electrons involves the wavevector k in reciprocal or k-space, with the electron momentum p given by $p = mv = \hbar k$, as explained

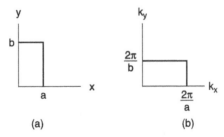

Figure 1.25. Sketch of (a) unit cell in two-dimensional x,y coordinate space; (b) corresponding Brillouin zone in k_x, k_y reciprocal space for a rectangular Bravais lattice.

above. We will consider three-dimensional crystals; in particular, we are interested in the properties of the III–V and the II–VI semiconducting compounds that have a cubic structure, so their three lattice constants are the same, namely, $a = b = c$. The electron motion expressed in the coordinates k_x, k_y, k_z of reciprocal space takes place in the Brillouin zone, and the shape of this zone for these cubic compounds is shown in Fig. 1.26. Points of high symmetry in the Brillouin zone are designated by capital (uppercase) Greek or Roman letters, as indicated.

The energy bands depend on the position in the Brillouin zone, and Fig. 1.27 presents these bands for the intrinsic (i.e., undoped) III–V compound GaAs. The figure plots energy versus the wavevector k in the following Brillouin zone directions: along Δ from point Γ to X, along Λ from Γ to L, along Σ from Γ to K, and along the path between points X and K. These points and paths are indicated in the sketch of the Brillouin zone in Fig. 1.30. We see from Fig. 1.27 that the various bands have prominent maxima and minima at the centerpoint Γ of the Brillouin zone. The energy gap or region where no band appears extends from the zero of energy at point Γ_8 to point Γ_6 directly above the gap at the energy $E_g = 1.35$ eV. The bands below point Γ_8 constitute the valence band, and those above point Γ_6 form the conduction band. Hence Γ_6 is the lowest-energy point of the conduction band, and Γ_8 is the highest point of the valence band.

At absolute zero all of the energy bands below the gap are filled with electrons, and all the bands above the gap are empty, so at $T = 0$ K the material is an insulator. At room temperature the gap is sufficiently small so that some electrons are thermally excited from the valence band to the conduction band, and these relatively few excited electrons gather in the region of the conduction band immediately above its minimum at Γ_6, a region that is referred to as a "valley." These electrons carry some electric current; hence the material is a semiconductor. Gallium arsenide is called a *direct-bandgap semiconductor* because the top of the valence band and the bottom of the conduction band are both at the same centerpoint (Γ) in the

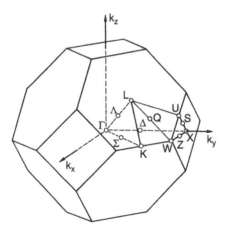

Figure 1.26. Brillouin zone of the gallium arsenide and zinc blende semiconductors showing the high-symmetry points Γ, K, L, U, W, and X, and the high-symmetry lines Δ, Λ, Σ, Q, S, and Z. (From G. Burns, *Solid State Physics*, Academic Press, Boston, 1985, p. 302.)

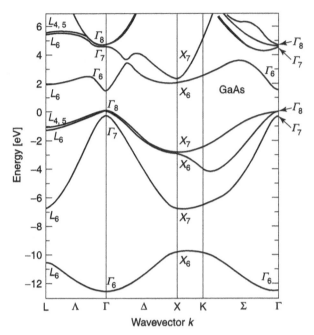

Figure 1.27. Band structure of the semiconductor GaAs calculated by the pseudopotential method. (From M. L. Cohen and J. Chelikowsky, *Electronic Structure and Electronic Properties of Semiconductors*, 2nd ed., Springer-Verlag, Solid State Science series, Vol. 75, Springer, Berlin, 1989, p. 65.)

Brillouin zone, as is clear from Fig. 1.27. Electrons in the valence band at point Γ_8 can become thermally excited to point Γ_6 in the conduction band with no change in the wavevector k. The compounds GaAs, GaSb, InP, InAs, and InSb and all the II–VI compounds included in Table B.6 have direct gaps. In some semiconductors such as Si and Ge the position of the top of the valence band in the Brillouin zone differs from the bottom of the conduction band, and these are called *indirect-gap semiconductors*.

Figure 1.28 depicts the situation at point Γ of a direct-gap semiconductor on an expanded scale, at temperatures above absolute zero, with energy bands approximated by parabolas. The conduction band valley at Γ_6 is shown occupied by electrons up to the Fermi level, which is defined as the energy of the highest occupied state. The excited electrons leave behind empty states near the top of the valence band, and these act like positive charges called *holes* in an otherwise full valence band. These hole levels exist above the energy $-E_F'$, as indicated in Fig. 1.28. Since an intrinsic or undoped semiconductor has just as many holes in the valence band as it has electrons in the conduction band, the corresponding volumes filled with these electrons and holes in k-space are equal to each other. These electrons and holes are the charge carriers of current, and the temperature dependence of their concentrations in GaAs, Si, and Ge is given in Fig. 1.29.

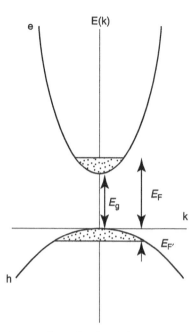

Figure 1.28. Sketch of the lower valence band and the upper conduction band of a semiconductor approximated by parabolas. The region in the valence band containing holes and that in the conduction band containing electrons are crosshatched. The Fermi energies E_F and $E_{F'}$ mark the highest occupied level of the conduction band and the lowest unoccupied level of the valence band, respectively. The zero of energy is taken as the top of the valence band, and the direct-bandgap energy E_g is indicated.

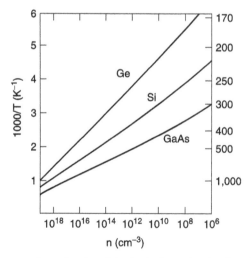

Figure 1.29. Temperature dependencies of the intrinsic carrier density of the semiconductors Ge, Si, and GaAs. (From G. Burns, *Solid State Physics*, Academic Press, Boston, 1985, p. 315.)

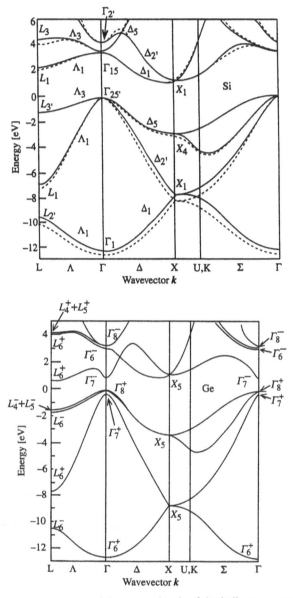

Figure 1.30. Band structure plots of the energy bands of the indirect-gap semiconductors Si and Ge. The bandgap (absence of bands) lies slightly above the Fermi energy $E = 0$ on both figures, with the conduction band above and the valence band below the gap. The figure shows that the lowest point or bottom of the conduction band of Ge is at the energy $E = 0.6$ at the symmetry point L (labeled L_6^+), and for Si it is 85% of the way along the direction Δ_1 from Γ_{15} to X_1. It is clear from Fig. 1.27 that the bottom of the conduction band of the direct-gap semiconductor GaAs is at the symmetry point Γ_6, The top of the valence band is at the centerpoint Γ of the Brillouin zone for all three materials. (From M. L. Cohen and J. Chelikowsky, *Electronic Structure and Electronic Properties of Semiconductors*, 2nd ed., Springer-Verlag, Solid State Science series, Vol. 75, Springer, Berlin, 1989, pp. 53, 64.)

In every semiconductor listed in Table B.6, including Si and Ge, the top of its valence band is located at the center of the Brillouin zone, but in the indirect-bandgap semiconductors Si, Ge, AlAs, AlSb, and GaP the lowest valley of their conduction bands appears at different locations in k-space and at point Γ. This is shown in Fig. 1.30 for the indirect-bandgap materials Si and Ge. We see from Fig. 1.30 that the conduction band of Ge is minimum at point L, which is in the middle of the hexagonal face of the Brillouin zone along the Λ or [111] direction depicted in Fig. 1.26. The Brillouin zone has eight such faces, with each point L shared by two zones, so the zone actually contains only four of these points proper to it, and we say that the valley degeneracy for Ge is 4. The semiconductor Si exhibits a lowest conduction band minimum along the Δ or [001] direction about 85% of the way to point X, as shown in Fig. 1.30. The corresponding valley of the compound GaP, not shown, also is located along Δ about 92% of the way to X. We see from Fig. 1.26 that there are six such Δ lines in the Brillouin zone, so this valley degeneracy is 6. Associated with each of the valleys that we have been discussing, at point L for Ge and along direction Δ for Si, there is a three-dimensional constant-energy surface in the shape of an ellipsoid that encloses the conduction electrons in the corresponding valleys, and these ellipsoids are sketched in Figs. 1.31a and 1.31b for Ge and Si, respectively.

Some interesting experiments such as cyclotron resonance have been carried out to map the configuration of these ellipsoid-type constant-energy surfaces. In a cyclotron resonance experiment conduction electrons are induced to move along

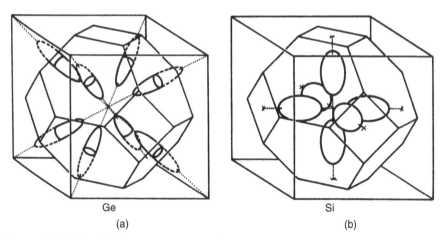

Ge	Si
(a)	(b)

Figure 1.31. Ellipsoidal constant energy surfaces in the conduction band of germanium (a) and silicon (b). The constant-energy surfaces of Ge are aligned along symmetry direction Λ and centered at symmetry point L. As a result, they lie half inside (solid lines) and half outside (dashed lines) the first Brillouin zone, so this zone contains the equivalent of four complete energy surfaces. The surfaces of Si lie along the six symmetry directions Δ (i.e., along $\pm k_x$, $\pm k_y$, $\pm k_z$), and are centered 85% of the way from the centerpoint Γ to symmetry point X. All six of them lie entirely within the Brillouin zone, as shown. Figure 1.26 shows the positions of symmetry points Γ, L, and X, and of symmetry lines Δ and Λ, in the Brillouin zone. (From G. Burns, *Solid State Physics*, Academic Press, Boston, 1985, p. 313.)

constant-energy surfaces at a velocity that always remains perpendicular to an applied magnetic field direction. By utilizing various orientations of applied magnetic fields relative to the ellipsoid, the electrons at the surface execute a variety of orbits, and by measuring the trajectories of these orbits, we can delineate the shape of the energy surface.

Table B.6 lists values of the energy gap E_g for types III–V and II–VI semiconductors at room temperature (left-hand value) and in several cases it also lists the value at absolute zero temperature (right-hand value). Table B.7 gives the temperature and pressure dependences, dE_g/dT and dE_g/dP, respectively, of the gap at room temperature.

1.3.4. Effective Mass

The kinetic energy of a conduction electron moving at a velocity v in a semiconductor is found to have the value $\frac{1}{2}m^*v^2 = p^2/2m^*$, where $\mathbf{p} = m^*\mathbf{v}$ is the momentum, and often the mass m^* in these expressions differs from the electron's fundamental or rest mass value m_0, which is a universal constant, so it is called an *effective mass*. This is analogous to the case of the kinetic energy of an object of mass m moving at a velocity v through a liquid. Its effective mass is reduced by the volume of the displaced liquid, and hence its measurable kinetic energy has a value lower than $\frac{1}{2}mv^2$.

On a simple one-dimensional model the energy E of a conduction electron has a quadratic dependence on the wavevector $\mathbf{k} = \mathbf{p}/h$ through the expression

$$E = h^2k^2/2m^* \tag{1.12}$$

The first derivative of this expression provides the velocity v

$$\frac{1}{h}\frac{dE}{dk} = \frac{hk}{m^*} = v \tag{1.13}$$

and the second derivative provides the effective mass m^*

$$\frac{1}{h^2}\frac{d^2E}{dk^2} = \frac{1}{m^*} \tag{1.14}$$

which, as we said above, differs in general from the free-electron mass m_0. These equations are rather trivial for the simple parabolic energy expression (1.12), but we see from the energy bands of Figs. 1.27 and 1.30 that the actual dependence of the energy E on k is much more complex than Eq. (1.12) indicates. This equation (1.14) is a general definition of the effective mass m^*, and the wavevector k dependence of m^* can be evaluated from the band structure plots by carrying out the differentiations. We see from a comparison of the slopes near the conduction band minimum and the valence band maximum of GaAs at point Γ on Fig. 1.27

(see also Fig. 1.28) that the upper electron bands have steeper slopes and hence lighter masses than do the lower hole bands with more gradual slopes.

1.3.5. Fermi Surfaces

At very low temperatures electrons fill the energy bands of solids up to the Fermi energy E_F, and the bands are empty for energies that exceed E_F. In three-dimensional k-space the set of values of k_x, k_y, k_z that satisfy the equation $\hbar^2(k_x^2 + k_y^2 + k_z^2)/2m = E_F$ form the *Fermi surface*. All k_x, k_y, k_z energy states that lie below this surface are full, and the states above the surface are empty. The Fermi surface encloses all the electrons in the conduction band that carry electric current. In the good conductors copper and silver the conduction electron density is 8.5×10^{22} and 5.86×10^{22} electrons/cm^3, respectively. From another viewpoint, the Fermi surface of a good conductor can fill the entire Brillouin zone. In the intrinsic semiconductors GaAs, Si, and Ge the carrier density at room temperature from Fig. 1.29 is approximately 10^6, 10^{10}, and 10^{13} carriers/cm^3, respectively, many orders of magnitude below that of metals, and semiconductors are seldom doped to concentrations above 10^{19} centers/cm^3. An *intrinsic semiconductor* is one with a full valence band and an empty conduction band at absolute zero of temperature. As we saw above, at ambient temperatures some electrons are thermally excited to the bottom of the conduction band, and an equal number of empty sites or holes are left behind near the top of the valence band. This means that only a small percentage of the Brillouin zone contains electrons in the conduction band, and the number of holes in the valence band is correspondingly small. In a one-dimensional representation this reflects the electron and hole occupancies depicted in Fig. 1.28.

If the conduction band minimum is at point Γ in the center of the Brillouin zone, as is the case with GaAs, then the Fermi surface will be very close to a sphere since the symmetry is cubic, and to a good approximation we can assume a quadratic dependence of the energy on the wavevector k, corresponding to Eq. (1.12). Therefore the Fermi surface in k-space is a small sphere given by the standard equation for a sphere

$$E_F = E_g + \frac{\hbar^2}{2m_e}(k_x^2 + k_y^2 + k_z^2)_F = E_g + \frac{\hbar^2}{2m_e}k_F^2 \qquad (1.15)$$

where E_F is the Fermi energy and the electron mass m_e relative to the free-electron mass has the values given in Table B.8 for various direct-gap semiconductors. Equations (1.15) are valid for direct-gap materials where the conduction band minimum is at point Γ. In all the semiconductor materials under consideration the valence band maxima are located at the Brillouin zone centerpoint Γ.

The situation for the conduction electron Fermi surface is more complicated for indirect gap semiconductors. We mentioned above that Si and GaP exhibit conduction band minima along the Δ direction of the Brillouin zone, and the corresponding six ellipsoidal energy surfaces are sketched in Fig. 1.31a. The longitudinal and transverse effective masses, m_L and m_T, respectively, have the following values

$$\frac{m_L}{m_0} = 0.92 \quad \frac{m_T}{m_0} = 0.19 \quad \text{for} \quad \text{Si} \tag{1.16a}$$

$$\frac{m_L}{m_0} = 7.25 \quad \frac{m_T}{m_0} = 0.21 \quad \text{for} \quad \text{GaP} \tag{1.16b}$$

for these two indirect-gap semiconductors. Germanium has its band minimum at points L of the Brillouin zone sketched in Fig. 1.26, and its Fermi surface is the set of ellipsoids centered at paints L with their axes along the Λ or [111] directions, as shown in Fig. 1.31a. The longitudinal and transverse effective masses for these ellipsoids in Ge are $m_L/m_0 = 1.58$ and $m_T/m_0 = 0.081$, respectively. Cyclotron resonance techniques together with the application of stress to the samples can be used to determine these effective masses for the indirect-bandgap semiconductors.

1.4. LOCALIZED PARTICLES

1.4.1. Donors, Acceptors, and Deep Traps

When an atom in column V of the periodic table, such as P, As, or Sb, which has five electrons in its outer or valence electron shell, is a substitutional impurity in Si, it uses four of these electrons to satisfy the valence requirements of the four nearest-neighbor silicons, and the one remaining electron remains weakly bound. The atom easily donates or passes on this electron to the conduction band, so it is called a *donor*, and the electron is called a *donor electron*. This occurs because the donor energy levels lie in the forbidden region close to the conduction band edge by the amount ΔE_d relative to the thermal energy value $k_B T$, as indicated in Fig. 1.24. A Si atom substituting for Ga plays the role of a donor in GaAs, Al substituting for Zn in ZnSe serves as a donor, and so on. The conductivity that is associated with the flow of electric current carried by these conduction band electrons is called *N-type*.

An atom from column III atom such as Al or Ga, called an *acceptor atom*, which has three electrons in its valence shell can serve as a substitutional defect in Si, and in this role it requires four valence electrons to bond with the tetrahedron of nearest-neighbor Si atoms. To accomplish this it draws or accepts an electron from the valence band, leaving behind a hole at the top of this band. The conductivity corresponding to electric current carried by the movement of these holes is called *P-type*. The transfer of an electron to Si occurs easily because the energy levels of the acceptor atoms are in the forbidden gap slightly above the valence band edge by the amount ΔE_A, which is small relative to $k_B T$, as indicated in Fig. 1.24. In other words, the excitation energies needed to ionize the donors and to add electrons to the acceptors are much less than the thermal energy at room temperature $T = 300$ K (i.e., $\Delta E_D, \Delta E_A \ll k_B T$), so virtually all donors are positively ionized and virtually all acceptors are negatively ionized at room temperature.

The donor and acceptor atoms that we have been discussing are known as "shallow centers," that is, shallow traps of electrons or holes, because their excitation energies are much less than that of the bandgap (ΔE_D, $\Delta E_A \ll E_g$). There are other centers with energy levels that lie deep within the forbidden gap, often closer to its center than to the top or bottom, in contrast to the case with shallow donors and acceptors. Since generally $E_g \gg k_B T$, these traps are not extensively ionized, and the energies involved in exciting or ionizing them are not small. Examples of deep centers are defects associated with broken bonds, or strain involving displacements of atoms.

1.4.2. Mobility

Another important parameter of a semiconductor is the mobility μ or charge carrier drift velocity v per unit electric field E, given by the expression $\mu = |v|/E$. This parameter is defined as positive for both electrons and holes. Table B.9 lists the mobilities μ_e and μ_h for electrons and holes, respectively, in the semiconductors under consideration. The electrical conductivity σ is the sum of contributions from the concentrations of electrons n and of holes p in accordance with the expression

$$\sigma = (ne\mu_e + pe\mu_h) \tag{1.17}$$

where e is the electronic charge. The mobilities have a weak power-law temperature dependence T^n, and the pronounced T dependence of the conductivity is due principally to the dependence of the electron and hole concentrations on the temperature. In doped semiconductors this generally arises mainly from the Boltzmann factor $\exp(-E_i/k_B T)$ associated with the ionization energies E_i of the donors or acceptors. Typical ionization energies for donors and acceptors in Si and Ge listed in Table B.10 range from 0.0096 to 0.16 eV, which is much less than the bandgap energies 1.12 eV and 0.67 eV of Si and Ge, respectively. Figure 1.24 shows the locations of donor and acceptor levels on an energy band plot, and makes clear that their respective ionization energies are much less than E_g. The thermal energy $k_B T = 0.026$ eV at room temperature ($T = 300$ K) is often comparable to the ionization energies. In intrinsic or undoped materials the main contribution is from the exponential factor $\exp(-E_g/2 k_B T)$ in the following expression from the law of mass action

$$n_i = p_i = 2K \left(\frac{k_B T}{2\pi h^2} \right)^{3/2} (m_e m_h)^{3/4} \exp \frac{-E_g}{2k_B T} \tag{1.18}$$

where K is a constant and the intrinsic concentrations of electrons n_i and holes p_i are equal to each other because the thermal excitation of n_i electrons to the conduction band leaves behind the same number p_i of holes in the valence band, that is, $n_i = p_i$. We see that the expression (1.18) contains the product $m_e m_h$ of the effective masses m_e and m_h of the electrons and holes, respectively, and the ratios m_e/m_0 and m_h/m_0 of these effective masses to the free-electron mass m_0 are presented in

Table B.8. These effective masses strongly influence the properties of excitons to be discussed next.

1.4.3. Excitons

An ordinary negative electron and a positive electron, called a *positron*, situated a distance *r* apart in free space experience an attractive *Coulomb force*, which has the value $-e^2/4\pi\varepsilon_0 r^2$, where *e* is their charge and ε_0 is the dielectric constant of free space. A quantum-mechanical calculation shows that the electron and positron interact to form an atom called *positronium*, which has bound state energies given by the Rydberg formula introduced by Niels Bohr in 1913 to explain the hydrogen atom

$$E = -\frac{e^2}{8\pi\varepsilon_0 a_0 n^2} = -\frac{6.8}{n^2}\text{eV} \qquad (1.19)$$

where a_0 is the Bohr radius given by $a_0 = 4\pi\varepsilon_0 h^2/m_0 e^2 = 0.0529$ nm, m_0 is the free-electron (and positron) mass, and the quantum number *n* takes on the values $n = 1,2,3,\ldots,4$. For the lowest-energy or ground state, which has $n = 1$, the energy is 6.8 V, which is exactly half the ground-state energy of a hydrogen atom, since the effective mass of the bound electron–positron pair is half that of the bound electron–proton pair in the hydrogen atom. Figure 1.32 shows the energy

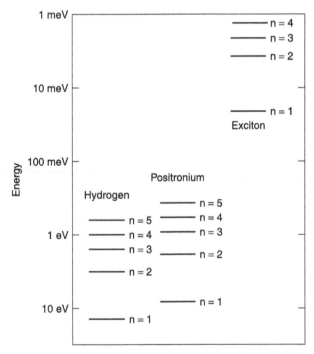

Figure 1.32. The first few energy levels in the Rydberg series of a hydrogen atom (a), positronium (b), and a typical exciton (c).

levels of positronium as a function of the quantum number n. This set of energy levels is often referred to as a *Rydberg series*. The continuum at the top of the figure is the region of positive energies where the electron and the hole are so far away from each other that the Coulomb interaction no longer has an appreciable effect, and the energy is all of the kinetic type, $1/2(mv^2) = p^2/2m$, or energy of motion, where v is the velocity and $p = mv$ is the momentum.

The analog of positronium in a solid such as a semiconductor is the bound state of an electron–hole pair called an *exciton*. For a semiconductor the electron is in the conduction band and the hole is in the valence band. The electron and hole both have effective masses m_e and m_h, respectively, which are less than that m_0 of a free electron, so the reduced mass m_R is given by $m_R = m_e m_h/(m_e + m_h)$. When the electron effective mass is appreciably less than the hole effective mass, $m_e \ll m_h$, the relationship between them is conveniently written in the form

$$m_R = \frac{m_e}{1 + (m_e/m_h)} \tag{1.20}$$

which shows that for this case m_R becomes comparable to the electron mass. For example, if $m_e/m_h = 0.2$, then $m^* = 0.83\, m_e$. A comparison of the data in Table B.8 shows that this situation is typical for GaAs-type semiconductors. We also see from Table B.12 that the relative dielectric constant $\varepsilon/\varepsilon_0$ has the range of values $5.6 < \varepsilon/\varepsilon_0 < 15.7$ for these materials, where ε_0 is the dielectric constant of free space. Both of these factors have the effect of decreasing the exciton energy E_{ex} from that of positronium, and as a result this energy is given by

$$E_{ex} = \frac{m^*/m_0}{(\varepsilon/\varepsilon_0)^2}\frac{e^2}{4\pi\varepsilon_0 a_0 n^2} = \frac{13.6\, m^*/m_0}{(\varepsilon/\varepsilon_0)^2 n^2}\quad \text{eV} \tag{1.21}$$

as shown plotted in Fig. 1.32. These same two factors also increase the effective Bohr radius of the electron orbit, and it becomes

$$a_{eff} = \frac{\varepsilon/\varepsilon_0}{m^*/m_0} a_0 = \frac{0.0529\varepsilon/\varepsilon_0}{m^*/m_0}\quad \text{nm} \tag{1.22}$$

Using the GaAs electron effective mass and the heavy-hole effective mass values from Table B.8, Eq. (1.22) gives $m^*/m_0 = 0.058$. Utilizing the dielectric constant value from Table B.11, we obtain, using Eqs. (1.21 and 1.22), for GaAs

$$E_0 = 4.5\,\text{meV}, \quad a_{eff} = 12.0\,\text{nm} \tag{1.23}$$

where E_0 is the ground-state ($n = 1$) energy. This demonstrates that an exciton extends over quite a few atoms of the lattice, and its radius in GaAs is comparable with the dimensions of a typical nanostructure. An exciton has the properties of a particle; it is mobile and able to move around the lattice. It also exhibits characteristic optical spectra. Figure 1.32 plots the energy levels for an exciton with the ground-state energy $E_0 = 18$ meV.

Technically speaking, the exciton that we have just discussed is a weakly bound electron–hole pair called a *Mott–Wannier exciton*. A strongly or tightly bound exciton, called a *Frenkel exciton*, is similar to a longlived excited state of an atom or molecule. It is also mobile, and can move around the lattice by the transfer of the excitation or excited-state charge between adjacent atoms or molecules. Almost all the excitons encountered in semiconductors and in nanostructures are of the Mott–Wannier type, except in organic nanocrystals, where they are Frenkel excitons.

PROBLEMS

1.1. Find the volume and the surface area of the cuboctahedron sketched in Fig. 1.6.

1.2. Consider a nanofilm formed from the two-dimensional close-packed lattice of Fig. 1.4. The smallest $n = 1$ nanofilm is, of course, one atom, and the $n = 2$ one is the seven-atom ($N = 7$) structure sketched on the left side of Fig. C.1 of Appendix C. The number of atoms around the edge is given by $N_{edge} = 6(n - 1)$ valid for $n > 1$, in analogy with Eq. (1.4). Make accurate sketches of the $n = 3$, $n = 4$, and $n = 5$ planar nanoparticles, and prepare the analog of Table 1.3 for this two-dimensional case for n values ranging from 1 to 5.

1.3. Evaluate the coefficients A, B, and C of the equation $N = An^2 + Bn + C$ for the number of atoms in a two-dimensional nanoparticle.

1.4. Find the distances between [111] planes of the zinc blende lattice in terms of the cubic unit cell dimension a for the following cases: (a) two nearest sulfur planes, (b) two nearest zinc planes, and (c) adjacent S and Zn planes. Why are there two answers for case (c)?

1.5. Consider three fcc nanoparticles containing approximately 1000 atoms: one in the shape of an octahedron, one in the shape of a cube, and one in the shape of the structure depicted in Fig. 1.9. Which has the largest, and which has the smallest value of surface energy per unit volume? What are their values?

1.6. A two-dimensional close packed lattice of circular atoms is illustrated in Fig. 1.4a. What is the radius of the interstitial sites of this lattice, and how numerous are they compared to the atoms?

1.7. A square of side a has zero-dimensional apices, four one-dimensional edges, and one two-dimensional face. If the square is projected out into the third dimension and the vertices of the two squares are connected with straight lines of length a, the resulting three-dimensional cube has 8 vertices, 12 edges, and 6 faces. Consider projecting the cube out into a fourth dimension and connecting the respective vertices with straight lines of length a, to form a four-dimensional "cube" called a *tesseract*. How many apices, edges, faces, and cubes does the tesseract have?

REFERENCES

1. N. W. Ashcroft and N. D. Mermin, *Solid State Physics*, Saunders College, Philadelphia, 1976.

2. G. Burns, *Solid State Physics*, Academic Press, San Diego, 1985.

3. C. Kittel, *Introduction to Solid State Physics*, 7th ed., Wiley, New York, 1996.

4. P. Y. Yu and M. Cardona, *Fundamentals of Semiconductors*, 3rd ed., Springer-Verlag, Berlin, 2001.

5. T. P. Martin, T. Bergmann, H. Gohlich, and T. Lange, *Chem. Phys. Lett.* **172**, 209 (1990).

6. S. Sugano and H. Koizumi, *Microcluster Physics*, Springer, Berlin, 1998, p. 90.

7. C. P. Poole, Jr. and H. A. Farach, *Chemical Bonding in Semiconductor Physics*, K. Boer, ed., Wiley, New York, 2001, Vol. 1, Chapter 2.

Methods of Measuring Properties of Nanostructures

2.1. INTRODUCTION

The current revolution in nanoscience was brought about by the concomitant development of several advances in technology. One of them has been the progressive ability to fabricate smaller and smaller structures, and another has been the continual improvement in the precision with which such structures are made. A widely used method for the fabrication of nanostructures is lithography, which makes use of a radiation-sensitive layer to form well-defined patterns on a surface. Molecular-beam epitaxy, the growth of one crystalline material on the surface of another, is a second technique that has been perfected. There are also chemical methods, and the utilization of self-assembly, ths spontaneous aggregation of molecular groups.

For the past three decades (i.e., since the mid-1970s) the continual advancement in technology has followed Moore's law,[1] whereby the number of transistors incorporated on a memory chip doubles every 18 months. This has resulted from continual improvements in design factors such as interconnectivity efficiency, as well as from continual decreases in size. Economic considerations driving this revolution are the need for more and greater information storage capacity and for faster and broader information dispersal through communication networks. Another major factor responsible for the nanotechnology revolution has been the improvement of old and the introduction of new instrumentation systems for evaluating and characterizing nanostructures. Many of these systems are very large and expensive, in the million-dollar price range, often requiring specialists to operate them. The aim of the present chapter is to explain the principles behind the operation of some of these systems, and to delineate their capabilities.

In the following sections we describe instruments for determining the positions of atoms in materials, instruments for observing and characterizing the surfaces of structures, and various spectroscopic-type devices for obtaining information about the properties of nanostructures.

The Physics and Chemistry of Nanosolids. By Frank J. Owens and Charles P. Poole, Jr.
Copyright © 2008 John Wiley & Sons, Inc.

2.2. STRUCTURE

2.2.1. Atomic Structures

To understand a nanomaterial, we must first learn about its structure, meaning that we must ascertain the types of atoms that constitute its building blocks, and how their atoms are arranged relative to each other. Most nanostructures are crystalline, meaning that their thousands of atoms have a regular arrangement in space on a *crystal lattice*, as explained in Section 1.1.2. This lattice can be described by assigning the positions of the atoms in a unit cell, so the overall lattice arises from the continual replication of this unit cell throughout space. Figure 1.1 presents sketches of the unit cells of the four crystal systems in two dimensions, and the characteristics of the parameters a, b, and θ for these systems are listed in the four top rows of Table 2.1. There are 17 possible types of crystal structures called *space groups*, meaning 17 possible arrangements of atoms in unit cells in two dimensions, and these are divided between the four crystal systems in the manner indicated in the column 4 of the table. Of particular interest is the most efficient way to arrange identical atoms on a surface, and this corresponds to the hexagonal system shown in Fig. 1.4a.

In three dimensions the situation is much more complicated, and some particular cases were described in Chapter 1. There are now three lattice constants, a,b,c, for the three dimensions x,y,z, with the respective angles α, β, and γ between them (α is between b and c, etc.). There are seven crystal systems in three dimensions with a

TABLE 2.1. Crystal Systems, and Associated Number of Space Groups, in Two and Three Dimensions[a]

Dimension	System	Conditions	Space Group
2	Oblique	$a \neq b$, $\gamma \neq 90°$	2
2	Rectangular	$a \neq b$, $\gamma = 90°$	7
2	Square	$a = b$, $\gamma = 90°$	3
2	Hexagonal	$a = b$, $\gamma = 120°$	5
3	Triclinic	$a \neq b \neq c$ $\alpha \neq \beta \neq \gamma$	2
3	Monoclinic	$a \neq b \neq c$ $\alpha = \gamma = 90° \neq \beta$	13
3	Orthorhombic	$a \neq b \neq c$ $\alpha = \beta = \gamma = 90°$	59
3	Tetragonal	$a = b \neq c$ $\alpha = \beta = \gamma = 90°$	68
3	Trigonal	$a = b = c$ $\alpha = \beta = \gamma < 120°$, $\neq 90°$	25
3	Hexagonal	$a = b \neq c$ $\alpha = \beta = 90° \gamma = 120°$	27
3	Cubic	$a = b = c$ $\alpha = \beta = \gamma = 90°$	36

[a]There are 17 two-dimensional space groups and 230 three-dimensional space groups.

total of 230 space groups divided among the systems in the manner indicated in column 4 of the table. The object of a crystal structure analysis is to distinguish the symmetry and space group, to determine the values of the lattice constants and angles, and to identify the positions of the atoms in the unit cell.

There are some special cases of crystal structures that are important for nanocrystals, such as those involving simple cubic (sc), body-centered cubic (bcc), and face-centered cubic (fcc) unit cells, as shown in Fig. 1.3. Another important structural arrangement is formed by stacking planar hexagonal layers in the manner sketched in Fig. 1.4b, which for a monatomic (single-atom) crystal provides the highest-density or closest-packed arrangement of identical spheres. If the third layer is placed directly above the first layer, the fourth directly above the second, and so on, in an $A-B-A-B-\ldots$-type sequence, the hexagonal close-packed (hcp) structure results. If, on the other hand, this stacking is carried out by placing the third layer in a third position and the fourth layer above the first, and so forth, the result is an $A-B-C-A-B-C-A-\ldots$ sequence, and the structure is fcc, as explained in Chapter 1. The latter arrangement is more commonly found in nanocrystals.

Some properties of nanostructures depend on their crystal structure, while other properties such as catalytic reactivity and adsorption energies depend on the type of exposed surface. Epitaxial films prepared from fcc or hcp crystals generally grow with the planar close-packed atomic arrangement just discussed. Face-centered cubic crystals tend to expose surfaces with this same hexagonal two-dimensional atomic array.

2.2.2. Crystallography

To determine the structure of a crystal, and thereby ascertain the positions of its atoms in the lattice, a collimated beam of X rays, electrons, or neutrons is directed at the crystal, and the angles at which the beam is diffracted are measured. We will explain the method in terms of X rays, but much of what we say carries over to the other two radiation sources. The wavelength λ of the X rays expressed in nanometers is related to the X-ray energy E expressed in the units kiloelectronvolts (keV) through the expression

$$\lambda = \frac{1.240}{E} \text{ nm} \tag{2.1}$$

Ordinarily the beam is fixed in direction and the crystal is rotated through a broad range of angles to record the X-ray spectrum, which is also called a *diffractometer recording* or *X-ray diffraction scan*. Each detected X-ray signal corresponds to a coherent reflection, called a *Bragg reflection*, from successive planes of the crystal for which Bragg's law is satisfied

$$2d \sin \theta = n\lambda \tag{2.2}$$

as shown in Fig. 2.1, where d is the spacing between the planes, θ is the angle that the X-ray beam makes with respect to the plane, λ is the wavelength of the X rays, and $n = 1,2,3,\ldots$ is an integer that usually has the value $n = 1$.

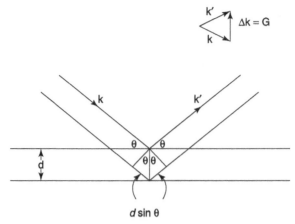

Figure 2.1. Reflection of X-ray beam incident at the angle θ off two parallel planes separated by the distance d. The difference in path length $2d \sin \theta$ for the two planes is indicated. (From C. P. Poole, Jr., *The Physics Handbook*, Wiley, New York, 1998, p. 333.)

Each crystallographic plane has three indices h,k,l, and for a cubic crystal they are ratios of the points at which the planes intercept the Cartesian coordinate axes x,y,z. The distance d between parallel crystallographic planes with indices hkl for a simple cubic lattice of lattice constant a has the particularly simple form

$$d = \frac{a}{(h^2 + k^2 + l^2)^{1/2}} \tag{2.3}$$

so higher-index planes have larger Bragg diffraction angles θ. Figure 2.2 shows the spacing d for (110) and (120) planes, where the index $l = 0$ corresponds to planes that are parallel to the z direction. It is clear from this figure that planes with higher indices

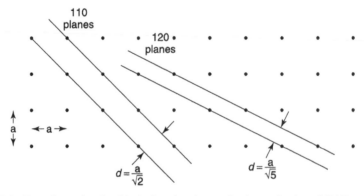

Figure 2.2. Two-dimensional cubic lattice showing projections of pairs of (110) and (120) planes (perpendicular to the surface) with the distances d between them indicated.

are closer together, in accordance with Eq. (2.3), so they have larger Bragg angles θ from Eq. (2.2). The amplitudes of the X-ray lines from different crystallographic planes also depend on the indices hkl, with some planes having zero amplitude, and these relative amplitudes help in identifying the structure type. For example, for a body-centered monatomic lattice the only planes that produce observed diffraction peaks are those for which $h + k + l = n$, an even integer, and for a FCC lattice the only observed diffraction lines either have all odd integers or all even integers.

To obtain a complete crystal structure, X-ray spectra are recorded for rotations around three mutually perpendicular planes of the crystal. This provides comprehensive information on the various crystallographic planes of the lattice. The next step in the analysis is to convert these data on the planes to knowledge of the positions of the atoms in the unit cell. This can be done by a mathematical procedure called *Fourier transformation*. Carrying out this procedure permits us to identify which one of the 230 crystallographic space groups corresponds to the structure, together with providing the lengths of the lattice constants edges a,b,c of the unit cell, and the values of the angles α,β,γ between them. In addition, the coordinates of the position of each atom in the unit cell can be deduced.

As an example of an X-ray diffraction structure determination, consider the case of nanocrystalline titanium nitride prepared by chemical vapor deposition with the grain size distribution shown in Fig. 2.3. The X-ray diffraction scan, with the various lines

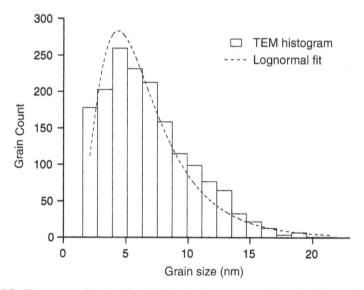

Figure 2.3. Histogram of grain size distribution in nanocrystalline TiN determined from a TEM micrograph. The fit parameters for the dashed curve are $D_0 = 5.8$ nm and $\sigma = 1.71$ nm. (From C. E. Krill, R. Haberkorn, and R. Birringer, in *Handbook of Nanostructured Materials and Nanotechnology*, H. S. Nalwa, ed., Academic Press, Boston, 2000, Vol. 2, Chapter 5, p. 207.)

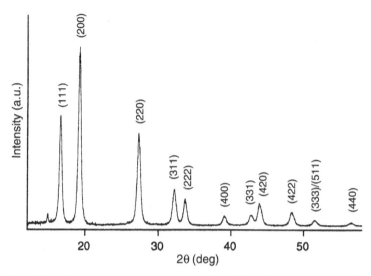

Figure 2.4. X-Ray diffraction scan of nanocrystalline TiN with the grain size distribution shown in Figure 2.3. Molybdenum K_α radiation was used with wavelength $\lambda = 0.07093$ nm calculated from Eq. (2.1). The X-ray lines are labeled with their respective crystallographic plane indices (*hkl*). Note that these indices are either all even or all odd, as expected for a fcc structure. The nonindexed weak line near $2\theta = 15°$ is due to an unidentified impurity. (From C. E. Krill et al., in *Handbook of Nanostructured Materials and Nanotechnology*, H. S. Nalwa, ed., Academic Press, Boston, 2000, Vol. 2, Chapter 3, p. 200.)

labeled by their crystallographic planes, is shown in Fig. 2.4. The fact that all the planes have either all odd or all even indices identifies the structure as FCC. The data show that TiN has the fcc NaCl structure sketched in Fig. 1.3c, with the lattice constant $a = 0.42417$ nm.

The widths of the Bragg peaks of the X-ray scan of Fig. 2.4 can be analyzed to provide information on the average grain size of the TiN sample. Since the widths arise from combinations or convolutions of grain size, microcrystalline strain, and instrumental broadening effects, it is necessary to correct for the instrumental broadening and to sort out the strain components to determine the average grain size. The assumption was made of spherical grains with the diameter D related to the volume V by the expression

$$D = \left(\frac{6V}{\pi}\right)^{1/3} \tag{2.4}$$

and several ways of making the linewidth corrections provided average grain size values between 10 and 12 nm, somewhat larger than expected from the transmission electron microscopic (TEM) histogram of Fig. 2.3. Thus X-ray diffraction can estimate average grain sizes, but a transmission electron microscope is needed to determine the actual distribution of grain sizes shown in Fig. 2.3.

A convenient way to estimate the average diameter D of particles in a powder sample from the broadening of the lines in their X-ray diffraction pattern (e.g., Fig. 2.4) is to make use of the Scherrer formula [2,3]

$$D = \frac{\lambda}{\cos \theta (B_N^2 - B_B^2)^{1/2}} \tag{2.5}$$

where B_N is the width of the lines at half-height in radians and B_B is the corresponding width from a pattern of the bulk material.

An alternate approach for obtaining the angles θ that satisfy the Bragg condition (2.2) in a powder sample is the Debye–Scherrer method sketched in Fig. 2.5. It employs a monochromatic X-ray beam incident on a powder sample generally contained in a very fine-walled glass tube. The tube can be rotated to smooth out the recorded diffraction pattern. The conical pattern of X rays emerging for each angle 2θ, with θ satisfying the Bragg condition (2.2), is incident on the film strip in arcs, as shown. It is clear from the figure that the Bragg angle θ has the value $\theta = S/4R$, where S is the distance between the two corresponding reflections on the film and R is the radius of the film cylinder. Thus a single exposure of the powder to the X-ray beam provides all of the Bragg angles at the same time. The Debye–Scherrer powder technique is often used for sample identification. To facilitate the identification, the Debye–Scherrer diffraction results for over 20,000 compounds are available to researchers in a JCPDS powder diffraction card file. This method has been widely used to obtain the structures of powders of nanoparticles.

A crystallographic lattice expands with temperature, and to measure this change, the differential distance Δd between crystallographic planes can be obtained by measuring the differential change $\Delta \theta$ in the angle of diffraction under backscattering

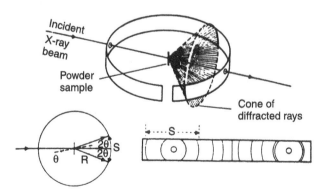

Figure 2.5. Debye–Scherrer powder diffraction technique, showing a sketch of the apparatus (top), an X-ray beam trajectory for the Bragg angle θ (lower left), and images of arcs of the diffraction beam cone on the film plate (lower right). (From G. Burns, *Solid State Physics*, Academic Press, Boston, 1985, p. 81.)

conditions. An evaluation of the differential $\Delta(d \sin \theta)$ from Eq. (2.2) provides the expression

$$\Delta\theta = -\left(\frac{\tan\theta}{d}\right)\Delta d \qquad (2.6)$$

which shows that $\Delta\theta$ becomes very large when the Bragg angle θ approaches $90°$, the condition for backscattering. Therefore, to measure the variation in the lattice parameters with temperature accurately, a backscattering geometry should be employed.

X-Ray crystallography is helpful for studying a series of isomorphic crystals, that is, crystals with the same crystal structure but different lattice constants, such as the solid solution series $Ga_{1-x}In_xAs$ or $GaAs_{1-x}Sb_x$, where x can take on the range of values $0 \leq x \leq 1$. For these cubic crystals the lattice constant a will depend on x since indium (In) is larger than gallium (Ga), and antimony (Sb) is larger than arsenic (As), as the data in Table B.1 indicate. For this case Vegard's law, Eq. (1.8) of Section 1.1.5, is a good approximation for estimating the value of a if x is known, or the value of x if a is known.

2.2.3. Particle Size Determination

In the previous section we discussed determination of the sizes of grains in polycrystalline materials via X-ray diffraction. These grains can range from nanoparticles with size distributions like that sketched in Fig. 2.3, to much larger micrometer-sized particles, held together tightly to form the polycrystalline material. This is the bulk or clustered grain limit. The opposite limit is that of grains or nanoparticles dispersed in a matrix so that the distances between them are larger than their average diameters or dimensions. It is of interest to know how to measure the sizes, or ranges of sizes, of these dispersed particles.

The most straightforward way to determine the size of a micrometer-sized grain is to look at it under a microscope, and for nanosized particles a transmission electron microscope (TEM), to be discussed in Section 2.3.1, serves this purpose. Figure 2.6 shows a TEM micrograph of polyaniline particles with diameters close to 100 nm dispersed in a polymer matrix.

Another method for determining the sizes of particles is by measuring how they scatter light. The extent of the scattering depends on the relationship between the particle size d and the wavelength λ of the light, and also on the polarization of the incident light beam. For example, the scattering of white light, which contains wavelengths in the range from 400 nm for blue to 750 nm for red, off the nitrogen and oxygen molecules in the atmosphere with respective sizes $d = 0.11$ and 0.12 nm, explains why the light reflected from the sky during the day appears blue, and that transmitted by the atmosphere at sunrise and sunset appears red.

Particle size determinations are made using a monochromatic (single-wavelength) laser beam scattered at a particular angle (usually $90°$) for parallel and perpendicular polarizations. The detected intensities can provide the particle size, the particle concentration, and the index of refraction. The Rayleigh–Gans theory is used to interpret the data for particles with sizes $d < 0.1\lambda$, which corresponds to the case for

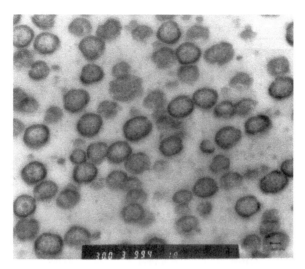

Figure 2.6. Transmission electron micrograph of polyaniline nanoparticles in a polymer matrix. (From B. Wessling, *Handbook of Organic Conductive Molecules and Polymers*, H. S. Nalwa, ed., Wiley, New York, 1997, Vol. 3, pp. 597–632.)

nanoparticles measured by optical wavelengths. The example of a laser-beam nano-particle determination shown in Fig. 2.7 shows that an organic solvent dispersion with sizes ranging from 9 to 30 nm, peaked at 12 nm. This method is applicable for use with nanoparticles that have diameters above 2 nm; for smaller nanoparticles other methods must be used.

The sizes of particles smaller than 2 nm can be conveniently determined by the method of mass spectrometry. The classical gas mass spectrometer sketched in Fig. 2.8 ionizes the nanoparticles to form positive ions by impact from electrons emitted by the heated filament (F) in the ionization chamber (I) of the ion source. The newly formed ions are accelerated through the potential drop in voltage V between the repeller (R) and accelerator (A) plates, then focused by lenses L, and collimated by slits S during their transit to mass analyzer. The magnetic field B of the mass analyzer, oriented normal to the page, exerts the force $F = qvB$, which bends the ion beam through an angle of 90° at the radius r, after which they are detected at the ion collector. The mass m:charge q ratio is given by the expression

$$\frac{m}{q} = \frac{B^2 r^2}{2V}.$$

(2.7)

The bending radius r is ordinarily fixed in a particular instrument, so either the magnetic field B or the accelerating voltage V can be scanned to focus the ions of various masses at the detector. The charge q of the nanoion is ordinarily known, so in practice it is the mass m that is determined. Generally the material forming the nanoparticle is known, so its density $\rho = m/V$ is also a known quantity, and its size or linear dimension d can be estimated or evaluated as the cube root of the volume: $d = (V)^{1/3} = (m/\rho)^{1/3}$.

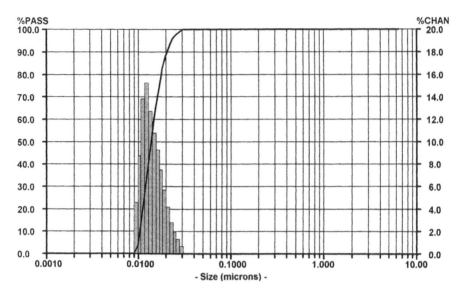

Figure 2.7. Laser light Doppler measurement of size distribution of conductive polymer particles dispersed in an organic solvent. The sizes range from 9 to 30 nm, with a peak at 12 nm. (From B. Wessling, in *Handbook of Nanostructured Materials and Nanotechnology*, H. S. Nalwa, ed., Academic Press, Boston, 2000, Vol. 5, Chapter 10, p. 508.)

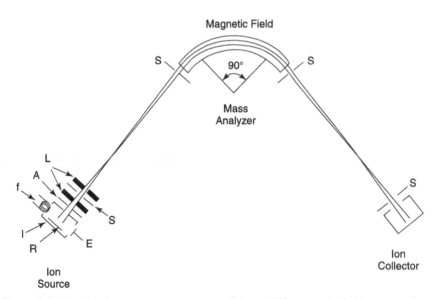

Figure 2.8. Sketch of a mass spectrometer utilizing a 90° magnetic field mass analyzer, showing details of the ion source: A—accelerator or extractor plate, E—electron trap; F—filament; I—ionization chamber; L—focusing lenses; r—ion bending radius; R—repeller; S—slits. The magnetic field of the mass analyzer is perpendicular to the plane of the page.

The mass spectrometer that has been described made use of the classic magnetic field mass analyzer. Modern mass spectrometers generally employ other types of mass analyzers, such as the quadrupole model, or the time-of-flight type, in which each ion acquires the same kinetic energy $\frac{1}{2}mv^2$ during its acceleration out of the ionization chamber, so the lighter-mass ions move faster and arrive at the detector before the heavier ions do, thereby providing a separation by mass.

Figure 2.9 gives an example of a time-of-flight mass spectrum obtained from soot produced by laser vaporization of a lanthanum–carbon target. The upper mass spectrum (a) of the figure, taken from the initial crude extract of the soot, shows

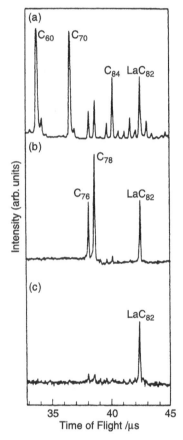

Figure 2.9. Time-of-flight mass spectra from soot produced by laser vaporization of a lanthanum–carbon target, showing the presence of the fullerene molecules C_{60}, C_{70}, C_{76}, C_{78}, C_{82}, C_{84}, and LaC_{84}. The spectra correspond, successively, to (a) the initial crude soot extract, (b) a fraction after passage through a chromatographic column, and (c) a second fraction after passage of the first fraction through another column to isolate and concentrate the endohedral compound LaC_{84}. [From K. Kikuchi, S. Suzuki, Y. Nakao, N. Yakahara, T. Wakabayashi, I. Ikemoto, and Y. Achiba, *Chem. Phys. Lett.* **216**, 67 (1993).]

lines from several fullerene molecules: C_{60}, C_{70}, C_{76}, C_{78}, C_{82}, C_{84}, and LaC_{84}. The last molecule in this group corresponds to an endohedral fullerene, namely, C_{84} with a lanthanum atom inside the fullerene cage. The second (b) and third (c) mass spectra were obtained by successively separating LaC_{84} from the other fullerenes using a procedure called *high-performance liquid chromatography* (HPLC).

2.2.4. Surface Structure

To obtain crystallographic information about the surface layers of a material, a technique called *low-energy electron diffraction* (LEED) can be employed because at low energies (10–100 eV) the electrons penetrate only very short distances into the surface, so their diffraction pattern reflects the atomic spacings in the surface layer. If the diffraction pattern arises from more than one surface layer, then the contribution of lower-lying crystallographic planes will be weaker in intensity. The electron beam behaves like a wave and reflects from crystallographic planes in analogy with an X-ray beam, and its wavelength λ, called the *de Broglie wavelength*, depends on the energy E expressed in the units of electron volts through the expression

$$\lambda = \frac{1.226}{E^{1/2}} \text{ nm} \tag{2.8}$$

which differs from Eq. (2.1) for X rays. Thus an electron energy of 25.2 eV gives a de Broglie wavelength λ equal to the Ga–As bond distance in gallium arsenide ($3^{1/2}a/4 = 0.244$ nm), where the lattice constant $a = 0.565$ nm, so we see that low energies are adequate for crystallographic electron diffraction measurements. Another technique for determining surface layer lattice constants is *reflection high-energy electron diffraction* (RHEED) carried out at grazing incidence angles where the surface penetration is minimal. When θ is small in the Bragg expression (2.2), then λ must be small, so the energy E of Eq. (2.8) must be large; hence the need for higher energies for applying RHEED diffraction at grazing incidence.

The thicknesses of surface layers can also be determined optically. A beam of light reflected from successive interfaces will be in-phase and reinforce if the interfaces are separated by an integral number of wavelengths. This property of reflected beams of light can be utilized to evaluate the distances between layers of atoms near the surface. Thicknesses down to <1 nm can be determined by this method.

The atomic constitution of surfaces can be probed by the Auger process, which is described at the end of Section 2.4.2. The energies of emitted Auger electrons are characteristic of the atoms from which they originate, so they identify which ones are in a surface layer.

2.3. MICROSCOPY

2.3.1. Transmission Electron Microscopy

Electron beams not only are capable of providing crystallographic information about nanoparticle surfaces but can also be used to produce images of the surface, and they play this role in electron microscopes. We will discuss several ways to use electron beams for the purpose of imaging, namely, several types of electron microscopes.

In a TEM the electrons from a source such as an electron gun enter the sample, are scattered as they pass through it, are focused by an objective lens, are amplified by a magnifying (projector) lens, and finally produce the desired image, in the manner reading from left to right (conventional TEM direction) in Fig. 2.10. The wavelength of the electrons in the incident beam is given by a modified form of Eq. (2.8)

$$\lambda = \frac{0.0388}{V^{1/2}} \text{ nm} \tag{2.9}$$

where the energy acquired by the electrons is $E = eV$, and V is the accelerating voltage expressed in the units kilovolt. If widely separated heavy atoms are present, they can dominate the scattering, with average scattering angles θ given by the expression $\theta \approx \lambda/d$, where d is the average atomic diameter. For an accelerating voltage of 100 kV and an average atomic diameter of 0.15 nm, we obtain $\theta \approx$ 0.026 radians or 1.5°. Images are formed because different atoms interact with and absorb electrons to a different extent. When individual atoms of heavy elements are further apart than several lattice parameters, they can sometimes be resolved by the TEM technique.

Electrons interact much more strongly with matter than do X rays or neutrons with comparable energies or wavelengths. For ordinary elastic scattering of 100-keV electrons the average distance traversed by electrons between scattering events, called the *mean free path*, varies from a few dozen nanometers for light elements to tens or

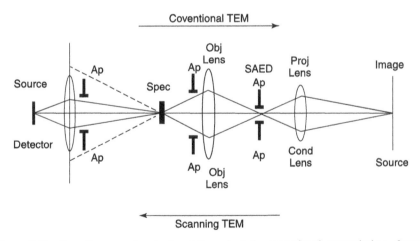

Figure 2.10. Ray diagram of a horizontally oriented conventional transmission electron microscope (top path) and of a scanning transmission electron microscope (bottom path). The selected area electron diffraction (SAED) aperture (Ap) and the sample or specimen (Spec) are indicated, as well as the objective (Obj) and projector (Proj) or condenser (Cond) lenses. (Adapted from P. R. Buseck, J. M. Cowley, and L. Eyring, *High-Resolution Transmission Electron Microscopy*, Oxford Univ. Press, New York, 1988, p. 6.)

perhaps hundreds of nanometers for heavy elements. The best results are obtained in electron microscopy by using film thicknesses comparable to the mean free path. Much thinner films exhibit too little scattering to provide useful images, and in thick films multiple scattering events predominate, blurring the image and rendering it difficult to interpret. Thick specimens can be studied by detecting backscattered electrons.

A TEM can form images by the use of the selected-area electron diffraction (SAED) aperture located between the objective and projector lenses in Fig. 2.10. The main part of the electron beam transmitted by the sample consists of electrons that have undergone no scattering. The beam also contains electrons that have lost energy through inelastic scattering with no deviation of their paths, and electrons that have been reflected by various *hkl* crystallographic planes. To produce a *bright-field image*, the aperture is inserted to allow only the main undeviated transmitted electron beam to pass, as shown in Fig. 2.11. Note that the electron-beam direction, which is depicted horizontally in Fig. 2.10, is directed from top to bottom in Fig. 2.11. We see that the brightfield image appears at the BF electron detector or viewing screen. If the aperture is positioned to select only one of the beams reflected from a particular *hkl* plane, the result is the generation of a darkfield image at one of the DF detectors or viewing screens. The details of the darkfield image that is formed

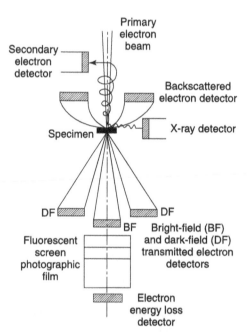

Figure 2.11. Positioning of signal detectors in vertically oriented electron microscope column. (From D. B. Williams, *Practical Analytical Electron Microscopy in Materials Science*, Phillips Electronic Instruments, Mahwah, NJ, 1984.)

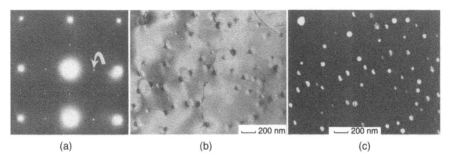

(a)   (b)   (c)

Figure 2.12. Transmission electron microscope figures from 2–3-nm-diameter γ' precipitates of $Ni_2(Ti,Al)$ in iron-base superalloy, showing (a) [100] fcc zone diffraction pattern displaying a regular arrangement of large and medium-sized bright spots from the superalloy, plus weak, very dim, spots from the γ' precipitates; (b) brightfield image showing 25-nm elastic strain fields around marginally visible γ' phase particles; (c) darkfield image obtained using an SAED aperture to pass the γ' diffraction spot marked with the arrow in (a). Image (c) resolves the γ' precipitate particles. (From T. J. Headley, cited in A. D. Romig, Jr., chapter in *Materials Characterization*, Vol. 10 of *Metals Handbook*, R. E. Whan, ed., American Society for Metals, Metals Park, OH, 1986, p. 442.)

can depend on the particular diffracted beam (particular *hkl* plane) that is selected for the imaging.

To illustrate this imaging technique, in Fig. 2.12 we present images of an iron base superalloy with a fcc austenite structure containing 2–3-nm γ' precipitates of $Ni_3(Ti, Al)$ with a fcc structure. The diffraction pattern in Fig. 2.12a obtained without using a filter exhibits regularly spaced large and medium-sized bright spots from the super-alloy, and very small, barely visible, dim spots from the γ' nanoparticles. In the brightfield image displayed in Fig. 2.12b the γ' particles are barely visible, but the 25-nm-diameter elastic strain fields generated by them are clearly seen. When the very dim γ' diffraction spot indicated by the arrow in Fig. 2.12a is selected by the SAED aperture for passage of the beam, the resulting darkfield image presented in Fig. 2.12c shows very clearly the positions of the various γ' precipitates.

A technique called *image processing* can be used to increase the information obtainable from a TEM image, and enhance some features that are close to the noise level. If the image is Fourier-transformed by a highly efficient technique called a *fast Fourier transform*, then it provides information similar to that in the direct-diffraction pattern. An example of the advantages of image processing is given by the sequence of images presented in Fig. 2.13 for a Ni nanoparticle supported on a SiO_2 substrate. Figure 2.13a shows the original image, and Fig. 2.13b presents the fast Fourier transform, which has the appearance of a diffraction pattern. Figures 2.13c–2.13e illustrate successive steps in the image processing, and Fig. 2.13f is an image of the SiO_2 substrate obtained by subtracting the particle image. Finally, Fig. 2.13g presents the nanoparticle reconstruction from the processed data. Note that the atom positions on the face at the upper right of the reconstructed nanoparticle correspond to a planar square lattice of the type found on the [100],

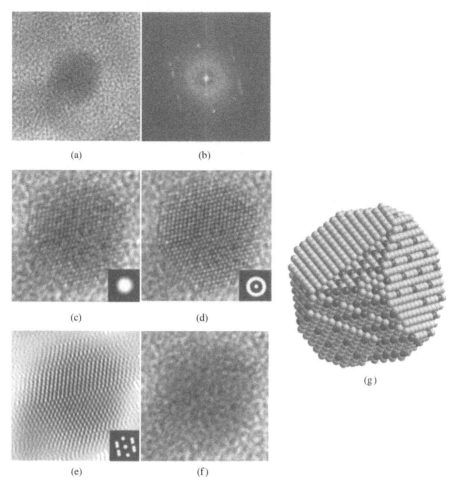

Figure 2.13. Transmission electron microscope image processing for a Ni particle on a SiO_2 substrate, showing (a) original brightfield image; (b) fast Fourier transform diffraction pattern type image; (c) processed image with aperture filter shown in inset; (d) image after further processing with aperture filter in the inset; (e) final processed image; (f) image of SiO_2 substrate obtained by subtracting out the particle image; (g) model of nanoparticle constructed from the processed data. (From M. José-Yacamán and J. A. Ascencio, in *Handbook of Nanostructured Materials and Nanotechnology*, H. S. Nalwa, ed., Academic Press, Boston, 2000, Vol. 2, Chapter 8, p. 405.)

[010], and [001] faces of the nanoparticle depicted in Fig. 1.9, and the triangular top face of the reconstructed nanopartilce is the hexagonal lattice arrangement present on the [111] and analogous faces of Fig. 1.9.

In addition to the directly transmitted and the diffracted electrons of the electron microscope, there are additional electrons in the beam that undergo inelastic scattering and lose energy by creating excitations in the specimen. This can occur by inducing vibrational motions in the atoms near their path, and thereby creating phonons or quantized lattice vibrations that will propagate through the crystal. If the sample is a metal, then the incoming electron can scatter inelastically by producing a *plasmon*, which is a collective excitation of the free-electron gas in the conduction band. A third, very important, source of inelastic scattering occurs when the incoming electron induces a single-electron excitation in an atom. This might involve inner-core atomic levels of atoms, such as inducing a transition from a *K* level ($n = 1$) or *L* level ($n = 2$) of the atom to a higher energy, which might be a discrete atomic level with a larger quantum number n, or the result might be the total removal (ionization) of the electron from the solid. Lower levels of energy loss occur when the excited electron is in the valence band of a semiconductor. This excitation can decay via the return of excited electrons to their ground states, thereby producing secondary radiation, and the nature of the secondary radiation can often give useful information about the sample. These types of transition are utilized in various branches of electron spectroscopy. They can be surface-sensitive because of the short penetration distance of the electrons into the material.

2.3.2. Field Ion Microscopy

Another instrumental technique in which the resolution approaches interatomic dimensions is field ion microscopy. In a field ion microscope a wire with a fine tip located in a high vacuum chamber is given a positive charge. The electric field and electric field gradient in the neighborhood of the tip are both quite high, and residual gas molecules that come close to the tip become ionized by them, transferring electrons to the tip, thereby acquiring a positive charge. These gaseous cations are repelled by the tip and move directly outward toward a photographic plate, where, upon impact, they create spots. Each spot on the plate corresponds to an atom on the tip, so the distribution of dots on the photographic plate represents a highly enlarged image of the distribution of atoms on the tip. Figure 2.14 presents a field ion micrograph from a tungsten tip, and Fig. 2.15 provides a stereographic projection of a cubic crystal with the orientation corresponding to the micrograph of Fig. 2.14. The locations of the various crystallographic planes identified on the projection are characteristic of the orientation and structure of the tungsten tip. The *International Tables for Crystallography*[4] provide stereographic projections for various point groups and crystal classes.

2.3.3. Scanning Microscopy

An efficient way to obtain images of the surface of a specimen is to scan the surface with an electron beam in a raster pattern, similar to the way an electron gun scans the screen in a television set. This is a systematic-type scan. Surface information is also obtained by a scanning probe in which the trajectory of the electron beam is directed across the regions of particular interest on the surface. The scanning can also be carried out by a probe that monitors electrons that tunnel between the surface and the probe tip, or by a probe that monitors the force exerted between

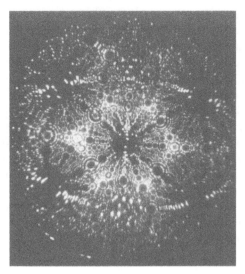

Figure 2.14. Field ion micrograph of a tungsten tip (T. J. Godfrey), explained by the stereographic projection of Fig. 2.15. (From G. D. W. Smith, chapter in *Materials Characterization*, Vol. 10 of *Metals Handbook*, R. E. Whan, ed., American Society for Metals, Metals Park, OH, 1986, p. 585.)

the surface and the tip. We shall describe in turn the instrumentation systems that carry out these three respective functions, namely, the scanning transmission electron microscope (SEM), the scanning tunneling microscope (STM), and the atomic force microscope (AFM).

It was mentioned above that the electron optics sketched in Fig. 2.10 for a conventional transmission electron microscope is similar to that of a scanning electron microscope, except that in the former TEM case the electrons travel from left to right, and in the latter SEM they move in the opposite direction, from right to left, as indicated in the figure. Since the workings of an electron microscope have been discussed widely, we will content ourselves with describing the electron deflection system of a scanning electron microscope, which is sketched in Fig. 2.16. The deflection is done magnetically through magnetic fields generated by electric currents flowing through coils, as occurs in many television sets. The magnetic field produced by a coil is proportional to the voltage V applied to the coil. We see from the upper inset on the left side of Fig. 2.16 that a sawtooth voltage is applied to the pairs of coils I_1, I_1 and I_2, I_2. The magnetic field produced by the coils exerts a force that deflects the electron beam from left to right along the direction of the line drawn at the bottom on the sample. The varying magnetic fields in the coil pairs f_1, f_1 and f_2, f_2 produce the smaller deflections from points 1 to 1' to 1'' shown in the detail A inset. Thus the electron beam scans repeatedly from left to right across the sample in a raster pattern that eventually covers the entire $r \times r$ frame area on the sample. Figure 2.17 shows 3-nm gold particles on a carbon substrate resolved by an SEM.

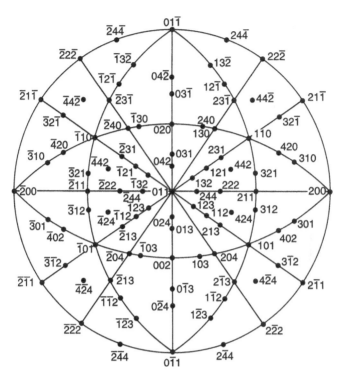

Figure 2.15. Stereographic [011] projection of a cubic crystal corresponding to the field ion tungsten micrograph in Fig. 2.14. (From G. D. W. Smith, chapter in *Materials Characterization*, Vol. 10 of *Metals Handbook*, R. E. Whan, ed., American Society for Metals, Metals Park, OH, 1986, p. 583.)

A scanning tunneling microscope (STM) utilizes a wire with a very fine point. This fine point is positively charged and acts as a probe when it is lowered to a distance of about 1 nm above the surface under study. Electrons at individual surface atoms are attracted to the positive charge of the probe wire and jump (tunnel) up to it, thereby generating a weak electric current. The probe wire is scanned back and forth across the surface in a raster pattern, in either a constant-height mode or a constant-current mode, as sketched in Fig. 2.18. In the constant-current mode a feedback loop maintains a constant probe height above the sample surface profile, and the up/down probe variations are recorded. This mode of operation assumes a constant tunneling barrier across the surface. In the constant-probe-height mode the tip is constantly changing its distance from the surface, and this is reflected in variations of the recorded tunneling current as the probe scans. The feedback loop establishes the initial probe height, and is then turned off during the scan. The scanning probe provides a mapping of the distribution of the atoms on the surface.

The STM often employs a piezoelectric tripod scanner, and an early design of this scanner built by Binning and Rohrer is depicted in Fig. 2.19. A *piezoelectric* is a

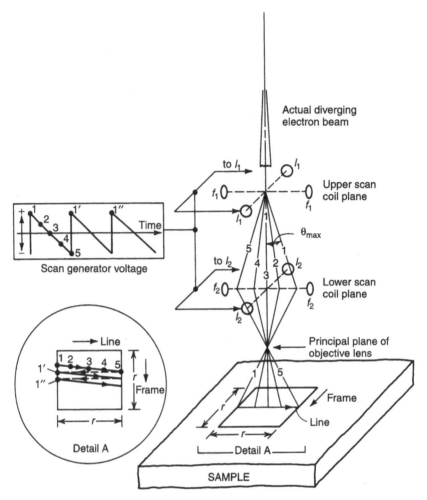

Figure 2.16. Double-deflection system of a scanning electron microscope. When the upper scan coils I_1–I_1 deflect the beam through an angle θ, the lower scan coils I_2–I_2 deflect it back through the angle 2θ, and the electrons sequentially strike the sample along the indicated line. The upper inset on the left shows the sawtooth voltage, which controls the current through the I_i scanning coils. The lower inset on the left shows the sequence of points on the sample for various electron trajectories 1,2,3,4,5 downward along the microscope axis. Scan coils f_1–f_1 and f_2–f_2 provide the beam deflection for the sequence of points 1, 1′, 1″ shown in detail A. (From J. D. Verhoeven, chapter in *Materials Characterization*, Vol. 10 of *Metals Handbook*, R. E. Whan, ed., American Society for Metals, Metals Park, OH, 1986, p. 493.)

material in which an applied voltage elicits a mechanical response, and the reverse. Applied voltages induce piezotransducers to move the scanning probe (or the sample) in nanometer increments along the x, y or z directions indicated on the arms (3) of the scanner in the figure. The initial setting is accomplished with

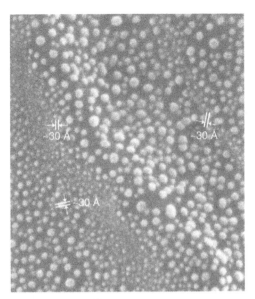

Figure 2.17. Micrograph of 3-nm (30-Å)-diameter gold particles on a carbon substrate taken with a scanning electron microscope. (From J. D. Verhoeven, chapter in *Materials Characterization*, Vol. 10 of *Metals Handbook*, R. E. Whan, ed., American Society for Metals, Metals Park, OH, 1986, p. 497.)

the aid of a stepper motor and micrometer screws. The tunneling current, which varies exponentially with the probe–surface atom separation, depends on the nature of the probe tip and the composition of the sample surface. From a quantum-mechanical perspective, the current depends on the dangling bond state of the tip apex atom and on the orbital states of the surface atoms.

The third technique for nanostructure surface studies is atomic force microscopy, and Fig. 2.20 is a diagram of a representative atomic force microscope (AFM). The fundamental difference between the STM and the AFM is that the former monitors the electric tunneling current between the surface and the probe tip, and the latter monitors the force exerted by the surface on the probe tip. The AFM, like the STM, has two modes of operation. The AFM can operate in a close contact mode in which the core–core repulsive forces with the surface predominate, or in a greater separation "noncontact" mode in which the relevant force is the gradient of the van der Waals potential. As in the STM case, a piezoelectric scanner can be used. The vertical motions of the tip during scanning may be monitored by the interference pattern of a light beam from an optical fiber, as shown in the upper diagram of the figure, or by the reflection of a laser beam, as shown in the enlarged view of the probe tip in the lower diagram of the figure. The AFM is sensitive to the vertical component of the surface forces. A related, more versatile, device called a *friction force microscope*, sometimes referred to as a *lateral force microscope*, simultaneously measures both normal and lateral forces of the surface on the tip.

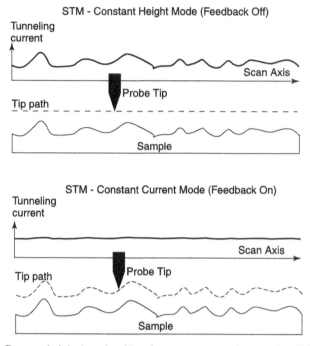

Figure 2.18. Constant-height (top sketch) and constant-current (bottom sketch) imaging mode of a scanning tunneling microscope. (From T. Bayburt, J. Carlson, B. Godfrey, M. Shank-Retzlaff, and S. G. Sligar, in *Handbook of Nanostructured Materials and Nanotechnology*, H. S. Nalwa, ed., Academic Press, Boston, 2000, Vol. 5, Chapter 12, p. 641.)

All three of these scanning microscopes are able to provide information on the topography and defect structure of a surface over distances close to an atomic scale. Figure 2.21 is a three-dimensional rendering of an AFM image of chromium deposited on a surface of SiO_2. The surface was prepared by the laser-focused

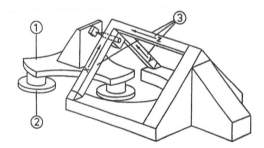

Figure 2.19. Scanning mechanism for scanning tunneling microscope, showing (1) the piezoelectric baseplate, (2) the three feet of the base plate, and (3) the piezoelectric tripod scanner holding the probe tip that points toward the sample. (From R. Wiesendanger, *Scanning Probe Microscopy and Spectroscopy*, Cambridge Univ. Press, 1994, p. 81.)

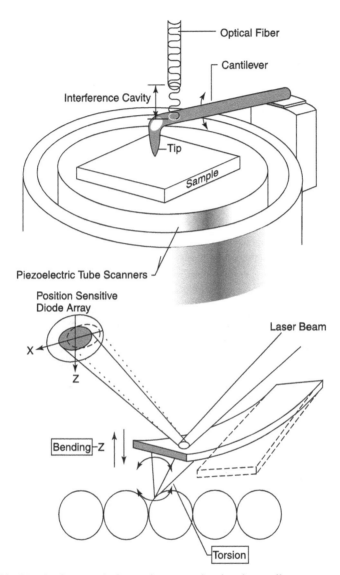

Figure 2.20. Sketch of an atomic force microscope showing the cantilever arm provided with a probe tip that traverses the sample surface through the action of the piezoelectric scanner. The upper figure shows an interference deflection sensor, and the lower enlarged view of the cantilever and tip is provided with a laser-beam deflection sensor. The sensors monitor the probe tip elevations upward from the surface during the scan.

deposition of atomic chromium in the presence of a Gaussian standing wave that reproduced the observed regular array of peaks and valleys on the surface. When the laser-focused chromium deposition was carried out in the presence of two plane waves displaced by 90° relative to each other, the two-dimensional arrangement

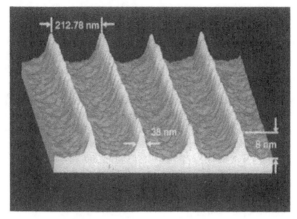

Figure 2.21. Three-dimensional rendering of an AFM image of nanostructure formed by laser-focused atomic Cr deposition in a Gaussian standing wave on an SiO_2 surface. [From J. J. McClelland, R. Gupta, Z. J. Jabbour, and R. L. Celotta, *Austral. J. Phys.* **49**, 555 (1996).]

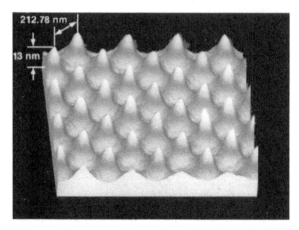

Figure 2.22. AFM image of nanostructure array formed when the laser-focused Cr deposition is carried out in two standing waves oriented at 90° relative to each other. [From R. Gupta, J. J. McClelland, Z. J. Jabbour, and R. L. Celotta, *Appl. Phys. Lett.* **67**, 1378 (1995).]

AFM image shown in Fig. 2.22 was obtained. Note that the separation between the peaks, 212.78 nm, is the same in both images. The peak heights are higher (13 nm) in the two-dimensional array than they are (8 nm) in the linear one.

2.4. SPECTROSCOPY

2.4.1. Infrared and Raman Spectroscopy

Vibrational spectroscopy involves photons that induce transitions between vibrational states in molecules and solids, typically in the infrared (IR) frequency range from 2 to

12×10^{13} Hz. Section 1.1.6 discusses the normal modes of vibration of molecules and solids. The energy gaps of many semiconductors are in this same frequency region and can be studied by IR techniques.

In IR spectroscopy an IR photon $h\nu$ is absorbed directly to induce a transition between two vibrational levels E_n and $E_{n'}$, where

$$E_n = \left(n + \frac{1}{2}\right) h\nu_0 \tag{2.10}$$

The vibrational quantum number $n = 0, 1, 2, \ldots$ is a positive integer, and ν_0 is the characteristic frequency for a particular normal mode. Owing to the selection rule $\Delta n = \pm 1$, IR transitions are observed only between adjacent vibrational energy levels, and hence have the frequency ν_0. In Raman spectroscopy a vibratonal transition is induced when an incident optical photon of frequency $h\nu_{\mathrm{inc}}$ is absorbed and another optical photon $h\nu_{\mathrm{emit}}$ is emitted:

$$E_n = |h\nu_{\mathrm{inc}} - h\nu_{\mathrm{emit}}| \tag{2.11}$$

From Eq. (2.10) the frequency difference is given by $|\nu_{\mathrm{inc}} - \nu_{\mathrm{emit}}| = |n' - n''|\nu_0 = \nu_0$ since the same IR selection rule $\Delta n = \pm 1$ is obeyed. Two cases are observed: $\nu_{\mathrm{inc}} > \nu_{\mathrm{emit}}$ corresponding to a *Stokes line* and $\nu_{\mathrm{inc}} < \nu_{\mathrm{emit}}$ for an *anti-Stokes line*. Infrared active vibrational modes arise from a change in the electric dipole moment μ of the molecule, while Raman active vibrational modes involve a change in the polarizability $P = \mu_{\mathrm{ind}}/E$, where the electric vector E of the incident light induces the dipole moment μ_{ind} in the sample. Some vibrational modes are IR active, that is, measurable by IR spectroscopy, and some are Raman active.

Infrared and optical spectroscopy is often carried out by reflection, and the measurements of nanostructures provide the reflectance (or reflectivity) R, which is the fraction of reflected light. For normal incidence we have

$$R = \frac{I_{\mathrm{r}}}{I_0} = \frac{|\varepsilon^{1/2} - 1|}{|\varepsilon^{1/2} + 1|} \tag{2.12}$$

where ε is the dimensionless dielectric constant of the material. The dielectric constant $\varepsilon(\nu)$ has real and imaginary parts, $\varepsilon = \varepsilon'(\nu) + \varepsilon''(\nu)$, where the real or dispersion part ε' provides the frequencies of the IR bands and the imaginary part ε'' measures the energy absorption or loss. A technique called *Kramers–Kronig analysis* is used to extract the frequency dependences of $\varepsilon'(\nu)$ and $\varepsilon''(\nu)$ from measured IR reflection spectra.

The classical way to carry out IR spectroscopy is to scan the frequency of the incoming light so that the detector can record changes in the light intensity for those frequencies at which the sample absorbs energy. A major disadvantage of this method is that the detector records meaningful information only while the scan is passing through absorption lines, while most of the time is spent scanning between lines when the detector has nothing to record. To overcome this deficiency, modern IR spectrometers irradiate the sample with a broad band of frequencies simultaneously, and then carry out a mathematical analysis of the resulting signal called a *Fourier transformation* to convert the detected signal back into the classical form of the spectrum. The resulting signal is called a *Fourier transform infrared* (FTIR) spectrum. The Fourier transform technique is also widely used in nuclear magnetic resonance, to be discussed below, and in other branches of spectroscopy. Its initial use was in X-ray diffraction crystallography.

Figure 2.23 presents an FTIR spectrum of silicon nitride (Si_3N_4) nanopowder showing the vibrational absorption lines corresponding to the presence of hydroxyl (Si—OH), amino (Si—NH_2), and imido (Si—NH—Si) groups on the surface. Figure 2.24 shows a similar FTIR spectrum of silicon carbonitride nanopowder (SiCN), revealing (curve a) the presence of the chemical species Si—OH, NH, and CH_x on the surface after activation at 873 K, and (curve b) their removal by heating for an hour under dry oxygen at 773 K. The latter treatment brings about the presence of carbonyl (C=O) on the surface. Infrared measurememts of nanostructures are discussed in more detail in Chapter 7.

Figure 2.25 shows how the width of the Raman spectral lines of germanium nanocrystals embedded in SiO_2 thin films exhibit a pronounced broadening when the particle size decreases below about 20 nm. The frequencies also shift at small particle sizes. The reasons for these effects are discussed in Chapter 7. Figure 2.26 shows a Raman spectrum recorded for germanium nanocrystals that arises from a distribution of particle sizes around an average value of 6.5 nm. These nanocrystals

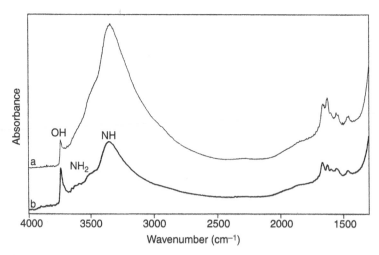

Figure 2.23. Fourier transform IR spectrum of silicon nitride nanopowder recorded (a) at room temperature under vacuum and (b) after activation at 773 K. (From G. Ramis, G. Busca, V. Lorenzelli, M.-I. Baraton, T. Merle-Méjean, and P. Quintard, in *Surfaces and Interfaces of Ceramic Materials*, L. C. Dufour et al., eds., Kluwer Academic Publishers, Dordrecht, 1989, p. 173.)

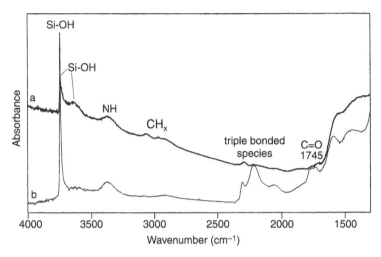

Figure 2.24. Fourier transform IR spectrum of silicon carbonitride nanopowder recorded (a) after activation at 873 K and (b) after subsequent heating in dry oxygen at 773 K for one hour. [From M.-I. Baraton, in *Handbook of Nanostructured Materials and Nanotechnology*, H. S. Nalwa, ed., Academic Press, Boston, 2000, Vol. 2, Chapter 2, p. 131; see also M.-I. Baraton, W. Chang, and B. H. Kear, *J. Phys. Chem.* **100**, 16647 (1996).]

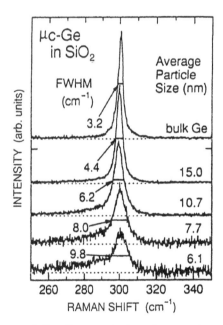

Figure 2.25. Dependence of the Raman spectra of germanium microcrystals (μ_C-Ge) embedded in SiO_2 thin films on the crystallite size. The average particle size and the full width at half-maximum height (FWHM) of the Raman line of each sample are indicated. [From M. Fujii, S. Hayashi, and K. Yamamoto, *Jpn. J. Appl. Phys.* **30**, 657 (1991).]

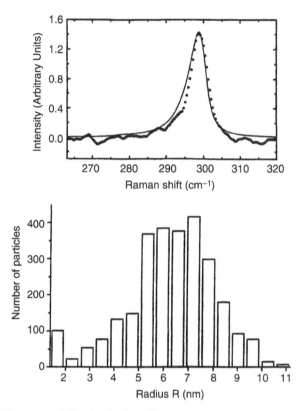

Figure 2.26. Histogram of the distribution of Ge nanocrystal particle sizes (lower figure) around the mean radius 6.5 nm, used to simulate the Raman spectrum depicted at the top. The size distribution was obtained by transmission electron microscopy. [From C. E. Bottani, C. Mantini, P. Milani, M. Manfredini, A. Stella, P. Tognini, P. Cheyssac, and R. Kofman, *Appl. Phys. Lett.* **69**, 2409 (1996).]

were prepared by chemical reduction, precipitation, and subsequent annealing of the phase $Si_xGe_yO_z$, and the histogram shown below the spectrum was determined from a scanning electron micrograph (SEM). Further Raman spectroscopy studies demonstrated that larger average particle sizes are obtained when the annealing is carried out for longer times and at higher temperatures.

We have been discussing what is traditionally considered as Raman scattering of light or Raman spectroscopy. This is spectroscopy in which the phonon vibration of Eq. (2.10), corresponding to the energy difference of Eq. (2.11), is an optical phonon of the type discussed in the previous section, namely, a phonon with a frequency of vibration in the IR region of the spectrum, corresponding to about $\approx 400\,\text{cm}^{-1}$, or a frequency of $\approx 1.2 \times 10^{13}$ Hz. When a low-frequency acoustical phonon is involved in the scattering of Eq. (2.11), then the process is referred to as *Brillouin scattering*. Acoustic phonons can have frequencies of vibration or energies that are a factor of 1000 less than those of optical phonons. Typical values are $\approx 1.5 \times 10^{10}$ Hz or

$\approx 0.5 \, \text{cm}^{-1}$. Brillouin spectroscopy, which involves both Stokes and anti-Stokes lines, as does Raman spectroscopy, is discussed in Section 7.3.

Since Chapter 7 covers the IR and optical spectroscopy of nanomaterials in some detail, in the present chapter we restrict ourselves to commenting on the representative optical spectra presented in Fig. 2.27, which were obtained from colloidal cadmium selenide semiconductor nanocrystals. These materials are transparent for photon energies below the bandgap, and absorb light above the gap. Colloidal nanocrystal synthesis provides a narrow range of size distributions, so optical spectra can accurately determine the dependence of the gap on the average particle radius, and this dependence is shown in Fig. 2.28 for CdSe. We see that the energy gap data for the colloidal samples plotted on this figure agree with the gaps evaluated from optical spectra of CdSe nanoparticle glass samples. The solid line in the figure is a theoretical plot made using a parabolic band model for the vibrational potential,

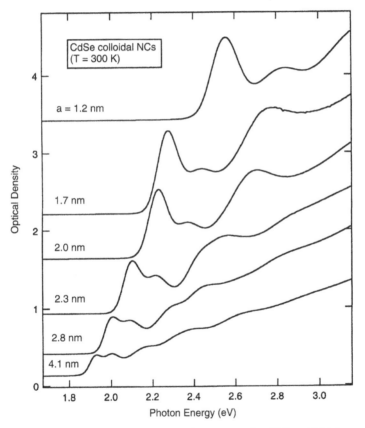

Figure 2.27. Room-temperature optical absorption spectra for CdSe colloidal nanocrystals (NCs) with mean radii in the range from 1.2 to 4.1 nm. (From V. I. Klimov, in *Handbook of Nanostructured Materials and Nanotechnology*, H. S. Nalwa, ed., Academic Press, Boston, 2000, Vol. 4, Chapter 7, p. 464.)

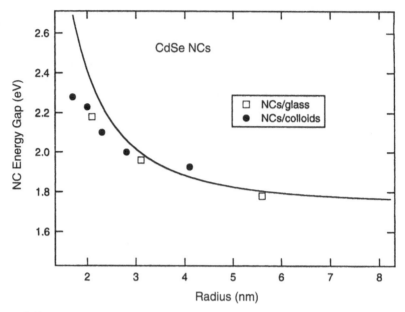

Figure 2.28. Dependence of the semiconductor energy gap of CdSe nanocrystals (NCs), in the form of a glass (□) and a colloidal suspension (•), on the mean particle radius. The colloidal nanocrystal spectral data are presented in Fig. 2.27. The solid curve gives the fit to the data using a parabolic band model for the vibrational potential. (From V. I. Klimov, in *Handbook of Nanostructured Materials and Nanotechnology*, H. S. Nalwa, ed., Academic Press, Boston, 2000, Vol. 4, Chapter 7, p. 465.)

and the measured energy is asymptotic to the bulk value for large particle sizes. The increase in the gap for small nanoparticle radii is due to quantum confinement effects, discussed in greater detail in Section 9.3.

2.4.2. Photoemission, X-Ray, and Auger Spectroscopy

Photoemission spectroscopy (PES) measures the energy distribution of electrons emitted by atoms and molecules in various charge and energy states. A material irradiated with ultraviolet light (UPS) or X rays (XPS) can emit electrons called *photoelectrons* from atomic energy levels with a kinetic energy equal to the difference between the photon energy $h\nu_{ph}$ and the ionization energy E_{ion}, which is the energy required to completely remove an electron from an atomic energy level

$$\mathrm{KE} = h\nu_{ph} - E_{ion} \tag{2.13}$$

The photoelectron spectrometer sketched in Fig. 2.29 shows, at the lower left side, X-ray photons incident on a specimen. The emitted photoelectrons then pass through a velocity analyzer that allows electrons within only a very narrow range of velocities

X-ray photoelectron spectrometer

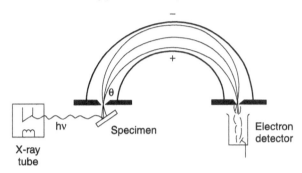

Figure 2.29. X-Ray photoelectron spectrometer showing photons *hv* generated by an X-ray tube incident on the specimen where they produce photoelectrons e characteristic of the specimen material, which then traverse a velocity analyzer, and are brought to focus at an electron detector that measures their kinetic energy.

to move along trajectories that take them from the entrance slit on the left, allowing them to pass through the exit slit on the right and then impinge on the detector. The detector measures the number of emitted electrons having a given kinetic energy, and this number is appreciable for kinetic energies for which Eq. (2.13) is satisfied.

The energy states of atoms or molecular ions in the valence band region have characteristic ionization energies that reflect perturbations by the surrounding lattice environment, so this environment is probed by the measurement. Related spectroscopic techniques such as inverse photoelectron spectroscopy (IPS), bremsstrahlung isochromat spectroscopy (BIS), electron energy loss spectroscopy (EELS), and Auger electron spectroscopy, provide similar information.

As an example of the usefulness of X-ray photoemission spectroscopy, the ratio of Ga to N in a GaN sample was determined by measuring the gallium $3d$ XPS peak at 1.1185 keV and the nitrogen $1s$ peak at 0.3975 keV, and the result gave the average composition $Ga_{0.95}N$. An XPS study of 10-nm, InP provided the indium $3d_{5/2}$ asymmetric line shown in Fig. 2.30a, and this was analyzed to reveal the presence of two superimposed lines, a main one at 444.6 eV arising from In in InP, and a weaker one at 442.7 attributed to In contained in the oxide In_2O_3. The phosphorus $2p$ XPS spectrum of Fig. 2.30b exhibits two well-resolved phosphorus peaks, one from InP and one from a phosphorus–oxygen species.

Inner electron or core-level transitions from levels n_1 to n_2 have frequencies v approximated by the well-known Rydberg formula from Eq. (1.19)

$$hv = - \left[\frac{me^4 Z^2}{32\pi^2 \varepsilon_0^2 h^2} \right] \left[\frac{1}{n_1^2} - \frac{1}{n_2^2} \right] \tag{2.14}$$

where Z is the atomic number and the other symbols have their usual meanings. The atomic number dependence of the frequency v of the K_α line, the innermost X-ray

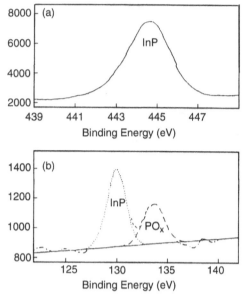

Figure 2.30. X-Ray photoelectron spectra (XPS) of nanocrystalline InP showing (a) $3d_{5/2}$ indium signal and (b) $2p$ phosphorus peaks. (From Q. Yitai, in *Handbook of Nanostructured Materials and Nanotechnology*, H. S. Nalwa, ed., Academic Press, Boston, 2000, Vol. 1, Chapter 9, p. 427.)

transition from $n_1 = 1$ to $n_2 = 2$, namely

$$\nu^{1/2} = a_K(Z - 1) \tag{2.15}$$

is called *Moseley's law*. The factor $(Z - 1)$ appears in Eq. (2.15) instead of Z to account for the shielding of the nucleus by the remaining $n_1 = 1$ electron, which lowers its apparent charge to $(Z - 1)$. A similar expression applies to the next-highest frequency L_α line in which $n_1 = 2$ and $n_2 = 3$. Figure 2.31 presents a plot of ν versus the atomic number Z for the experimentally measured K_α and L_α lines of the elements in the periodic table from $Z \approx 15$ to $Z = 60$. Measurements based on Moseley's law can provide information on the atom content of nanomaterial for all except the lightest elements. The ionization energies of the outer electrons of atoms are more dependent on the number of electrons outside closed shells than they are on the atomic number, as shown by the data in Fig. 2.32. These ionization energies are in the visible or near-ultraviolet (near-UV) region.

An energetic photon is capable of removing electrons from all occupied atomic energy levels that have ionization energies less than the incoming energy. When the photon energy drops below the largest ionization energy corresponding to the

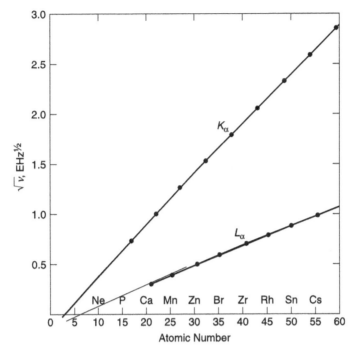

Figure 2.31. Moseley plot of the K_α and L_α characteristic X-ray frequencies versus the atomic number Z. (From C. P. Poole, Jr., H. A. Farach, and R. J. Creswick, *Superconductivity*, Academic Press, Boston, 1995, p. 515.)

K level, then the $n = 1$ electron can no longer be removed, and the X-ray absorption coefficient abruptly drops. It does not, however, drop to zero because the incoming energy is still sufficient to raise the $n = 1$ electron to a higher unoccupied level, such as a $3d$ or a $4p$ level, or to knock out electrons in the L ($n = 2$), M ($n = 3$), and further levels. The abrupt drop in absorption coefficient is referred to as an *absorption edge*; in this case it is a K absorption edge. It is clear from the relative spacings of the energy levels of Fig. 2.33 that transitions of this type are close in energy to the ionization energy, and they provide so-called fine structure on the absorption edge. They give information on the bonding states of the atom in question. The resolution of individual fine-structure transitions can be improved with the use of polarized X-ray beams. Several related X-ray absorption spectroscopy techniques are available for resolving fine structure.

Information on absorption edges can also be obtained via electron energy loss spectroscopy (EELS). This involves irradiating a thin film with a beam of monoenergetic electrons that have energies of, perhaps, 170 keV. As the electrons traverse the film, they exchange momentum with the lattice and lose energy by exciting or ionizing atoms, and an electron energy analyzer is employed to measure the amount of energy E_{abs} that is absorbed. This energy corresponds to a transition of the type

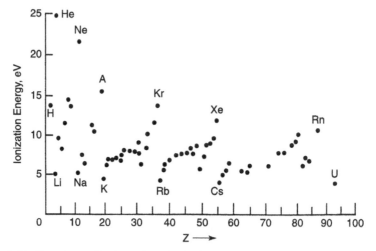

Figure 2.32. Experimentally determined ionization energy of the outer electron in various elements. (From R. Eisberg and R. Resnick, *Quantum Physics*, Wiley, New York, 1974, p. 364.)

indicated in Fig. 2.33, and is equal to the difference between the kinetic energy KE_0 of the incident electrons and that KE_{SC} of the scattered electrons

$$E_{abs} = KE_0 - KE_{SC} \qquad (2.16)$$

A plot of the measured electron intensity as a function of the absorbed energy contains peaks at the binding energies of the various electrons in the sample. The analog

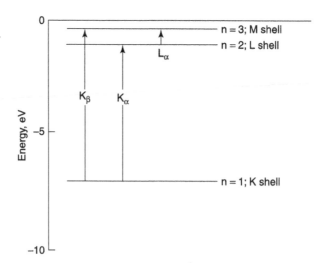

Figure 2.33. Energy-level diagram of molybdenum showing the transitions of the *K* and *L* series of X-ray lines.

of optical and X-ray polarization experiments can be obtained with EELS by varying the direction of the momentum transfer $\Delta \mathbf{p}$ between the incoming electron and the lattice relative to the crystallographic c axis. This vector $\Delta \mathbf{p}$ plays the role of the electric polarization vector $\mathbf{E}$ in photon spectroscopy. This procedure can increase the resolution of the absorption peaks.

Another procedure that measures electrons emitted from atoms in characteristic charge and energy states is the Auger process. In the photoemission experiment discussed above, when the incoming X-ray photon removed an electron from the lowest $(n-1)$ level or K shell of the atom, it left behind a vacancy or missing electron there, as illustrated in Fig. 2.34a. If an electron from the next-higher $(n=2)$ level or L shell drops down to fill this vacancy, then energy is conserved by the emission of a photon $h\nu$ with the energy ΔE corresponding to the difference in energy $\Delta E = E_K - E_L$ between that of the of the K shell and that of the L shell. This process is depicted in Fig. 2.34b. Another way to conserve energy when the L-level electron drops down is for an electron to be emitted from an even higher level such as the M shell $(n=3)$, as depicted in Fig. 2.34c. To conserve energy, the emitted electron carries away a kinetic energy E_{KE} given by

$$E_{KE} = \Delta E - |E_M| \tag{2.17}$$

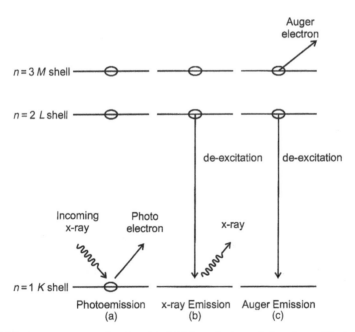

Figure 2.34. Auger process showing (a) initial excitation of the atom, (b) deexcitation accompanied by emission of an X ray, and (c) deexcitation accompanied by emission of an Auger electron from the M shell.

which is characteristic of the atom under consideration. Therefore a measurement of this Auger process kinetic energy E_{KE} provides the identity of the atom. This determination is independent of how the L-state electron had been initially removed to ionize the atom and establish the vacancy there.

The Auger electrons that are emitted from deep inside the sample ordinarily undergo secondary emission processes before they reach the surface, so they are rarely detected. This means that the Auger electrons that do emerge from the sample and are detected have come from atoms located within the first few surface layers. Hence Auger spectroscopy is a tool for determining the identity of atoms within these surface layers.

2.4.3. Magnetic Resonance

A further branch of spectroscopy that has provided information on nanostructures is magnetic resonance, which involves the study of microwave (radar frequency) and radiofrequency transitions. Most magnetic resonance measurements are made in fairly strong magnetic fields, typically $B \approx 0.33$ T (3300 G) for electron spin resonance (ESR), also called *electron paramagnetic resonance* (EPR), and $B \approx 10$ T for nuclear magnetic resonance (NMR). Several types of magnetic resonance are discussed here.

Nuclear magnetic resonance involves the interaction of a nucleus possessing a nonzero nuclear spin I with an applied magnetic field B_{app} to give the energy level splitting into $2I + 1$ lines with the energies

$$E_m = h\gamma B_{app}m \qquad (2.18)$$

where γ is the gyromagnetic ratio, sometimes called the *magnetogyric ratio*, characteristic of the nucleus, and m assumes integer or half-integer values in the range $-I \leqslant m \leqslant +I$ depending on whether I is an integer or a half-integer. The value of γ is sensitive to the local chemical environment of the nucleus, and it is customary to report the chemical shift δ of γ relative to a reference value γ_R:

$$\delta = \frac{\gamma - \gamma_R}{\gamma_R} \qquad (2.19)$$

Chemical shifts are very small, and are usually reported in parts per million (ppm). The most favorable nuclei for study are those with $I = \frac{1}{2}$, such as ^{1}H, ^{19}F, ^{31}P, and ^{13}C. The first three are dominant isotopes, with the latter only 1.1% abundant.

Fullerene molecules such as C_{60} and C_{70} are discussed in Chapter 10. The well-known buckyball C_{60} has the shape of a soccer ball with 12 regular pentagons and 20 hexagons. The fact that all of its carbon atoms are equivalent was determined unequivocally by the ^{13}C NMR spectrum, which contains only a single narrow line. In contrast to this, the rugby-ball-shaped C_{70} fullerene molecule, which contains 12 pentagons (2 regular) and 25 hexagons, has five types of carbon, and this is confirmed by the ^{13}C NMR spectrum presented at the top of Fig. 2.35. The five NMR

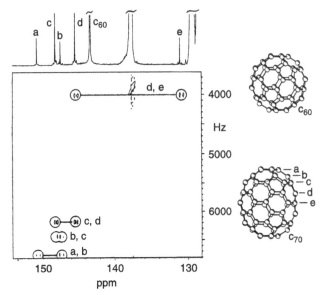

Figure 2.35. NMR spectrum from ^{13}C isotope–enriched C_{70} fullerene molecules. The five NMR lines from the a, b, c, d, and e carbons, indicated in the C_{70} molecule on the lower right, have the expected intensity ratios 10 : 10 : 20 : 20 : 10 in the upper spectrum corresponding to the number of each carbon type in the molecule. The so-called two-dimensional NMR spectrum at the lower left, which is a plot of the spin–spin coupling constant frequencies versus the chemical shift in parts per million (ppm), shows doublets corresponding to the indicated carbon bond pairs. The ppm chemical shift scale at the bottom applies to both the lower two-dimensional spectrum and the upper conventional NMR spectrum. One of the NMR lines arises from the fullerene species C_{60}, which is present as an impurity. [From K. Kikuchi, S. Suzuki, Y. Nakao, N. Nakahara, T. Wakabayashi, H. Shiromaru, I. Ikemoto, and Y. Achiba, *Chem. Phys. Lett.* **216**, 67 (1993).]

lines from the a, b, c, d and e carbons, indicated in the upper figure, have the intensity ratios 10 : 10 : 20 : 20 : 10 corresponding to the number of each carbon type in the molecule. Thus NMR has provided a confirmation of the structures of these two fullerene molecules.

Electron spin resonance, detects unpaired electrons in transition ions, especially those with odd numbers of electrons such as Cu^{2+} ($3d^9$) and Gd^{3+} ($4f^7$). Free radicals like those associated with defects or radiation damage can also be detected. The energies or resonant frequencies are three orders of magnitude higher than NMR for the same magnetic field. A different notation is employed for the energy $E_m = g\mu_B B_{app} m$ of state m, where μ_B is the Bohr magneton and g is the dimensionless g factor, which has the value 2.0023 for a free electron. For the unpaired electron with spin $S = \frac{1}{2}$ on a free radical, EPR measures the energy difference $\Delta E = E_{1/2} - E_{-1/2}$ between the levels $m = \pm\frac{1}{2}$, to give a single line spectrum at the energy

$$\Delta E = g\mu_B B_{app} \qquad (2.20)$$

Equations (2.18) and (2.20) are related through the expression $g\mu_B = h\gamma$. If the unpaired electron interacts with a nuclear spin of magnitude I then $2I + 1$ hyperfine structure lines appear at the energies

$$\Delta E(m_I) = g\mu_B B_{app} + Am_I \qquad (2.21)$$

where A is the hyperfine coupling constant and m_I takes on the $2I + 1$ values in the range $I \leqslant m_I \leqslant I$. Figure 2.36 shows an example of such a spectrum arising from the endohedral fullerene compound LaC_{84}, which was detected in the mass spectrum shown in Fig. 2.9. The lanthanum atom inside the C_{84} cage has the nuclear spin the $I = \frac{7}{2}$, and the unpaired electron delocalized throughout the C_{84} fullerene cage interacts with this nuclear spin to produce the eight-line hyperfine multiplet shown in the figure.

Electron paramagnetic resonance has been utilized to study conduction electrons in metal nanoparticles, and to detect the presence of conduction electrons in nanotubes to determine whether the tubes are metals or very narrowband semiconductors. This technique has been employed to identify trapped oxygen holes in colloidal TiO_2 semiconductor nanoclusters. It has also been helpful in clarifying spin–flip resonance transitions and Landau bands in quantum dots.

The construction of new nanostructured biomaterials is being investigated by studying the structure and organization of supermolecular assemblies on the basis of the interaction of proteins with phospholipid bilayers. This can be conveniently studied by attaching spin labels (e.g., paramagnetic nitroxides) to the lipids and using EPR to probe the restriction of the spin label motions arising from phospholipids associated with membrane-inserted domains of proteins.

Microwaves or radar waves can also provide useful information about materials when employed under nonresonant conditions in the absence of an applied magnetic field. For example, energy gaps that occur in the microwave region can be estimated

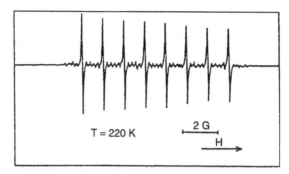

T = 220 K   2 G   H

Figure 2.36. EPR spectrum of the endohedral fullerene molecule LaC_{84} dissolved in toluene at 220 K. The unpaired electron delocalized on the carbon cage interacts with the $I = \frac{7}{2}$ nuclear spin of the lanthanum atom located inside the cage to produce the observed eight-line hyperfine multiplet. [From R. D. Johnson, D. S. Bethune, and C. S. Yannoni, *Acc. Chem. Res.* **25**, 169 (1992).]

by the frequency dependence of the microwave absorption signal. Microwaves have been used to study photon-assisted single-electron tunneling and Coulomb blockades in quantum dots.

2.5. VARIOUS BULK PROPERTIES

A number of properties of bulk materials that are measured by standard laboratory instrumentation undergo systematic changes when their dimensions become very small. We will comment on several of them without discussing the associated measurement techniques.

2.5.1. Mechanical Properties

When bulk structures decrease to the nano range of sizes, their mechanical properties commonly exhibit a considerable improvement. The hardness, mechanical strength, and yield strength often approach their theoretical limiting values for a perfect crystal in nanowires such as whiskers. The main reason for this is the decrease in the number of defects in the materials, and the fact that grain boundaries block ion movement. The hardness H and the yield strength Y_S generally depend inversely on the square root of the grain size d in accordance with the Hall–Petch expressions

$$H = H_0 + \frac{H'}{d^{1/2}} \tag{2.22}$$

$$Y = Y_0 + \frac{Y'}{d^{1/2}} \tag{2.23}$$

where the constants H_0 and Y_0 are characteristic of the material, and the factors H' and Y' depend on the frictional stress of the lattice. It has been known for many decades that carefully prepared whiskers in the low-micrometer range can be stronger than bulk crystals by more than two orders of magnitude. Heat treatments can cause defects to move to the surface and enhance the strength further. More recently carbon nanotubes have also been found to exhibit great strength. This will be discussed in greater detail in Chapter 10.

2.5.2. Electrical Properties

The electrical conductivity σ sometimes increases, and of course its reciprocal the resistivity $\rho = 1/\sigma$ correspondingly decreases with size, because of the improved order and the lesser amount of defects in the lattice. The increase in surface scattering sometimes has the opposite effect on conductivity. The bandgap E_g of a semiconductor increases, an effect confirmed by the shift of the optical absorption peak to shorter wavelengths. This increase in E_g decreases the extent to which donor electrons can be thermally activated to the conduction band and holes can thermally populate the top of the valence band. As a result, the semiconductor can carry less current, and σ decreases. The Curie temperature of a ferroelectric material such as $BaTiO_3$ can decrease significantly with size.

2.5.3. Magnetic Properties

When a ferromagnet decomposes into nanosized grains, the result is a collection of single-domain particles that orient as large individual magnetic moments μ_m in an applied magnetic field, a phenomenon called *superparamagnetism*. The magnetic susceptibility of a granular sample composed of ρ single-domain particles per unit volume depends inversely on the temperature in accordance with the Curie law

$$\chi = \frac{C}{T} \tag{2.24}$$

where the Curie constant C has the value $C = \mu_m{}^2 \rho/3k_B$, where k_B is Boltzmann's constant. The magnetic moment μ_m of a 25-nm single domain could exceed a million Bohr magnetons. The effect of nanosizing on magnetic properties is examined more thoroughly in Chapter 13.

2.5.4. Other Properties

Decreasing the narrow dimensions of films, whiskers, and particles to nanometers generally has the effect of decreasing the lattice constants and the distance between crystallographic planes. Melting points and other phase transition points are often lowered. For example, the melting point of a 1.5-nm-diameter CdS nanoparticle drops by a factor of 2 from its bulk value, which exceeds 1600 K.

PROBLEMS

2.1. For a simple cubic lattice, find the spacing, in terms of the lattice constant a, between adjacent planes oriented perpendicular to the [111] direction. Do the same for the planes in the [110] and [001] directions.

2.2. A material has an optical reflectivity $R = 0.7$; what its dimensionless dielectric constant $\varepsilon/\varepsilon_0$? What is its index of refraction n?

2.3. Consider the Bragg reflection of X rays of energy 50 keV from the [111] planes of a simple cubic crystal with a lattice constant $a = 0.52$ nm. How many orders of reflection can be detected? Find the Bragg angle θ for the first three orders.

2.4. Equation (2.4) applies for spherical grains. Find the analogous expression for a cylindrical grain with a length L equal to its diameter D.

2.5. What would be the energy E of an electron beam with a de Broglie wavelength equal to the spacing between the [111] planes of a ZnSe crystal?

2.6. Consider a mass spectrometer in which a magnetic field with strength $B = 8$ T focuses singly charged fullerence ions C_{60} at the ion detector. What B field values will focus C_{84} and LaC_{82} ions at the ion detector?

2.7. Moseley's law for K_α X-ray lines is given by Eq. (2.15), and the analogous law for L_α lines is $v^{1/2} = a_L (Z - 1)$. Find the ratio a_K/a_L of the slopes.

2.8. The lines of an X-ray diffraction pattern from a bulk material have a width of 0.3 radians. Find the linewidths for a granular sample containing 25-nm grains.

2.9. The mass-to-charge ratio m/q of a singly charged fullerence molecule (C_{60}) is measured in a mass spectrometer using an accelerating voltage V of 3×10^3 and a magnetic field B of 12 T. What is its bending radius r?

REFERENCES

1. G. E. Moore. *IEEE IEDM Tech. Digest* 11–13 (1975).
2. J. C. Wilson. *Proc. Phys. Soc.* (London). **81**, 41 (1963).
3. H. P. Klug, and L. E. Alexander *X-Diffraction Procedures*, Wiley, New York, 1974, pp. 655, 656, 660, 661.
4. T. Hahn, ed., *International Tables for Crystallography*, 4th ed., Kluwer Academic Publishers, Dordrecht, 1996.

CHAPTER 3

Properties of Individual Nanoparticles

3.1. INTRODUCTION

The purpose of this chapter is to describe the unique properties of individual nano-particles. It is essentially a survey of the properties of various kinds of nanoparticles and how their properties are different from those of the bulk material. The reasons for the differences are discussed in later chapters. Various methods of making different kinds of nanoparticles are discussed. Because nanoparticles have 10^6 atoms or less, their properties differ from those of the same atoms bonded together to form bulk materials. First it is necessary to define what we mean by a *nanoparticle*. The terms *nanoparticle* and *nanotechnology* are new. However, nanoparticles themselves were around and studied long before these words were coined. For example, many of the beautiful colors of stained glass windows of medieval churches resulted from the presence of small metal oxide clusters in the glass, having a size comparable to the wavelength of light. Particles of different sizes scatter different wavelengths of light, imparting different colors to the glass. Small colloidal particles of silver are a part of the process of image formation in photography. Water at ambient tempera-ture consists of clusters of hydrogen-bonded water molecules. Nanoparticles are gen-erally considered to be a number of atoms or molecules bonded together in a cluster with a radius less than 100 nm. A nanometer is 10^{-9} meters (m) or 10 angstroms (Å), so particles having a radius of ~ 1000 Å or less can be considered to be nanoparticles. Figure 3.1 gives a somewhat arbitrary classification of atomic clusters according to their size showing the relationship between the number of atoms in the cluster and its radius. For example, a cluster of 1 nm radius has approximately 25 atoms with most of the atoms on the surface of the cluster. This definition based on size is not totally satisfactory because it does not really distinguish between molecules and nanoparticles. There are many molecules with more than 25 atoms, particularly bio-logical molecules. For example, the heme molecule, $FeC_{34}H_{32}O_4N_4$, which is incor-porated in the hemoglobin molecule in our blood and transports oxygen to our cells, contains 75 atoms. Nanoparticles can be built by assembling individual atoms or by subdividing bulk materials. What makes nanoparticles very interesting and endows them with their unique properties is that their size can be smaller than critical

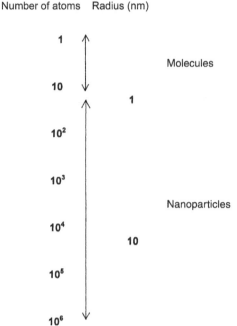

Number of atoms   Radius (nm)

Figure 3.1. Distinction between molecules, nanoparticles, fine particles, and bulk according to the size and number of atoms in the cluster.

lengths that characterize many physical phenomena. Generally the physical properties of materials can be characterized by some critical length, such as a thermal diffusion length or a scattering length. The electrical conductivity of a metal is strongly determined by the distance the electrons travel between collisions with the vibrating atoms or impurities of the solid. This distance is called the *mean free path* or the *scattering length*. If the sizes of particles are less than these characteristic lengths, it is possible that new physical or chemical phenomena or reactions may occur.

Perhaps a working definition of a nanoparticle is an aggregate of atoms with at least one dimension between 1 and 100 nm viewed as a subdivision of a bulk material, and of dimension less than the characteristic length of some phenomena. Since molecules are not subdivisions of bulk materials, they are not considered nanoparticles.

3.2. METAL NANOCLUSTERS

3.2.1. Magic Numbers

Figure 3.2 illustrates a device used to make clusters of metal atoms. A high-intensity laser beam is incident on a metal rod contained in a vacuum chamber,

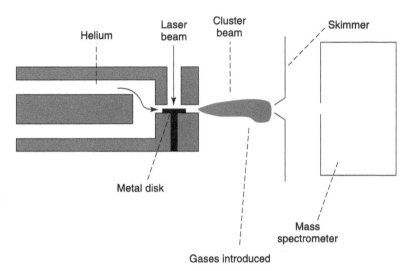

Figure 3.2. Apparatus to form metal nanoparticles by laser-induced evaporation of atoms from the surface of a metal. Various gases such as oxygen can be introduced to study the chemical interaction between the nanoparticles and the gases. (With permission from F. J. Owens and C. P. Poole, the *New Superconductors*, Plenum Press, 1999.)

causing evaporation of atoms from the surface of the metal. The atoms are then swept away by a burst of helium and passed through an orifice into a vacuum where the expansion of the gas causes cooling and formation of clusters of the metal atoms. These clusters are then ionized by UV radiation and passed into a mass spectrometer that measures their mass-to-charge ratio (m/e; also expressed as m/z). Figure 3.3 shows the mass spectrum data of lead clusters formed in such an experiment where the number of ions (counts) is plotted as a function of the number of atoms in the cluster. (Usually mass spectra data are plotted as counts vs. mass over charge.) The data show that clusters of 7 and 10 atoms are more likely than other clusters, which means that these clusters are more stable than are clusters of other sizes. Figure 3.4a is a plot of the ionization potential of atoms as a function of their atomic number Z, which is the number of electrons in the atom. The ionization potential is the energy necessary to remove the outer electron from the atom. The maximum ionization potentials occur for the rare-gas atoms, ^{2}He, ^{10}Ne, and ^{18}Ar, because their outermost s and p orbitals are filled. More energy is required to remove electrons from filled orbitals than unfilled orbitals. Figure 3.4b shows the ionization potential of sodium clusters as a function of the number of atoms in a cluster. Peaks are observed at clusters having two and eight atoms. These numbers are referred to as *electronic magic numbers*. Their existence suggests that clusters can be viewed as superatoms, and this result has motivated the development of the jellium model of clusters. In the case of larger clusters stability, as discussed in Chapter 1, is determined by structure. Certain structures have a specific numbers of atoms that are more stable than others. These numbers are referred to as *structural magic numbers*.

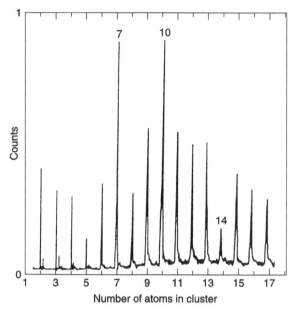

Figure 3.3. Mass spectrum data of Pb clusters. [With permission from M. A. Duncan and D. H. Rouvray, *Sci. Am.* 110 (Dec. 1989).]

3.2.2. Theoretical Modeling of Nanoparticles

The jellium model envisions a cluster of atoms as a large atom. The positive nuclear charge of each atom of the cluster is assumed to be uniformly distributed over a sphere the size of the cluster. A spherically symmetric potential well is used to represent the potential describing the interaction of the electron with the positive spherical charge distribution. Thus the energy levels can be obtained by solving the Schrödinger equation for this system in a fashion analogous to that for the hydrogen atom. Figure 3.5 compares the energy-level scheme for the hydrogen atom and the energy-level scheme for a spherical positive charge distribution. The superscripts refer to the number of electrons that fill a particular energy level. The electronic magic number corresponds to the total number of electrons on the superatom when the top level is filled. Notice that the order of the levels in the jellium model differs from that of the hydrogen atom. In this model the electronic magic numbers correspond to those clusters of a size that will allow all the energy levels to be filled.

An alternative model that has been used to calculate the properties of small clusters is to treat them as molecules and use existing molecular orbital theories such as *density functional theory* to calculate their properties. This approach can be used to calculate the actual geometric–electronic structure of small metal clusters. In the quantum theory of the hydrogen atom the electron circulating about the nucleus is described by a wave. The mathematical form for this wave, called the *wavefunction* ψ, is obtained by solving the Schrödinger equation, which includes the electrostatic potential between the electron and the positively charged nucleus. The square of

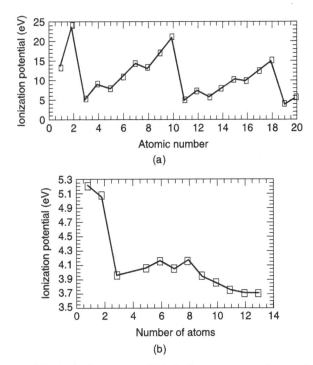

Figure 3.4. Plots of the ionization energy of (a) single atoms versus the atomic number. (ionization energy of the sodium atom at atomic number 11 = 5.14 eV); (b) sodium nanoparticles versus the number of atoms in the cluster. [Adapted from Herman et al., *J. Chem. Phys.* **80**, 1780 (1984).]

the amplitude of the wavefunction represents the probability of finding the electron at some position relative to the nucleus. The wavefunction of the lowest level of the hydrogen atom, designated the $1s$ level, has the form

$$\psi(1s) = A \exp - r/\rho \tag{3.1}$$

where r is the distance of the electron from the nucleus and ρ is the radius of the first Bohr orbit. This comes from solving the Schrödinger equation for an electron having an electrostatic interaction with a positive nucleus given by e/r. The equation of the hydrogen atom is one of the few exactly solvable problems in physics, and is one of the best understood systems in the universe. In the case of a molecule such as the H_2^+ ion, molecular orbital theory assumes that the wavefunction of the electron around the two hydrogen nuclei can be described as a linear combination of the wavefunctions of the isolated H atoms. Thus the wavefunction of the electrons in the ground state will have the form

$$\psi = a\psi(1)_{1s} + a\psi(2)_{1s} \tag{3.2}$$

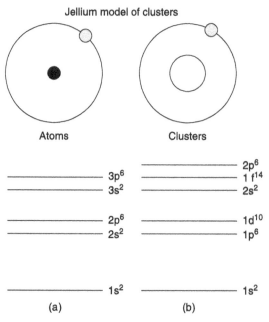

Figure 3.5. A comparison of the energy levels of the hydrogen atom and those of the jellium model of a cluster. The electronic magic numbers of the atoms are 2, 10, 18, and 36 for He, Ne, Ar, and Kr (Kr energy levels are not shown) and 2, 18, and 40 for the clusters. [With permission from B. K. Rao et al., *J. Cluster Sci.* **10**, 477 (1999).]

The Schrödinger equation for the molecular ion is

$$\left[\frac{-h^2}{8\pi^2 m}\nabla^2 - \frac{e^2}{r_a} - \frac{e^2}{r_b}\right]\psi = E\psi \tag{3.3}$$

The Laplacian symbol ∇^2 denotes a double differentiation operation. The last two terms in the brackets are the electrostatic attractions of the electron to the two positive nuclei, which are at distances r_a and r_b, respectively, from the electron. For the hydrogen molecule, which has two electrons, a term, e^2/r_{ij}, for the electrostatic repulsion of the two electrons would be added. The Schrödinger equation is solved with this linear combination, Eq. (3.2), of wavefunctions. When there are many atoms in the molecule and many electrons the problem becomes complex, and many approximations can be used to obtain the solution. Density functional theory represents one approximation. With the development of large fast computer capability and new theoretical approaches, it is possible using molecular orbital theory to determine the geometric–electronic structure of large molecules with a high degree of accuracy. The calculations can find the structure with the lowest energy, which will be the equilibrium geometry. These molecular orbital methods with some modification have been applied to metal nanoparticles.

3.2.3. Geometric Structure

Generally the crystal structure of large nanoparticles is the same as the bulk structure with somewhat different lattice parameters. X-Ray diffraction studies of 80-nm aluminum particles have shown that they have the face-centered cubic (fcc) unit cell shown in Fig. 3.6a, which is the structure of the unit cell of bulk aluminum. However in some instances it has been shown that small particles having diameters of <5 nm may have different structures. For example, it has been shown that gold particles of 3–5 nm have an icosahedral structure rather than the bulk fcc structure. It is of interest to consider an aluminum cluster of 13 atoms because this is a magic number. Figure 3.6b shows three possible arrangements of atoms for the cluster. From the criteria of maximizing the number of bonds and minimizing the number of atoms on the surface, as well as the fact that the structure of bulk aluminum is fcc, one might expect the structure of the particle to be fcc. However, molecular orbital calculations based on the density functional method predict that the icosahedral form has a lower energy than do the other forms, suggesting the possibility of a structural change. There are no experimental measurements of the Al_{13} structure to verify this prediction. The experimental determination of the structure of small metal nanoparticles is difficult, and not much structural data are available. In the late 1970s and early 1980s G. D. Stien was able to determine the structures of Bi_N, Pb_N, In_N, and Ag_N nanoparticles. The particles were made using an oven to vaporize the metal and a supersonic expansion of an inert gas to promote cluster formation. Deviations from the fcc structure were observed for clusters smaller than 8 nm in diameter. Indium clusters undergo a change of structure when the size is smaller than 5.5 nm. Above 6.5 nm, a diameter corresponding to about 6000 atoms, the clusters have a face-centered tetragonal structure with a c/a ratio of 1.075. In a tetragonal unit cell the edges of the cell are perpendicular, the long axis is denoted by c and the two short axes, by a. Below $\sim$6.5 nm the c/a ratio

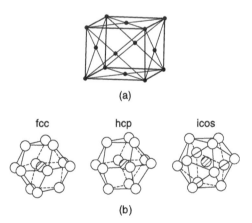

(a)

(b)

Figure 3.6. (a) The fcc unit cell of bulk aluminum; (b) three possible structures of Al_{13}—a fcc structure (fcc), an hexagonal close-packed structure (hcp), and an icosahedral (icos) structure.

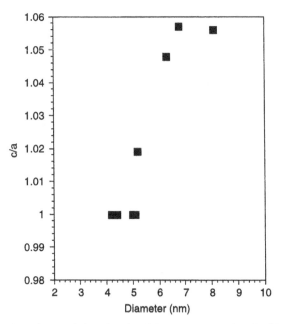

Figure 3.7. Plot of c-axis : a-axis length ratio of the tetragonal unit cell of indium nanoparticles versus diameter of the nanoparticles. [Plotted from data in A. Yokozeki and G. D. Stein, *J. Appl. Phys.* **49**, 224 (1978).]

begins to decrease, and at 5 nm $c/a = 1$, meaning that the structure is FCC. Figure 3.7 is a plot of c/a versus the diameter of indium nanoparticles. Figure 3.8 is a plot of the measured lattice parameters of aluminum nanoparticles versus particle size showing that the lattice parameter decreases slightly with reduced size. The structure of isolated nanoparticles may be different from the structures of particles in which molecules are bonded to their surfaces, referred to as *ligand-stabilized nanoparticles*. One way to accomplish this bonding is by self-assembly, which is discussed in the next chapter.

A different structure can result in a change in many properties. One obvious property that will be different is the electronic structure. In Chapter 8 the effect of nanosizing on the electronic structure will be discussed in more detail. Table 3.1 gives the result of the density functional calculations of some of the electronic properties of Al_{13}. Notice that binding energy per atom in Al_{13} is less than that in the bulk aluminum. The Al_{13} cluster has an unpaired electron in the outer shell. The addition of an electron to form $Al_{13}(-)$ closes the shell with a significant increase in the binding energy. The molecular orbital approach is also able to account for the dependence of the binding energy and ionization energy on the number of atoms in the cluster. Figure 3.9 shows some examples of the structure of boron nanoparticles of different sizes calculated by density functional theory. Figure 3.6 illustrates another important property of metal nanoparticles. For these small particles, all of the atoms that make up the particle except one are on the surface. This has important implications for many of the properties of the nanoparticles such as their vibrational structure,

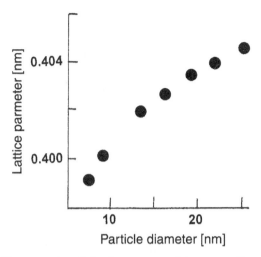

Figure 3.8. Lattice parameter of aluminum nanoparticles versus diameter of particles.

stability, and reactivity. In this chapter, although we are discussing metal nanoparticles as though they can exist as isolated entities, this is not always the case. Some nanoparticles such as aluminum are highly reactive. If one were to have an isolated aluminum nanoparticle exposed to air, it would immediately react with oxygen, resulting in an oxide coating of Al_2O_3 on the surface. X-Ray photoelectron spectroscopy of oxygen-passivated, 80-nm, aluminum nanoparticles indicates that they have a 3–5-nm layer of Al_2O_3 on the surface and that this thickness is relatively independent of the diameter of the particle. Figure 3.10 shows a high-resolution transmission electron microscope (TEM) image of an aluminum nanoparticle where the oxide layer is evident. As we will see later, nanoparticles can be made in solution without exposure to air. For example, aluminum nanoparticles can be made by decomposing aluminum hydride in certain heated solutions. In this case the molecules of the solvent may be bonded to the surface of the nanoparticle, or a surfactant (surface-active agent) such as oleic acid can be added to the solution. The surfactant will coat the particles and prevent them from aggregating. Such metal nanoparticles are said to be *passived*, that is, coated with some other chemical to which they are exposed. The chemical nature of this layer will significantly influence the properties of the nanoparticle.

TABLE 3.1. Calculated Binding Energy per Atom and Atomic Separation in Some Aluminum Nanoparticles Compared with Bulk Aluminum

Cluster	Binding Energy, eV	Al Separation, Å
Al_{13}	2.77	2.814
Al_{13}^-	3.10	2.75
Bulk Al	3.39	2.86

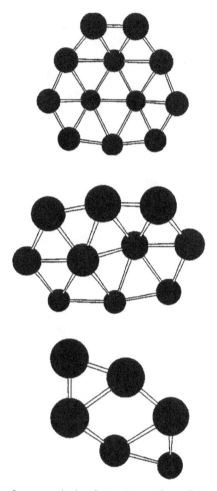

Figure 3.9. Illustration of some calculated structures of small boron nanoparticles. (F. J. Owens, unpublished.)

Self-assembly has been used to coat gold nanoparticles. The self-assembly concept is discussed in the next chapter. Gold nanoparticles have been passivied by self-assembly using octadecylthiol, which produces a self-assembled monolayer (SAM), $C_{18}H_{37}S$—Au. Here the long hydrocarbon chain molecule is tethered at its end to the gold particle Au by the thio headgroup SH, which forms a strong S—Au bond. Attractive interactions between the molecules produce a symmetric ordered arrangement of these molecules about the particle. This symmetric arrangement of the molecules around the particle is a key characteristic of the self-assembled monolayers.

3.2.4. Electronic Structure

When atoms form a lattice, the discrete energy levels of the atoms are smudged out into energy bands. The term *density of states* refers to the number of energy levels in a

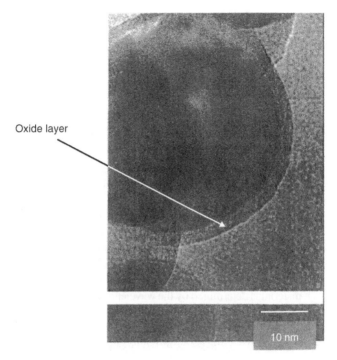

Figure 3.10. High-resolution TEM image of a 40-nm Al nanoparticle showing the aluminum oxide layer on the surface. [Adapted from R. Scheffian et al., *J. Energ. Mater.* **21**, 1441 (2006).]

given interval of energy. For a metal the top band is not totally filled. In the case of a semiconductor, the top occupied band, called the *valence band*, is filled, and there is a small energy separation referred to as the *bandgap* between it and the next-higher unfilled band. When a metal particle having bulk properties is reduced in size to a few hundred atoms, the density of states in the conduction band, the top band containing electrons, changes dramatically. The continuous density of states in the band is replaced by a set of discrete energy levels, which may have energy-level spacings exceeding the thermal energy k_BT, and a gap opens up. The changes in the electronic structure during the transition of a bulk metal to a large cluster, and then down to a small cluster of less than 15 atoms, are illustrated in Fig. 3.11. The small cluster is analogous to a molecule having discrete energy levels with bonding and antibonding orbitals. Eventually a size is reached where the dmensions of the particles are in the order of the wavelengths of the electrons. In this situation the energy levels can be modeled by the quantum-mechanical treatment of a particle in a box. This is referred to as the *quantum size effect*. The emergence of new electronic properties can be understood in terms of the *Heisenberg uncertainty principle*, which states that the more an electron is spatially confined, the broader will be its range of momentum. The average energy will not be determined as much by the chemical nature of the atoms as by the dimension of the particle. It is interesting to note that the quantum size effect occurs in semiconductors at larger sizes because of the longer wavelength

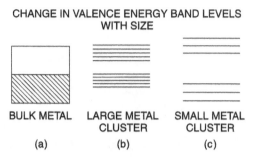

CHANGE IN VALENCE ENERGY BAND LEVELS
WITH SIZE

BULK METAL   LARGE METAL   SMALL METAL
CLUSTER   CLUSTER

(a)   (b)   (c)

Figure 3.11. Illustration of how energy levels of a metal change when the number of atoms of the material is reduced: (a) Valence band of bulk metal; (b) metal cluster of 100 atoms showing opening of a bandgap; (c) metal cluster having three atoms.

of conduction electrons and holes in semiconductors owing to the larger effective mass. In a semiconductor the wavelength can approach 1 μm, whereas in a metal it is in the order of 0.5 nm. In Chapter 8 we will explain the reasons why the electronic structure changes with size in the nanometer regime.

The color of a material is determined by the wavelength of light that is absorbed by it. The absorption occurs because electrons are induced by the photons of the incident light to make transitions between the lower-lying occupied levels and higher unoccupied energy levels of the materials. Clusters of different sizes will have different electronic structures and different energy-level separations. Figure 3.12 compares the calculated energy levels of some excited states of boron clusters B_6, B_{10}, and B_{12} showing the difference in the energy-level separations. Light-induced transitions between these levels

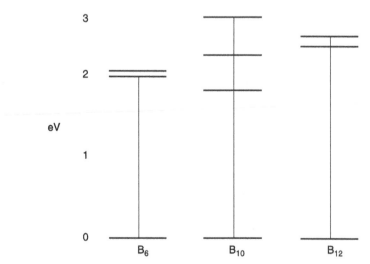

Figure 3.12. Density functional calculation of excited-state energy levels of B_6, B_{10}, and B_{12} nanoparticles. Photon-induced transitions between the lowest level and the upper levels determine the color of the particles. (F. J. Owens, unpublished.)

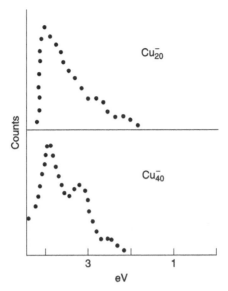

Figure 3.13. UV photoelectron spectrum in the valence band region of singly ionized copper nanoparticles having 20 and 40 atoms. [Adapted from Pettiete et al., *J. Chem. Phys.* **88**, 5377 (1988).]

determine the color of the materials. This means that clusters of different sizes can have different colors, and the size of the cluster can be used to engineer the color of a material. We will come back to this when we discuss semiconducting clusters.

One method of studying the electronic structure of nanoparticles is UV photo-electron spectroscopy, which is described in more detail in Chapter 2. An incident UV photon removes electrons from the outer valence levels of the atom. The electrons are counted and their energy measured. The data obtained from the measurement are the number of electrons (counts) versus energy. Because the clusters have discrete energy levels, the data will be a series of peaks with separations corresponding to the separations of the energy levels of the cluster. Figure 3.13 shows the UV photo-electron spectrum of the outer levels of copper clusters having 20 and 40 atoms. It is clear the electronic structure in the valence region is different, depending on the size of the cluster. The energy of the lowest peak is a measure of the electron affinity of the cluster. The *electron affinity* is defined as the increase in electronic energy of the cluster when an electron is added to it. Figure 3.14 is a plot of the measured electron affinities versus the size of Cu clusters, again showing peaks at certain cluster sizes.

3.2.5. Reactivity

Since the electronic structure of a nanoparticle depends on the size of the particle, the ability of the cluster to react with other species should depend on cluster size. This has important implications for the design of catalytic agents.

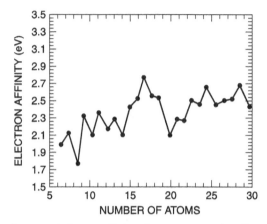

Figure 3.14. Plot of measured electron affinity of copper versus size of nanoparticle. [Adapted from Pettiete et al., *J. Chem. Phys.* **88**, 5377 (1988).]

There is experimental evidence for the effect of size on the reactivity of nanopar-ticles. Their reactivity with various gases can be studied by the apparatus sketched in Fig. 3.2, in which gases such as oxygen are introduced into the region of the cluster beam. A laser beam aimed at a metal disk dislodges metallic particles, which are carried along to a mass spectrometer by a flow of helium gas. Downstream before the particles in the cluster beam enter the mass spectrometer, various gases are intro-duced, as shown in Fig. 3.2. Figure 3.15 shows the results of studies of the reaction of oxygen with aluminum particles. Figure 3.15a is the mass spectrum of the aluminum particles before the oxygen is introduced. Figure 3.15b shows the spectrum after oxygen enters the chamber. The results show that two peaks have increased substantially and certain peaks (12, 14, 19, and 20) have disappeared. The Al_{13} and Al_{23} peaks have increased substantially, and peaks from Al_{15} to Al_{22} have decreased.

These data are clear evidence for the dependence of the reactivity of aluminum clusters on the number of atoms in the cluster. Similar size dependences have been observed for the reactivity of other metals. Figure 3.16 presents a plot of the reaction rate of iron with hydrogen as a function of the size of the iron nanopar-ticles. The data show that particles of certain sizes such as the one with 10 atoms and sizes greater than 18 atoms are more reactive with hydrogen than others. Notice that reactivity does not simply scale inversely with the surface area of the particles when the particles have nanometer dimensions. This has important impli-cations for the use of nanoparticles as catalysts. Catalysts are materials that modify the rate of a chemical reaction, usually speeding up the reaction. Ordinarily the cata-lyst participates in the reaction by combining with one or more of the reactants and at the end of the process is regenerated. In other words, the catalyst is constantly recycled as the reaction proceeds. This will be discussed in more detail in the next chapter.

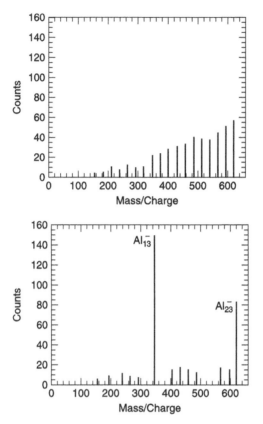

Figure 3.15. Mass spectrum of Al nanoparticles before (a) and after (b) exposure to oxygen gas. [Adapted from Leuchtner et al., *J. Chem. Phys.* **91**, 2753 (1989).]

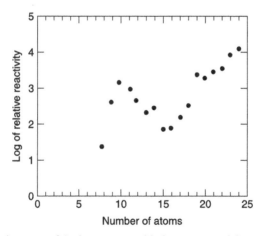

Figure 3.16. Reaction rate of hydrogen gas with iron nanoparticles versus particle size. [Adapted from R. L. Whetten et al., *Phys. Rev. Lett.* **54**, 1494 (1985).]

3.2.6. Fluctuations

Very small nanoparticles have all or almost all of their atoms on the surface. Figure 3.17 presents a plot of the percentage of atoms of gold nanoparticles that are on the surface of the particle versus the diameter of the particle obtained by using the results of Table 1.1. The plot shows clearly that the fraction of surface atoms increases sharply below ~10 nm. Surface atoms are less restricted in their ability to vibrate than are those in the interior, and they are able to make larger excursions from their equilibrium positions. This happens because on the surface each atom has less nearest neighbors to bond with than in the bulk, and thus the binding energy is lower. This can lead to changes in the structure of the particle. Changes in the geometry of gold clusters with time have been observed using an electron microscope. The gold clusters of 10–100 Å radii are prepared in vacuum and deposited on a silicon substrate, which is then covered with an SiO_2 film. The electron microscope pictures of gold nanoparticles presented in Fig. 3.18, which were taken at different times, show a number of fluctuation induced changes in the structure brought about by the particles transforming between different structural arrangements. At higher temperatures these fluctuations can cause breakdown in the symmetry of the nanoparticle, resulting in the formation of a liquidlike droplet of atoms.

3.2.7. Magnetic Clusters

Although this theory is not rigorously correct and leads to incorrect predictions of properties, an electron in an atom can be viewed as a point charge orbiting the nucleus. Its motion around the nucleus gives it orbital angular momentum and produces a magnetic field (except for s states). The magnetic field pattern arising from this movement resembles that of a bar magnet. The electron is said to have an orbital

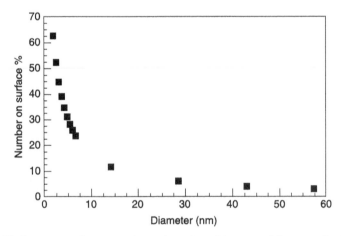

Figure 3.17. Percentage of atoms on the surface of a gold nanoparticle versus the diameter of the particle.

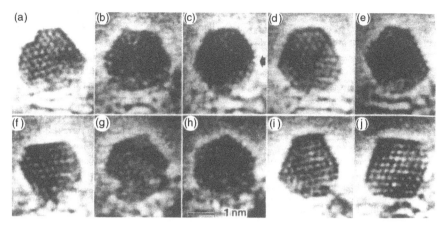

Figure 3.18. A series of electron microscope pictures of gold nanoparticles containing approximately 460 atoms taken at various times showing fluctuation-induced changes in structure. (With permission from S. Sugano and H. Koizumi, in *Microcluster Physics*, Springer, Berlin, 1998, p. 18.)

magnetic moment. There is also another contribution to the magnetic moment arising from the fact that the electron possesses a spin. Classically one can think of the electron as a spherical charge rotating about some axis. Thus there is both a spin and an orbital magnetic moment, which can be added to give the total magnetic moment of the electron. The total moment of the atom will be the vector sum of the moments of each electron of the atom. In energy levels filled with an even number of electrons the electron magnetic moments are paired oppositely, and the net magnetic moment is zero. Thus few atoms in solids have a net magnetic moment. There are, however, some transition ion atoms such as iron, manganese, and cobalt that have partially filled inner *d*-orbital levels, and hence they possess a net magnetic moment. Crystals of these atoms can become ferromagnetic when the magnetic moments of all the atoms are aligned in the same direction. In this section we discuss the magnetic properties of nanoclusters of metal atoms that have a net magnet moment. In a cluster the magnetic moment of each atom will interact with the moments of the other atoms, and can force all the moments to align in one direction with respect to some symmetry axis of the cluster. The cluster will have a net moment, and is said to be magnetized. The magnetic moments of such clusters can be measured by means of the Stern–Gerlach experiment illustrated in Fig. 3.19. The cluster particles are sent into a region where there is an inhomogeneous magnetic field, which separates the particles according to whether their magnetic moment is up or down. From the beam separation, and knowledge of the strength and gradient of the magnetic field, the magnetic moment can be determined. However, for magnetic nanoparticles the measured magnetic moment is found to be less than the value for a perfect parallel alignment of the moments in the cluster. The atoms of the cluster vibrate, and this vibrational

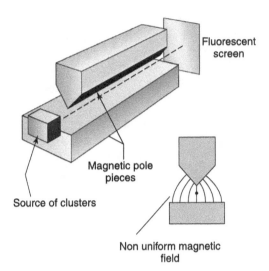

Figure 3.19. Illustration of the Stern–Gerlach experiment used to measure the magnetic moments of nanoparticles. A beam of metal clusters from a source is sent between the poles of permanent magnets shaped to produce a uniform dc magnetic field gradient, which produces a net force on the magnetic dipole moments of the clusters, thereby deflecting the beam and separating atoms with different magnetic moments. The magnetic moment can be determined by the extent of the deflection, which is measured on a photographic plate or fluorescent screen.

energy increases with temperature. These vibrations cause some misalignment of the magnetic moments of the individual atoms of the cluster so that the net magnetic moment of the cluster is less than what it would be if the moments of all the atoms were aligned in the same direction. The component of the magnetic moment of an individual cluster will interact with an applied dc magnetic field, and is more likely to align parallel to the field than antiparallel to it. The overall net moment will be lower at higher temperatures; more precisely, it is inversely proportional to the temperature, an effect called *superparamagnetism*. When the interaction energy between the magnetic moment of a cluster and the applied magnetic field is greater than the vibrational energy, there is no vibrational averaging, but because the clusters rotate, there is some averaging. This is called *locked moment magnetism*.

One of the most interesting observed properties of nanoparticles is that clusters made up of nonmagnetic atoms can have a net magnetic moment. For example, clusters of rhenium show a pronounced increase in their magnetic moment when they have less than 20 atoms. Figure 3.20 is a plot of the magnetic moment versus the size of the rhenium cluster. The magnetic moment is large when $n < 15$. In Chapter 13 the role of nanosizing on the properties of magnetic materials will be discussed in more detail.

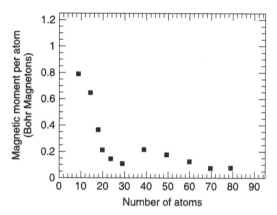

Figure 3.20. Plot of the magnetic moment per atom of rhenium nanoparticles versus the number of atoms in the particle. [Adapted from A. J. Cox et al., *Phys. Rev.* **B49**, 12295 (1994).]

3.2.8. Bulk-to-Nano Transition

At what number of atoms does a cluster assume the properties of the bulk material? In a cluster of less than 100 atoms, the amount of energy needed to ionize it (i.e., to remove an electron from the cluster) differs from the workfunction. The workfunction is the amount of energy needed to remove an electron from the bulk solid. Clusters of gold have been found to have the same melting point as bulk gold only when they contain 1000 atoms or more. Figure 3.21 is a plot of the melting temperature of gold nanoparticles versus the diameter of the particle. Possible reasons for the

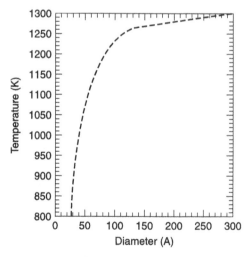

Figure 3.21. Melting temperature of gold nanoparticles versus particle diameter (10 Å = 1 nm). [Adapted from J. P. Borel et al., *Surf. Sci.* **106**, I (1981).]

reduction of the melting temperature will be discussed in later chapters. The average separation of copper atoms in a copper cluster approaches the value of the bulk material when the clusters have 100 atoms or more. In general it appears that different physical properties of clusters reach the characteristic values of the solid at different cluster sizes. The size of the cluster where the transition to bulk behavior occurs appears to depend on the property being measured.

3.3. SEMICONDUCTING NANOPARTICLES

3.3.1. Optical Properties

Because of their role in quantum dots, nanoparticles consisting of the elements, which are normal constituents of semiconductors, have been the subject of much study, with particular emphasis on their electronic properties. A nanoparticle of Si_n can be made by laser evaporation of a Si substrate in the region of a helium gas pulse. The beam of neutral clusters is photolyzed by a UV laser producing ionized clusters whose m/e ratio is then measured in a mass spectrometer. The most striking property of nanoparticles made of semiconducting elements is the pronounced changes in their optical properties compared to those of the bulk material. There is a significant shift in the optical absorption spectra toward the shorter wavelengths referred to as the blueshift as the particle size is reduced.

In a bulk semiconductor a bound electron–hole pair, called an *exciton*, can be produced by a photon having energy greater than the bandgap of the material. The bandgap is the energy separation between the top filled energy level of the valence band and the nearest unoccupied level in the conduction band above it. The photon excites an electron from the filled band to the unoccupied band above. The result is an electron vacancy or hole in the otherwise filled valence band, which corresponds to an electron with an effective positive charge. Because of the Coulomb attraction between the positive hole and the negative electron, a bound pair, called an *exciton*, is formed that can move through the lattice. The separation between the hole and the electron is many lattice parameters. The existence of the exciton has a strong influence on the electronic properties of the semiconductor and its optical absorption. The exciton can be modeled as a hydrogenlike atom and has energy levels with relative spacings analogous to the energy levels of the hydrogen atom but with lower actual energies, as explained in Section 1.4.3. Light-induced transitions between these hydrogenlike energy levels produce a series of optical absorptions that can be labeled by the principle quantum numbers of the corresponding hydrogen energy levels. Figure 3.22 presents an optical absorption spectrum of cuprous oxide (Cu_2O), showing the absorption due to the exciton. We are particularly interested in what happens when the size of the nanoparticle becomes smaller than or comparable to the radius of the orbit of the electron–hole pair. There are two situations, called the *weak-confinement* and the *strong-confinement regimes*. In the weak regime the particle radius is larger than the radius of the electron–hole pair, but the range of motion of the exciton is limited, which causes a blueshift of the

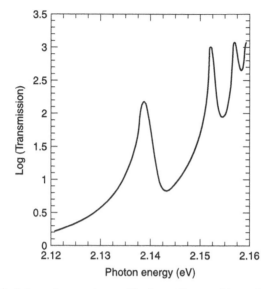

Figure 3.22. Optical absorption spectrum of hydrogenlike transitions of weakly bound excitons in Cu_2O. [Adapted from P. W. Baumeister, *Phys. Rev.* **121**, 359 (1961).]

absorption spectrum. When the radius of the particle is smaller than the orbital radius of the electron–hole pair, the motion of the electron and the hole become independent, and the exciton does not exist. The hole and the electron have their own set of energy levels. Here there is also a blueshift, and the emergence of a new set of absorption lines. Figure 3.23 shows the optical absorption spectra of a CdSe

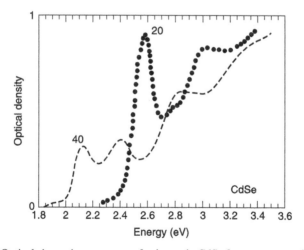

Figure 3.23. Optical absorption spectrum of exictons in CdSe for two nanoparticles having 20 and 40 of atoms, respectively. [Adapted from D. M. Mittleman, *Phys. Rev.* **B49**, 14435 (1994).]

nanoparticle at two different sizes measured at 10 K. One can see that the lowest-energy absorption region, referred to as the *absorption edge*, is shifted to higher energy as the particle size decreases. Since the absorption edge is due to the bandgap, this means that the bandgap is increasing as the particle size decreases. Notice also that the intensity of the absorption increases as the particle size is reduced. The higher-energy peaks are associated with the exciton, and they shift to higher energies with the decrease in particle size. These effects are a result of the confinement of the exciton that was discussed above. Essentially, as the particle size is reduced, the hole and the electron are forced closer together, and the separation between the energy levels changes. This subject will be discussed in greater detail in Section 9.4.

3.3.2. Photofragmentation

It has been observed that nanoparticles of silicon and germanium can undergo fragmentation when subjected to laser light from a Q-switched Nd : YAG laser. The products depend on the size of the cluster, the intensity of the laser light, and the wavelength. Figure 3.24 shows the dependence of the cross section for photofragmentation (a measure of the probability for breakup of the cluster) with 532 nm laser light, versus the size of a Si fragment. One can see that certain-sized fragments are more likely to dissociate than others. Some of the fissions that have been observed are

$$Si_{12^+} + h\nu \rightarrow Si_{6^+} + Si_6 \qquad (3.5)$$

$$Si_{20^+} + h\nu \rightarrow Si_{10^+} + Si_{10} \qquad (3.6)$$

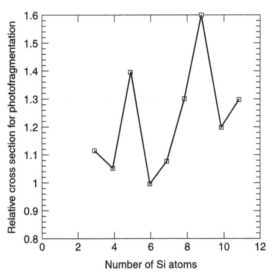

Figure 3.24. Photodissociation cross section of silicon nanoparticles versus number of atoms in particle. [Adapted from L. Bloomfield et al., *Phys. Rev. Lett.* **54**, 2266 (1985).]

TABLE 3.2. Examples of Smallest Obtainable Multiply Charged Clusters of Different Kinds[a]

Atom	Charge		
	$+2$	$+3$	$+4$
Kr	Kr_{73}	—	—
Xe	Xe_{52}	Xe_{114}	Xe_{206}
CO_2	$(CO_2)_{44}$	$(CO_2)_{106}$	$(CO_2)_{216}$
Si	Si_3	—	—
Au	Au_3	—	—
Pb	Pb_7	—	—

where $h\nu$ is a photon of light energy. Similar results have been obtained for germanium nanoparticles. When the cluster size is greater than 30 atoms, the fragmentation has been observed to occur explosively.

3.3.3. Coulomb Explosion

Multiple ionization of clusters causes them to become unstable, resulting in very rapid high-energy dissociation or explosion. The fragment velocities from this process are very high. The phenomenon is called *Coulombic explosion*. Multiple ionizations of a cluster cause a rapid redistribution of the charges on the atoms of the cluster, making each atom more positive. If the strength of the electrostatic repulsion between the atoms is greater than the binding energy between the atoms, the atoms will rapidly fly apart from each other with high velocities. The minimum number of atoms N required for a cluster of charge Q to be stable depends on the kinds of atoms, and on the nature of the bonding between the atoms of the cluster. Table 3.2 gives the smallest size that is stable for doubly charged clusters of different types of atoms and molecules. The table also shows that larger clusters are more readily stabilized at higher degrees of ionization. The attractive forces between the atoms of the cluster can be overcome by the electrostatic repulsion between the atoms when they become positively charged as a result of photoionization.

3.4. RARE-GAS AND MOLECULAR CLUSTERS

3.4.1. Inert-Gas Clusters

Table 3.2 lists a number of different kinds of nanoparticles. Besides metal atoms and semiconducting atoms, nanoparticles can be assembled from rare gases such as krypton and xenon, and molecules such as water. Xenon clusters are formed by adiabatic expansion of a supersonic jet of the gas through a small capillary into a vacuum. The gas is then collected by a mass spectrometer, where it is ionized by an electron beam, and its m/e ratio measured. As in the case of metals, there are magic numbers, meaning that clusters having a certain number of atoms are more stable than others.

For the case of xenon, the most stable clusters occur at particles having 13, 19, 25, 55, 71, 87, and 147 atoms. Argon clusters have similar structural magic numbers. Since the inert-gas atoms have filled electronic shells, their magic numbers are structural magic numbers as discussed in Chapter 1. The forces that bond rare-gas atoms into clusters are weaker than those that bond metallics and semiconducting atoms. Even though rare-gas atoms have filled electron shells, because of the movement of the electrons about the atoms, they can have an instantaneous electric dipole moment P_1. An electric dipole moment occurs when a positive charge and a negative charge are separated by some distance. This dipole produces an electric field $2P_1/R^3$ at another atom a distance R away. This in turn induces a dipole moment P_2 on the second atom, $2\alpha P_1/R^3$, where α is the electronic polarizability. Thus two rare-gas atoms will experience an attractive potential

$$U(R) = \frac{2P_1 P_2}{R^3} = \frac{-4\alpha P_1^2}{R^6} \tag{3.7}$$

This is known as the *van der Waals potential*. As the two atoms get much closer together, there will be repulsion between the electronic cores of each atom. Experimentally this has been shown to have the form B/R^{12}. Thus the overall interaction potential between two inert-gas atoms has the form

$$U(R) = \frac{B}{R^{12}} - \frac{C}{R^6} \tag{3.8}$$

This is known as the Lennard-Jones potential, and it is used to calculate the structure of inert-gas clusters. The force between the atoms arising from this potential is a minimum for the equilibrium distance $R_{min} = \left(\frac{2B}{C}\right)^{1/6}$, and it is attractive for larger separations and repulsive for smaller separations of the atoms. More generally, it is weaker than the forces that bind metal and semiconducting atoms into clusters.

3.4.2. Superfluid Clusters

Clusters of ^{4}He and ^{3}He atoms formed by supersonic free jet expansion of helium gas have been studied by mass spectrometry, and magic numbers are found at cluster sizes of $N = 7,10,14,23,30$ for ^{4}He and $N = 7,10,14,21,30$ for ^{3}He. One of the more unusual properties displayed by clusters is the observation of superfluidity in He clusters having 64 and 128 atoms. *Superfluidity* is the result of the difference in the behavior of atomic particles having half-integer spin, called fermions, and particles having integer spin, called *bosons*. The difference between them lies in the rules that determine how they occupy the energy levels of a system. Fermions such as electrons are allowed to have only two particles in each energy level with their spins oppositely aligned. Bosons, on the other hand, do not have this restriction. This means that as the temperature is lowered and more of the lower levels become occupied, bosons can all occupy the lowest level, whereas fermions will distributed

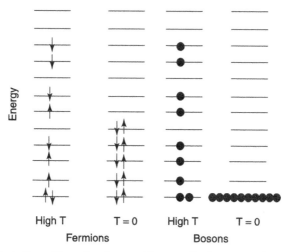

Figure 3.25. Illustration of how fermions and bosons distribute over the energy levels of a system at high and low temperatures.

in pairs at the lowest sequence of levels. Figure 3.25 illustrates this difference. The case where all the bosons are in the lowest level is referred to as *Bose–Einstien condensation*. When this occurs, the wavelength of each boson is the same as every other, and all of the waves are in phase.

When Bose–Einstein condensation occurs in liquid ^{4}He at temperature 2.2 K, called the *lambda point* (λ-point), the liquid helium becomes a superfluid, and its viscosity drops to zero. Normally when a liquid is forced through a small thin tube, it moves slowly because of friction with the walls, and increasing the pressure at one end increases the velocity. In the superfluid state the liquid moves quickly through the tube, and increasing the pressure at one end does not change the velocity. The transition to the superfluid state at 2.2 K is marked by a discontinuity in the specific heat known as the *lambda transition*. The *specific heat* is the amount of heat energy necessary to raise the temperature of one gram of the material by one degree Kelvin. Figure 3.26 presents a plot of the specific heat versus temperature for bulk liquid helium, and for a helium cluster of 64 atoms, showing that clusters become superfluid at a temperature lower than that of the bulk liquid of He atoms.

3.4.3. Molecular Clusters

Individual molecules can form clusters. One of the most common examples of this is the water molecule. It has been known for almost 30 years, long before the invention of the word *nanoparticle*, that water does not consist of isolated H_2O molecules. The broad Raman spectra of the O—H stretch of the water molecule in the liquid phase from 3200 to 3600 cm^{-1} has been shown to be due to a number of overlapping peaks arising from both isolated water molecules and water molecules hydrogen-bonded into clusters. The H atom of one molecule forms a bond with the oxygen

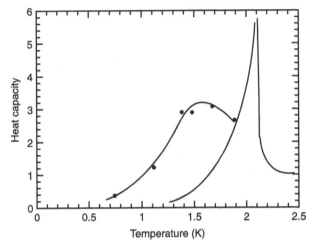

Figure 3.26. Heat capacity versus temperature for liquid helium (solid line) and a liquid consisting of clusters of 64 helium atoms (dark circles). The peak corresponds to the transition to the superfluid state. [Adapted from P. Sindzingre, *Phys. Rev. Lett.* **63**, 1601 (1989).]

atom of another. Figure 3.27 shows the structure of one such water cluster. At ambient conditions 80% of water molecules are bonded into clusters, and as the temperature is raised, the clusters dissociate into isolated H_2O molecules. There are other examples of molecular clusters such as $(NH_3)_n^+$, $(CO_2)_{44}$, and $(C_4H_8)_{30}$.

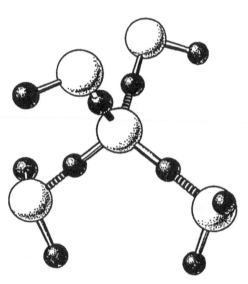

Figure 3.27. A hydrogen-bonded cluster of five water molecules. The large spheres are oxygen and the small spheres are hydrogen atoms.

3.4.4. Nanosized Organic Crystals

Nanosized organic crystals have a number of interesting applications. Many π-conjugated organics exhibit third-order nonlinear optical properties, which means that their refractive indices depend on the intensity of the incident light. In the case of these nonlinear optical materials the refractive index is given by

$$n = n_0 + n_2 I \tag{3.9}$$

where I is the intensity of the light. Nonlinear optical effects can be the basis of optical switches and used to convert visible-light frequencies to higher frequencies in the ultraviolet region of the spectrum. Organic nanocrystals such as perylene, pyrene, and anthracene display quantum confinement effects, discussed in Section 9.3.5. In bulk crystals of perylene a luminescence is observed at 560 nm from the self-trapped exciton. However, in nanocrystals the luminescence occurs from 470 to 482 nm for a size change from 50 to 200 nm. Nanoorganic crystals also have applications in the pharmaceutical industry, where pills made of nanosized crystals should enter the bloodstream faster than tablets consisting of micro sized components.

3.5. METHODS OF SYNTHESIS

Earlier in the chapter we described one method of making nanoparticles using laser evaporation, in which a high-intensity laser beam is incident on a metal rod, causing atoms to be evaporated from the surface of the metal. These metal atoms are then cooled into nanoparticles. There are, however, other ways to make nanoparticles, and we will describe several of them.

3.5.1. RF Plasma

Figure 3.28 illustrates a method of nanoparitcle synthesis, which utilizes a plasma generated by radio frequency (RF) heating coils. The starting metal is contained in a pestle in an evacuated chamber. The metal is heated above its evaporation point using high-voltage RF coils wrapped around the evacuated system in the vicinity of the pestle. Helium gas is then allowed to enter the system, forming a high temperature plasma in the region of the coils. The metal vapor nucleates on the He gas atoms and diffuses up to a colder collector rod where nanoparticles are formed. The particles are generally passivated by the introduction of some gas such as oxygen. In the case of aluminum nanoparticles, the oxygen forms a layer of aluminum oxide about the particles.

3.5.2. Chemical Methods

Probably the most useful methods of synthesis in terms of their potential to be scaled up are chemical methods. There are a number of different chemical methods that can

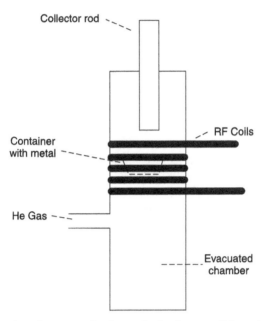

Figure 3.28. Illustration of apparatus for the synthesis of nanoparticles using an RF-produced plasma.

be used to make nanoparticles of metals, and we will give some examples. Several types of reducing agents can be used to produce nanoparticles such as $NaBEt_3H$, $LiBEt_3H$, and $NaBH_4$, where Et denotes the ethyl ($\cdot C_2H_5$) radical. For example, nanoparticles of molybdenum (Mo) can be produced in toluene solution with $NaBEt_3H$ at room temperature, providing a high yield of Mo nanoparticles having dimensions from 1 to 5 nm. The equation for the reaction is

$$MoCl_3 + 3NaBEt_3H \Rightarrow Mo + 3NaCl + 3BEt_3 + \tfrac{3}{2}H_2 \qquad (3.10)$$

Nanoparticles of aluminum have been made by decomposing $Me_2EtNAlH_3$ in toluene and heating the solution to 105°C for 2 h (Me is methyl, $\cdot CH_3$). Titanium iso-propoxide is added to the solution and acts as a catalyst for the reaction. The choice of catalyst determines the size of the particles produced. For instance, 80-nm particles were made using titanium. A surfactant such as oleic acid can be added to the solution to coat the particles, and prevent aggregation.

3.5.3. Thermolysis

Nanoparticles can be made by decomposing solids at high temperature having metal cations, molecular anions, or metal organic compounds. The process is called *thermolysis*. For example, small lithium particles can be made by decomposing lithium

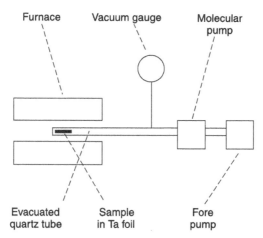

Furnace   Vacuum gauge   Molecular pump

Evacuated quartz tube   Sample in Ta foil   Fore pump

Figure 3.29. Apparatus used to form metal nanoparticles by thermally decomposing solids consisting of metal cations and molecular anions, or metal organic solids located in a furnace. (F. J. Owens, unpublished.)

azide (LiN_3). The material is placed in tantalum foil located in an evacuated quartz tube and heated in a furnace to 400°C in the apparatus shown in Fig. 3.29. At $\sim$370°C the LiN_3 decomposes, releasing N_2 gas, which is observed by an increase in the pressure on the vacuum gauge. In a few minutes the pressure drops back to its original low value, indicating that all the N_2 has been removed. The remaining lithium atoms coalesce in the oven to form small colloidal metal particles. Particles of <5 nm can be made by this method. Passivation can be achieved by introducing an appropriate gas.

The presence of these nanoparticles can be detected by electron paramagnetic resonance (EPR) of the conduction electrons of the metal particles. Electron paramagnetic resonance, which is described in more detail in Section 2.4.3, measures the energy absorbed when electromagnetic radiation such as microwaves induce a transition between the spin states m_s split by a dc magnetic field. Generally the experiment measures the derivative of the absorption as a function of an increasing dc magnetic field. Normally, because of the low penetration depth of the microwaves into a metal, it is not possible to observe the EPR of the conduction electrons. However, in a collection of nanoparticles there is a large increase in surface area, and the size is of the order of the penetration depth, so it is possible to detect the EPR of the conduction electrons. Generally EPR derivative signals are quite symmetric, but for the case of conduction electrons relaxation effects make the lines very asymmetric, and the extent of the asymmetry is related to the small dimensions of the particles. This asymmetry is quite temperature-dependent, as indicated in Fig. 3.30, which shows the EPR spectra of lithium particles at 300 and 77 K which had been made by the process described above. It is possible to estimate the size of the particles from the g-factor shift, and linewidth of the spectra.

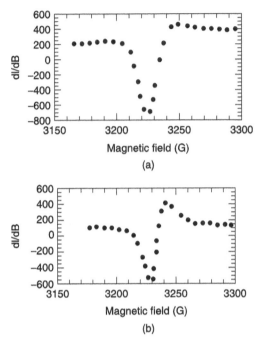

Figure 3.30. Electron paramagnetic resonance spectra at 300 K (a) and 77 K (b) arising from conduction electrons in lithium nanoparticles made from the thermal decomposition of LiN_3. (F. J. Owens, unpublished.)

3.5.4. Pulsed-Laser Methods

Pulsed lasers have been used in the synthesis of nanoparticles of silver. Silver nitrate solution and a reducing agent are flowed through a blenderlike device. In the blender there is a solid disk that rotates in the solution. The solid disk is subjected to pulses from a laser beam creating hotspots on the surface of the disk. The apparatus is illustrated in Fig. 3.31. Silver nitrate and the reducing agent react at these hotspots, resulting in the formation of small silver particles, which can be separated from the solution using a centrifuge. The size of the particles is controlled by the energy of the laser and the rotation speed of the disk. This method is capable of a high rate of production of $2-3$ g/min.

3.5.5. Synthesis of Nanosized Organic Crystals

To grow large crystals from solution, a saturated solution of the material at some higher temperature is allowed to slowly cool, or a saturated solution at some constant temperature is allowed to slowly evaporate. On the other hand, to grow small crystals, it is necessary to get the materials out of solution as fast as possible. One way to do this is illustrated in Fig. 3.32. The material is dissolved in a solvent in which it is very soluble. The solution is then placed in a syringe and injected into a liquid in which it

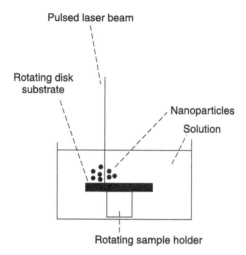

Figure 3.31. Apparatus to make silver nanoparticles using a pulsed-laser beam that creates hotspots on the surface of a rotating disk. [Adapted from J. Singh, *Mater. Today* **2**, 10 (2001).]

is not soluble. Sonication is applied to prevent aggregation of the crystals, which come out of solution rapidly. This method can produce crystals of low-micrometer dimensions. To get the material out of solution faster, more complex equipment is needed. One approach draws on the fact that the solubility of some organics in super-critical fluids such as liquid CO_2 depends on the pressure applied to the liquid. Above $31.1°C$ and 73.8 bars of pressure, CO_2 is a liquid. Figure 3.33 is a plot of the

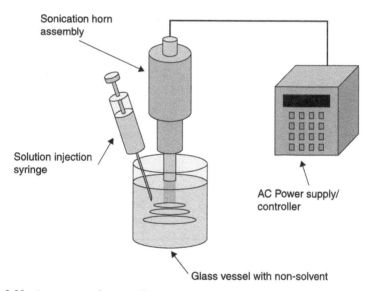

Figure 3.32. Apparatus to form small organic crystals by squirting a solution of the material into a liquid in which it is not soluble while applying sonication.

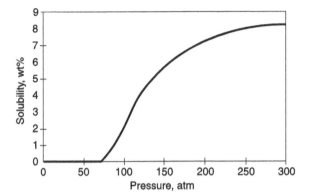

Figure 3.33. Solubility of naphthalene in liquid carbon dioxide at 45°C versus pressure. (V. Stepanov, private communication.)

solubility of naphthalene in liquid CO_2 at 45°C as a function of pressure. In this method the organic material is dissolved in liquid CO_2 under pressure. The apparatus used for this is shown in Fig. 3.34. The system consists of a container called a *saturation vessel* that holds an appropriate amount of the organic material. The CO_2 is introduced into the vessel, and its pressure and the temperature are raised to above the point where the CO_2 liquefies. To achieve optimum solubility (i.e., saturation), the vessel is packed with 3-mm glass beads coated with the organic material that is be nanosized. The pressure and the temperature are maintained for a number of hours to make sure that all the material dissolves. Sometimes a small amount of

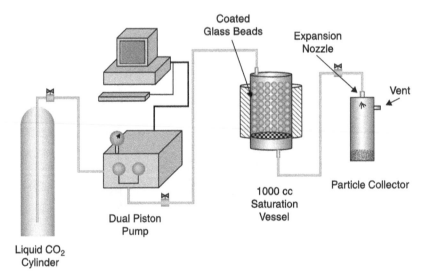

Figure 3.34. Apparatus to produce organic nanocrystals by rapid reduction of pressure over liquid carbon dioxide in which the organic material is dissolved. [V. Stepanov et al., *Propel., Explos. Pyrotech.* **30**, 178 (2005).]

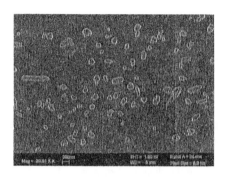
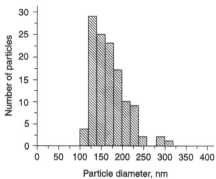

Figure 3.35. Electron microscope picture of crystals made by rapid expansion of saturated superfluids (a) and their size distribution (b). [V. Stepanov et al., *Propel. Explos. Pyrotech.* **30**, 178 (2005).]

another solvent such as acetone is added to the liquid CO_2. The valve from the saturation vessel to the collector cell is opened and there is an abrupt release of pressure, which causes the material to precipitate out of solution very rapidly, accumulating at the bottom of the collector cell. This process can produce nanosized crystals of organic materials. Figure 3.35 shows an electron microscope picture of crystals of cyclotrimethylenenitramine (an energetic material also known as RDX) made by this method. The distribution of particles sizes is shown in Fig. 3.35b. The particle size can be controlled by the magnitude of pressure applied to the liquid CO_2 before the pressure is released. Figure 3.36 is a plot of the particle size versus this pressure.

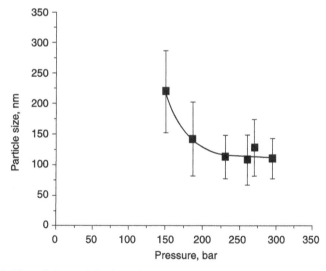

Figure 3.36. Plot of the particle size of crystals versus the pressure applied to liquid CO_2 before it is suddenly reduced. (V. Stepanov, private communication.)

3.6. SUMMARY

In this chapter a number of examples have been presented showing that the physical, chemical, and electronic properties of nanoparticles depend strongly on the number and kind of atoms that make up the particle. We have seen that the color, the reactivity, the stability, and the magnetic behavior all depend on the particle size. In some instances entirely new behavior not seen in the bulk has been observed such as magnetism in clusters that are constituted from nonmagnetic atoms. In the following chapters many of the reasons for these changes will be discussed. Besides providing new research challenges for scientists to understand the new behavior, the results have enormous potential for new applications, allowing the design of properties by control of particle size. It is clear that nanoscale materials can form the basis of a new class of atomically engineered materials.

PROBLEMS

3.1. Using Table 1.1 (of Chapter 1), plot the percentage of atoms of copper nanoparticles that are on the surface versus the particle diameter in nanometers. What properties are affected when a large percentage of the atoms are on the surface?

3.2. The aluminum oxide layer on aluminum nanoparticles is about 1.5 nm thick and is relatively independent of diameter. Assuming a spherical nanoparticle, calculate the volume fraction of the particle that is aluminum oxide from a radius of 4–20 nm at 2-nm intervals. Plot the volume fraction of the material that is aluminum oxide.

3.3. Consider a neon dimer consisting of two neon atoms separated by 3.14 Å and bonded by the Lennard-Jones potential, where $\varepsilon = 0.0013$ eV and $\sigma = 2.74$. If the dimer is doubly ionized by a pulsed laser such that each neon becomes positively charged, what will happen? Calculate the maximum kinetic energy of the neon ions and their initial velocities. In the form of the Lennard-Jones potential in the text, Equation (3.8) $C = 4\,\varepsilon\,\sigma^6$ and $B = 4\,\varepsilon\,\sigma^{12}$.

3.4. It was reported in the journal *Nature* that deuterium clusters subjected to femtosecond laser pulses underwent a Coulombic explosion (a femtosecond is 10^{-15} seconds). The fragments of the dissociation were reported to have energies up to one million electonvolts (MeV). When the deuterium fragments collided, they had sufficient energy to undergo unclear fusion by the reaction $D + D \Rightarrow$ ^3He + neutron. Do you think that this is theoretically possible? In order to evaluate this possibility, consider a linear cluster of 10 deuterium atoms with interatomic separations of 0.7416 Å. The binding energy of the end atom of the chain is 4.5 eV. Assuming that the cluster is subjected to a pulse laser that removes all the electrons of the chain, calculate the energy that the end atom carries off. Is this enough to produce fusion if it collidies head-on with another deuterium atom of the same energy?

3.5. Figure 3.22 shows the optical absorption peaks arising from excitons in Cu_2O. Fit the peaks to the energy levels for a hydrogen type atom energy-level scheme, which has the form $E_n = E_g - C/n^2$. The bandgap is 2.169 eV, and the peaks from $n = 2$ to $n = 5$ occur at 2.154, 2.158, 2.163, and 2.165 eV. Calculate the radius of the $n = 5$ orbit. How is the spectrum changed if the particle size becomes smaller than the radius of this orbit?.

3.6. Figure 3.21 plots the melting temperature of gold particles versus their diameters in the range from 2.7 to 30 nm. Calculate the melting temperatures as a function of the fraction of atoms on the surface for several particles with diameters below 15 nm, and plot their melting temperature versus the fraction of surface atoms. What do the results say about the role of surface atoms in the melting of nanoparticles?

CHAPTER 4

The Chemistry of Nanostructures

In this chapter we consider how composition, structure, and size of nanomaterials affect the chemical properties of the materials such as their ability to react with other materials. Chemical processes for producing nanostructures will also be discussed.

4.1. CHEMICAL SYNTHESIS OF NANOSTRUCTURES

4.1.1. Solution Synthesis

In Section 3.5.2 we presented one example of a chemical method to form Mo nanoparticles of diameters 1–5 nm. Many chemical methods are used to produce different kinds of nanoparticles. Here we present a few representative examples.

A colloidal dispersion of rhodium nanoparticles can be made by refluxing a solution of rhodium chloride and poly(vinyl alcohol) (PVA) in a 1 : 1 volume mixture of methanol and water near 80°C. This is done in an argon atmosphere. *Refluxing* means continuous evaporation and condensation. The methanol serves as a reducing agent according to the chemical reaction

$$RhCl_3 + \tfrac{3}{2}CH_3OH \rightarrow Rh + \tfrac{3}{2}HCHO + 3HCl \tag{4.1}$$

The presence of the polymer serves to stabilize the particles and prevents aggregation.

Aluminum nanoparticles can be produced by adding stoichiometric proportions of $AlCl_3$ and $LiAlH_4$ to trimethylbenzene. The mixture is contained in a flask under a nitrogen gas atmosphere. The mixture is then heated to the boiling point of the solvent, which is 162°C, and refluxed for 20 h. Aluminum particles are formed by the reaction

$$AlCl_3 + 3LiAlH_4 \rightarrow 3LiCl + 4AlH_3 \tag{4.2}$$

followed by

$$4AlH_3 \rightarrow 4Al + 6H_2 \tag{4.3}$$

The Physics and Chemistry of Nanosolids. By Frank J. Owens and Charles P. Poole, Jr.
Copyright © 2008 John Wiley & Sons, Inc.

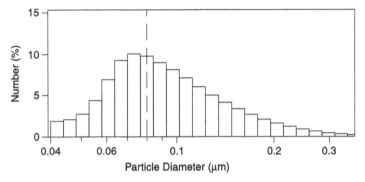

Figure 4.1. Distribution of particle sizes of Al nanoparticles made by decomposing a solution of dimethylethylamine. (V. Stepanov, private communication.)

After cooling the aluminum particles are passivated by gradually introducing air into the flask, forming the aluminum oxide layer on the surface as shown in Fig. 3.10. Particles of average size 1000 nm can be made by this method.

Aluminum nanoparticles of smaller size can be formed by decomposing a 0.5 M solution of dimethylethylamine, $Me_2EtNAlH_3$, in toluene in the presence of titanium isoproproxide (Me denotes methyl and Et is ethyl). The titanium isopropoxide serves as a substrate for the decomposition, and the particle size may be controlled by varying the concentration of the titanium isopropoxide. The solution under nitrogen is heated to 105°C for 2 h while being subjected to sound waves, a process referred to as *sonication*. The sonication helps prevent aggregation of the particles by keeping them in motion. Figure 4.1 shows the distribution of particle sizes obtained by laser light scattering for this process. The median particle size is 87 nm. In both of these methods the particles can be coated with polymers by dissolving the appropriate polymer in the solution.

Because of applications to data storage in computer systems, the synthesis of magnetic nanoparticles is a particularly important area of research. Methods for large-scale chemical synthesis of Fe_2O_3 magnetic nanoparticles have been developed. Iron chloride, $FeCl_3\ 6H_2O$, and sodium oleate, $Na[CH_3(CH_2)_7CH=CH(CH_2)_7CO_2]$, are reacted to form an iron oleate complex and NaCl. The iron oleate complex is then slowly heated in 1-octadecene, [$CH_3(CH_2)_{15}CH=CH_2$], to 320°C and held at that temperature for 30 min. The size of the particles can be controlled by the temperature of heating and the time held at the elevated temperature. Figure 4.2 shows a high-resolution transmission electron microscope (TEM) image of two different-sized particles that can be produced. It is interesting that the particles are uniform in size and spherical in shape.

4.1.2. Capped Nanoclusters

In the chemical synthesis of metal nanoparticles it is often necessary to stop the size expansion of the particles. This can be accomplished by introducing into the solution,

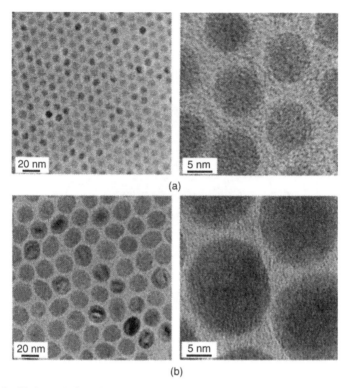

Figure 4.2. High-resolution electron microscope images of two different-sized Fe_2O_3 nano-particles (a) and (b) that can be made from ferrous chloride and sodium oleate. [Adapted from J. Park et al., *Nature Mater.* 3, 891 (2004).]

at some point during the reaction, a coating polymer or surfactant. An example of this is the capping of gold nanoparticles by attaching alkanethiolate molecules to the surface as illustrated in Fig. 4.3. The capped gold particles can be made by mixing an aqueous solution of $HAuCl_4$ with a solution of Oct_4NBr in diethyl ether (Oct denotes octyl) that contains the RSH-capping material and the reducing agent $NaBH_4$. The governing chemical reactions are

$$HAuCl_4 + Oct_4NBr(Et_2O) \rightarrow Oct_4NAuCl_4(Et_2O) + HBr \qquad (4.4)$$

followed by

$$nOct_4NAuCl_4(Et_2O) + mRSH + 3nNaBH_4 \rightarrow Au(SR)_m + \text{others} \qquad (4.5)$$

The alkanthiolate (SR)-capped gold particles can be precipitated, filtered, and washed with acetone.

Water is the most common solvent used for chemical reactions. Metal nanoparticles can be made water-soluble. For example, the gold nanoparticles capped with

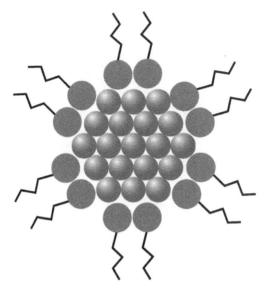

Figure 4.3. Illustration of an alkanthiolate capped gold nanoparticle.

thiols having a carboxyl group on the end will be soluble in water, forming an ionized carboxylate. This approach can be generalized to render different kinds of nanoparticles soluble by adding to whatever chemical coats the particle a chemical group that makes it soluble in water.

4.1.3. Solgel Processing

A *gel* is a semirigid colloidal dispersion of solid particles in a liquid. Typically the rigidity of the gel is such that it will not flow under the influence of gravity. Solgel processing that can be done at room temperature involves the formation of a dispersion of nanoparticles in a gel. Most commonly used are metal alkoxide precursors that readily undergo catalyzed hydrolysis and condensation to form a gel of metal oxide nanoparticles. For example, a gel of Al_2O_3 nanoparticles can be made by hydrolysis of aluminum *sec*-butoxide, (ASB), $Al(OC_4H_9)_3$ by the reaction

$$Al(OC_4H_9)_3 + H_2O \rightarrow Al(OC_4H_9)_2(OH) + C_4H_9OH \qquad (4.6)$$

A catalyst is used to start the reaction and control it. This reaction is followed by a condensation polymerization

$$Al(OC_4H_9)_2(OH) + H_2O \rightarrow AlO(OH) + 2C_4H_9(OH) \qquad (4.7)$$

which is followed by

$$2Al(OH)_3 \rightarrow Al_2O_3 + 3H_2O \qquad (4.8)$$

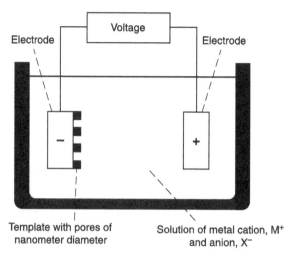

Figure 4.4. Illustration of an electrochemical cell having a nanoporous tempelate over the negative electrode that allows production of metal nanorods.

4.1.4. Electrochemical Synthesis of Nanostructures

If two electrodes are placed in a water solution of $Cu(SO)_4$ and a potential is applied between the electrodes, then copper will be deposited on the negative electrode. Electrochemical deposition can be used to synthesize metal nanorods by placing over the negative electrodes a template such as a polycarbonate containing channels of nanometer-scale diameter. Figure 4.4 illustrates the electrochemical cell for doing this. When a voltage is applied, the cations such as Cu_2^+ diffuse toward the negative electrode resulting in the growth of nanowires inside the pores of the template. After completion the template can be chemically removed. Wires of Ni, Co, Cu, and Au having diameters from 10 to 200 nm have been made by this process.

4.2. REACTIVITY OF NANOSTRUCTURES

It has long been known that as particle size decreases, the inherent reactivity of the materials increases because of the increased surface area of the particles. However, when particle size is very small (~ 1 nm), the reactivity no longer scales inversely with the particle size. We saw in Chapter 3 (Fig. 3.16) that the rate of reaction of hydrogen with iron nanoparticles having less than 25 atoms showed an oscillatory dependence on the number of atoms in the nanoparticle. These effects can be explained in terms of the jellium model of small nanoparticles, which treats the particle as a large atom. Because certain particle sizes will have filled outer shells of electrons, they will be less reactive than those having unfilled outer shells. Another way to look at this is to examine the electron affinity of particles versus size. *Electron affinity* is a measure of the ability of the nanoparticle to attract an electron. Because a reaction

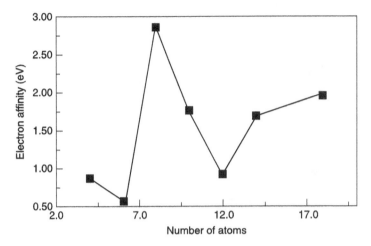

Figure 4.5. Plot of the calculated electron affinity of boron nanoparticles versus the number of atoms in the particle.

between particles involves sharing of the electrons of the particles, the electron affinity of the particle is a good index of how reactive the particle will be. For small particles the electron affinity can be calculated by molecular orbital methods such as density functional theory. The electron affinity is the difference between the calculated total energy of a neutral particle and a particle of the same structure having a negative charge. Figure 4.5 shows the calculated electron affinity of small boron nanoparticles versus the number of atoms in the particle. This clearly shows that at small sizes the reactivity of the particle does not scale inversely as the particle size.

The coating on a particle will also affect its reactivity. For example, Al nanoparticles synthesized under an inert atmosphere when exposed to oxygen rapidly react with oxygen, forming a 3-nm Al_2O_3 layer on their surface as shown in Fig. 3.10. However, once the particles are coated with Al_2O_3, they do not react with other materials provided the temperature is kept below 600°C, the melting temperature of aluminum. If they are heated above the melting temperature, the Al_2O_3 coating may crack and the bare aluminum within can react with whatever gas it is exposed to.

Chemical reactions can also be produced in materials by mechanical means such as impact or shock. The best-known examples of this are organic explosives such as trinitrotoluene (TNT). If TNT is impacted by dropping a weight on it, the material can undergo an exothermic reaction called *detonation*. The mechanism of reaction is believed to involve the movement of dislocations in the material due to the mechanical deformation. The moving dislocations generate heat, which induces chemical reactions. We will discuss dislocations in more detail in the Chapter 12 on mechanical properties. A *dislocation* is a defect in the lattice structure, which could be a line in the lattice where the adjacent rows of the lattice do not line up in the normal lattice structure. It has been found that when a powder consisting of nanosized particles is subjected to impact, it requires a much stronger impact to initiate a reaction. This is because the boundaries between the nanosized grains block dislocation movement.

4.3. CATALYSIS

4.3.1. Nature of Catalysis

Catalysis involves the modification of the rate of a chemical reaction, usually a speeding up or acceleration of the reaction rate, by the addition of a substance, called a *catalyst*, that is not consumed during the reaction. Ordinarily the catalyst participates in the reaction by combining with one or more of the reactants, and at the end of the process it is regenerated without change. In other words, the catalyst is being constantly recycled as the reaction progresses. When two or more chemical reactions are proceeding in sequence or in parallel, a catalyst can play the role of selectively accelerating one reaction relative to the others.

There are two main types of catalyst. *Homogeneous catalysts* are dispersed in the same phase as the reactants; the dispersal ordinarily is in a gas or a liquid solution. *Heterogeneous catalysts* are in a different phase than the reactants, separated from them by a phase boundary. Heterogeneous catalytic reactions usually take place on the surface of a solid catalyst, such as silica or alumina, which has a very high surface area that typically arises from their porous or spongelike structure. The surfaces of these catalysts are impregnated with acid sites, or coated with a catalytically active material such as platinum, and the rate of the reaction tends to be proportional to the accessible area of a platinum coated surface. Many reactions in biology are catalyzed by biological catalysts called *enzymes*. For example, particular enzymes can decompose large molecules into groups of smaller ones, add functional groups to molecules, or induce oxidation–reduction reactions. Enzymes are ordinarily specific for particular reactions.

Catalysis can play two principal roles in nanoscience: (1) catalysts are involved in some methods for the preparation of quantum dots, nanotubes, and a variety of other nanostructures; and (2) some nanostructures themselves can serve as catalysts for additional chemical reactions.

4.3.2. Surface Area of Nanoparticles

Nanoparticles have an appreciable fraction of their atoms at the surface, as the data in Table 1.1 demonstrate. A number of properties of materials composed of micrometer-sized grains, as well as those composed of nanometer-sized particles, depend strongly on the surface area. For example, the electrical resistivity of a granular material is expected to scale with the total area of the grain boundaries. The chemical activity of a conventional heterogeneous catalyst is proportional to the overall surface area per unit volume, so the high areas of nanoparticles provide them with the possibility of functioning as efficient catalysts. It does not follow, however, that catalytic activity will necessarily scale with the surface area in the nanoparticle range of sizes.

This is particularly true when the particle size is in the order of a nanometer or less. Figure 3.16, which is a plot of the reaction rate of H_2 with Fe particles as a function of the particle size, does not show any trend in this direction; nor does the dissociation rate plotted in Fig. 4.6 for atomic carbon formed on rhodium aggregates deposited on an

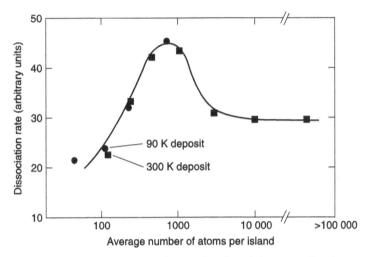

Figure 4.6. Effect of catalytic particle size on the dissociation rate of carbon monoxide. Rhodium aggregates of various sizes, characterized by the number of Rh atoms per aggregate, were deposited on alumina (Al_2O_3) films. The rhodium was given a saturation carbon monoxide (CO) coverage, then the material was heated from 90 K to 500 K (circles), or from 300 to 500 K (squares), and the amount of atomic carbon formed on the rhodium provided a measure of the dissociation rate for each aggregate (island) size.

alumina film. This result shows that catalytic activity can be optimized by an appropriate choice of size of the catalyst particle. Figure 4.7 shows that the activity or turnover frequency (TOF) of the cyclohexene hydrogenation reaction (frequency of converting cyclohexene C_6H_4 to cyclohexane C_6H_6) normalized to the concentration of surface Rh metal atoms decreases with increasing particle size from 1.5 to 3.5 nm, and then begins leveling off. The Rh particle size had been established by the particular alcohol $C_nH_{2n+1}OH$ (inset of Fig 4.7) used in the catalyst preparation, where $n = 1$ for methanol, 2 for ethanol, 3 for 1-propanol, and 4 for 1-butanol.

The specific surface area of a catalyst is customarily reported in units of square meters per gram, denoted by the symbol S, with typical values for commercial catalysts in the range from 100 to $400\,m^2/g$. The general expression for this specific surface area per gram S is

$$S = \frac{(\text{area})}{\rho\,(\text{volume})} = \frac{A}{\rho V} \qquad (4.9)$$

where ρ is the density, which is expressed in the units g/cm^3. A sphere of diameter d has area $A = \pi d^2$ and volume $V = \pi d^3/6$, to give $A/V = 6/d$. A cylinder of diameter d and length L has volume $V = \pi d^2 L/4$. The limit $L \ll d$ corresponds to the shape of a disk with the area $A \sim \pi d^2/2$, including both sides, to give $A/V \sim 2/d$. In like manner, a long cylinder or wire of diameter d and length $L \gg d$ has $A \sim 2\pi r L$ and

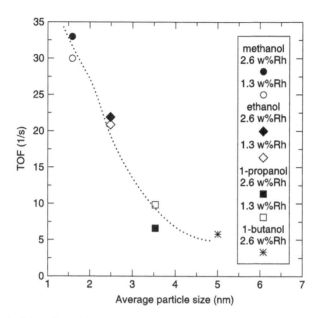

Figure 4.7. Activity of cyclohexene hydrogenation, measured by the turnover frequency (TOF) or rate of conversion of cyclohexene to cyclohexane, plotted as a function of the rhodium (Rh) metal particle size on the surface. The inset gives the alcohols (alkanols) used for the preparation of each particle size. (From G. W. Busser, J. G. van Ommen, and J. A. Lercher, Preparation and characterization of polymer-stabilized rhodium particles, in Advanced Catalysts and Nanostructured Materials Elsevier, W. R. Moser, ed., 1996, Chapter 9, p. 225.)

$A/V \sim 4/d$. Using the units square meters per gram (m^2/g) for these various geometries, we obtain the expressions

$$S(r) = \frac{6 \times 10^3}{\rho d} \qquad \text{sphere of diameter } d \qquad (4.10)$$

$$S(r) = \frac{6 \times 10^3}{\rho a} \qquad \text{cube of side } a \qquad (4.11)$$

$$S(r) \sim \frac{2 \times 10^3}{\rho L} \qquad \text{thin disk, } L \ll d \qquad (4.12)$$

$$S(r) \sim \frac{4 \times 10^3}{\rho d} \qquad \text{long cylinder or wire, } d \ll L \qquad (4.13)$$

where the length parameters a, d, and L are expressed in nanometers, and the density ρ has the units g/cm^3. In Eq. (4.12) the area of the side of the disk is neglected, and in

Eq. (4.13) the areas of the two ends of the wire are disregarded. Similar expressions can be written for distortions of the cube (Eq. 4.11) into the quantum well and quantum wire configurations.

The densities of types III/V and II/VI semiconductors, from Table B.5 (of Appendix B), are in the range from 2.42 to 8.27 g/cm³, with GaAs having the typical value $\rho = 5.32$ g/cm³. Using this density we calculated the specific surface areas of the nanostructures represented by Eqs. (4.10), (4.12), and (4.13) for various values of the size parameters d and L, and the results are presented in Table 4.1. The specific surface areas for the smallest structures listed in the table correspond to quantum dots (column 2, sphere), quantum wires (column 3, cylinder), and quantum wells (column 4, disk). Their specific surface areas are within the range typical of commercial catalysts.

The data tabulated in Table 4.1 represent minimum specific surface areas in the sense that for a particular mass, or for a particular volume, a spherical shape has the lowest possible area, and for a particular linear mass density, or mass per unit length, a wire of circular cross section has the minimum possible area. It is of interest to examine how the specific surface area depends on shape. Consider a cube of side a with the same volume as a sphere of radius r

$$\frac{4\pi r^3}{3} = a^3 \tag{4.14}$$

so $a = (4\pi/3)^{1/3}r$. With the aid of Eqs. (4.10) and (4.11), we obtain for this case $S_{cub} = 1.24\, S_{sph}$, so a cube has 24% more specific surface than a sphere with the same volume.

TABLE 4.1. Specific Surface Areas of GaAs Spheres, Long Cylinders (Wires) and Thin Disks as a Function of Size[a]

Size, nm	Surface Area, m²/g		
	Sphere	Wire	Disk
2	281	187	187
3	187	125	125
5	112	76	76
10	57	38	38
15	38	26	26
20	29	19	19
30	19	13	13
50	11	8	8
100	6	4	4

[a]The smaller sizes represent spherical quantum dots of diameter d, cylindrical quantum wires of diameter d, and planar quantum wells of thickness L, respectively.

4.3.3. Porous Materials

In the previous section we saw that an efficient way to increase the surface area of a material is to decrease its grain size or its particle size. Another way to increase the surface area is to fill the material with voids or empty spaces. Some substances, such as zeolites, crystallize in structures in which there are regularly spaced cavities where atoms or small molecules can lodge, or they can move in and out during changes in environmental conditions. A molecular sieve, which is a material suitable for filtering out molecules of particular sizes, ordinarily has a controlled narrow range of pore diameters. There are also other materials, such as silicas and aluminas, which can be prepared so that they have a porous structure of a generally random type, that is, sponge like on a mesoscopic or micrometer scale. It is quite common for these materials to have pores with diameters in the nanometer range. Pore surface areas are sometimes determined by the Brunauer–Emmett–Teller (BET) adsorption isotherm method in which the uptake of a gas such as nitrogen (N_2) by the pores is measured.

Most commercial heterogeneous catalysts have a very porous structure, with surface areas of several hundred square meters per gram. Ordinarily a heterogeneous catalyst consists of a high-surface-area material that serves as a catalyst support or substrate, and the surface linings of its pores contain a dispersed active component, such as acid sites or platinum atoms, which bring about or accelerate the catalytic reaction. Examples of substrates are the oxides silica (SiO_2), gamma-alumina (γ-Al_2O_3), titania (TiO_2 in its tetragonal anatase form), and zirconia (ZrO_2). Mixed oxides are also in common use, such as high-surface-area silica–alumina. A porous material ordinarily has a range of pore sizes, and this is illustrated by the upper right spectrum in Fig. 4.8 for the organosilicate molecular sieve MCM-41,

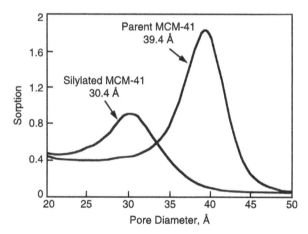

Figure 4.8. Distribution of pore diameters in two molecular sieves with mean pore diameters of 3.04 and 3.94 nm, determined by the physisorption of argon gas. [From J. S. Beck, J. C. Vartuli, W. J. Roth, M. E. Leonowicz, C. T. Kresge, K. D. Schmitt, C. T.-W. Chu, D. H. Olson, E. W. Sheppard, S. B. McCullen, J. B. Higgins, and J. L. Schenkler, *J. Am. Chem. Soc.* **114**, 10834 (1962).]

which has a mean pore diameter of 3.94 nm (39.4 Å). The introduction of relatively large trimethylsilyl groups $(CH_3)_3Si$ to replace protons of silanols SiH_3OH in the pores occludes the pore volume, and shifts the distribution of pores to a smaller range of sizes, as shown in the lower left spectrum of the figure. The detection of the nuclear magnetic resonance (NMR) signal from the ^{29}Si isotope of the trimethylsilyl groups in these molecular sieves, with its $+12$ ppm chemical shift shown in Fig. 4.9, confirmed its presence in the pores after the trimethylsilation treatment.

The active component of a heterogeneous catalyst can be a transition ion, and traditionally over the years the most important active component has been platinum dispersed on the surface. Examples of some metal oxides that serve as catalysts, either by themselves or distributed on a supporting material, are NiO, Cr_2O_3, Fe_2O_3, Fe_3O_4, Co_3O_4, and β-$Bi_2Mo_2O_9$. Preparing oxides and other catalytic materials for use ordinarily involves calcination, which is a heat treatment at several hundred degrees Celsius. This treatment can change the structure of the bulk and the surface, and Fig. 4.10 illustrates this for the catalytically active material β-$Bi_2Mo_2O_9$. We deduce from the figure that for calcination in air the grain sizes grow rapidly between 300°C and 350°C, reaching $\cong 20$ nm, with very little additional change up to 500°C. Sometimes a heat treatment induces a phase change of catalytic importance, as in the case of hydrous zirconia (ZrO_2), which transforms from a

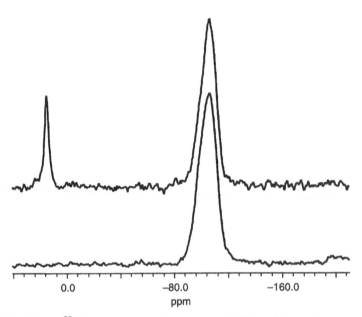

Figure 4.9. Silicon (^{29}Si) nuclear magnetic resonance (NMR) spectrum of an organosilicate molecular sieve before (lower spectrum) and after (upper spectrum) the introduction of the large trimethylsilyl groups $(CH_3)_3Si$ to replace the protons of the silanols SiH_3OH in the pores. The signal on the left with a chemical shift of $+12$ ppm arises from ^{29}Si of trimethylsilyl, and the strong signal on the right at -124 ppm is due to ^{29}Si in silanol SiH_3OH. [From J. C. Vartuli et al., in Moser, op. cit. (see Fig. 4.7 legend), Chapter 1, p. 13.]

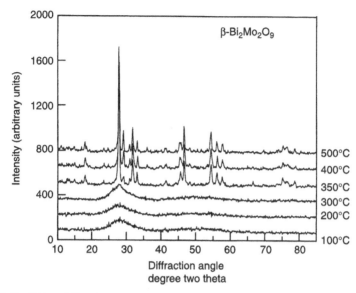

Figure 4.10. X-Ray diffraction patterns of the catalyst ß-Bi$_2$Mo$_2$O$_9$ taken at 100°C intervals during calcination in air over the range 100–500°C. [From W. R. Moser, J. E. Sunstrom, and B. Marshik-Guerts, in Moser, op. cit. (see Fig. 4.7 legend), Chapter 12, p. 296.]

high-surface-area amorphous state to a low-surface-area tetragonal phase at 450°C, as shown in Fig. 4.11. The change is exothermic, that is, a change accompanied by the emission of heat, as shown by the exotherm peak at 450°C in the differential thermal analysis (DTA) curve of Fig. 4.12.

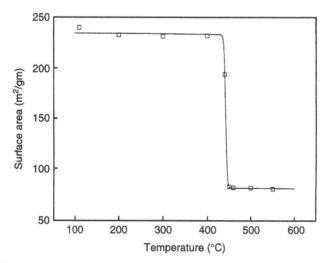

Figure 4.11. Temperature dependence of the surface area of hydrous zirconia, showing the phase transition at 450°C. [From D. R. Milburn et al., in Moser, op. cit. (see Fig. 4.7 legend), Chapter 6A, p. 138.]

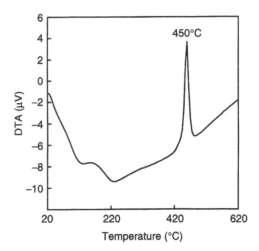

Figure 4.12. Differential thermal analysis curve of a hydrous zirconia catalyst. The heat emission peak at 450°C arises from the exothermic phase transition shown in Fig. 4.11. [From D. R. Milburn et al., in Moser, op. cit., (see Fig. 4.7 legend), Chapter. 6A, p. 137.]

For some reactions the catalytic activity arises from the presence of acid sites on the surface. These sites can correspond to either Brønsted acids, which are proton donors, or Lewis acids, which are electron pair acceptors. Figure 4.13 sketches the structures of a Lewis acid site on the left and a Brønsted acid site on the right located on the surface of a zirconia sulfate catalyst, perhaps containing platinum (Pt/ZrO_2SO_4). The figure shows a surface sulfate group SO_4 bonded to adjacent Zr atoms in a way that creates the acid sites. In mixed-oxide catalysts the surface acidity depends on the mixing ratio. For example, the addition of ≤ 20 wt% ZrO_2 to γ-Al_2O_3 does not significantly alter the activity of the Brønsted acid sites, but it has a pronounced effect in reducing the strength of the Lewis acid sites from very strong for pure γ-alumina to a rather moderate value after the addition of the 20 wt% ZrO_2. The infrared stretching frequency of the —CN vibration of adsorbed CD_3CN is a measure of the Lewis acid site strength, and Fig. 4.14 shows the progressive decrease in this frequency with increasing incorporation of ZrO_2 in the γ-Al_2O_3.

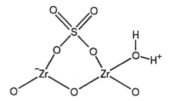

Figure 4.13. Configuration of sulfate group on the surface of a zirconia–sulfate catalyst, showing the Lewis acid site Zr⁻ at the left and the Brønsted acid site H⁺ at the right. [From G. Strukel et al., in Moser, op. cit. (see Fig. 4.7 legend), Chapter 6B, p. 147.]

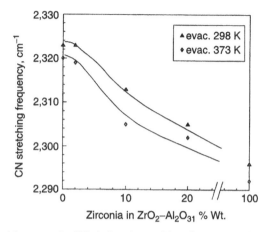

Figure 4.14. Cyanide group (—CN) infrared stretching frequency of CD_3CN adsorbed on zirconia–alumina catalysts as a function of zirconia (ZrO_2) content after evacuation at the temperatures 25°C (upper curve) and 100°C (lower curve). The stretching frequency is a measure of the Lewis acid site strength. [From G. Centi et al., in Moser, op. cit. (see Fig. 4.7 legend), Chapter 4, p. 81.]

4.4. SELF-ASSEMBLY

4.4.1. The Self-Assembly Process

Proteins are large molecules, with molecular weights in the tens of thousands, found in virtually all the cells and tissues of the body, and they are essential for life. They are formed by the successive addition of hundreds of amino acids, each brought to its site of attachment by a transfer RNA molecule, in an order prescribed by a messenger RNA molecule. Each amino acid readily bonds to the previous one when it arrives in place. Thus sequences of amino acids assemble into a polypeptide chain, which continually increases in length to eventually become a protein. This type of self-assembly process, which occurs naturally in all living systems, has its analog in nanoscience. Here we also encounter the spontaneous organization of small molecules into larger, well-defined, stable, ordered molecular complexes or aggregates, and we encounter the spontaneous adsorption of atoms or molecules onto a substrate in a systematic, ordered manner. This process involves the use of weak, reversible interactions between parts of molecules without any central control, and the result is a configuration that is in equilibrium. The procedure is automatically error-checking, so faulty or improperly attached subunits can be replaced during the growth.

   The traditional organic synthesis of very large molecules called *macromolecules* comprises a number of time-consuming steps that involve breaking and remaking strong covalent bonds, and these steps are carried out under kinetic control. The yields are small, and errors are not readily recognized or corrected. In contrast to this, the self-assembly variety of synthesis makes use of weak, noncovalent bonding interactions such as those involving hydrogen bonds and van der Waals forces, which permit the reactions to proceed under thermodynamic control, with the continual correction of errors. The initial individual molecules or subunits are usually small in size and number and easy to synthesize, and the final product is produced in a thermodynamic equilibrium state.

4.4.2. Semiconductor Islands

One type of self-assembly involves the preparation of semiconductor islands, and it can be carried out by a technique called *heteroepitaxy*, which involves the placement or deposition of the material that forms the island on a supporting substance called a *substrate* consisting of a different material with a closely matched interface between them. Heteroepitaxy has been widely used for research, as well as for the fabrication of many semiconductor devices, so it is a well-developed technique. It involves bringing atoms or molecules to the surface of the substrate where they do one of three things. They are either adsorbed and diffuse about on the surface until they join or nucleate with another adatom to form an island, they attach themselves to or aggregate into an existing island, or they desorb and thereby leave the surface. Small islands can continue to grow, migrate to other positions, or evaporate. There is a critical size at which they become stable, and no longer experience much evaporation. Thus there is an initial nucleation stage when the number of islands increases with the coverage. This is followed by an aggregation stage when the number of islands levels off and the existing ones grow in size. Finally there is the coalescence stage when the main events that take place involve the merger of existing islands with each other to form larger clusters.

The various stages can be described analytically or mathematically in terms of the rates of change dn_i/dt of the concentrations of individual adatoms n_1, pairs of adatoms n_2, clusters of size 3 n_3, and so on. An example of a kinetic equation that is applicable at the initial or nucleation stage is the following expression for isolated atoms

$$\frac{(dn_1)}{(dt)} = (R_{ads} + R_{det} + 2R_1) - (R_{evap} + R_{cap} + 2R_1') \qquad (4.15)$$

where R_{ads} is the rate of adsorption, R_{det} is the rate of detachment of atoms from clusters larger than pairs, and R_1 is the rate of breakup of adatom pairs. The negative terms correspond to the rate of evaporation R_{evap}, the rate of capture of adatoms by clusters R_{cap}, and the rate of formation of pairs of adatoms $2R_1'$. The factor of 2 associated with R_1 and R_1' accounts for the participation of two atoms in each pair process. Analogous expressions can be written for the rate of change of the number of pairs dn_2/dt, for the rate of change of the number of triplet clusters dn_3/dt, and so on. Some of the terms for the various rates R_i depend on the extent of the coverage of the surface, and the equation itself is applicable mainly during the nucleation stage.

At the second or aggregation stage the percentage of isolated adatoms becomes negligible, and a free-energy approach can provide some insight into the island formation process. Consider the Gibbs free-energy density $g_{sur-vac}$ between the bare surface and the vacuum outside, the free-energy density $g_{sur-lay}$ between the surface and the layers of adatoms, and the free-energy density $g_{lay-vac}$ between these layers and the vacuum. These are related to the overall Gibbs free-energy density g through the expression

$$g = g_{sur-vac}(1 - \varepsilon) + (g_{sur-lay} + g_{lay-vac})\varepsilon \qquad (4.16)$$

where ε is the fraction of the surface covered. As the islands form and grow the relative contributions arising from these terms gradually change, and the growth process evolves to maintain the lowest thermodynamic free energy. These free energies can be used to define a spreading pressure $P_S = g_{sur-vac} - (g_{sur-lay} + g_{lay-vac})$, which involves the difference between the bare surface free energy $g_{sur-vac}$ and that of the layers $(g_{sur-lay} + g_{lay-vac})$, and it is associated with the spreading of adatoms over the surface For the condition $(g_{sur-lay} + g_{lay-vac}) < g_{sur-vac}$, the addition of adatoms increases ε, and thereby causes the free energy to decrease. Thus the adatoms that adsorb will tend to remain directly on the bare surface, leading to a horizontal growth of islands and the eventual formation of a monolayer. The spreading pressure P_S is positive and contributes to the dispersal of the adatoms. This is referred to as the *Franck–van der Merwe growth mode*.

For the opposite condition $(g_{sur-lay} + g_{lay-vac}) > g_{sur-vac}$, growth of the fractional surface coverage ε increases the free energy, so it is thermodynamically unfavorable for the adsorbed layer to be thin and flat. The newly added adatoms tend to keep the free energy low by aggregating on the top of existing islands, leading to a vertical rather than a horizontal growth of islands. This is called the *Volmer–Weber growth mode*.

We mentioned above that heteroepitaxy involves islands or a film with a nearly matched interface with the substrate. The fraction f of mismatch between the islands and the surface is given by the expression

$$f = \frac{|a_f - a_s|}{a_s} \qquad (4.17)$$

where a_f is the lattice constant of the island or film and a_s is the lattice constant of the substrate. For small mismatches ($<2\%$), very little strain develops at the growth of a film consisting of many successive layers on top of each other. If the mismatch exceeds 3%, then the first layer is appreciably strained, and the extent of the strain builds up as several additional layers are added. Eventually, beyond a transition region the strain subsides, and thick films are strained only in the transition region near the substrate. This mismatch complicates the free-energy discussion following Eq. (4.16), and favors the growth of three-dimensional islands to compensate for the strain and thereby minimize the free energy. This is called the *Stranski–Krastanov growth mode*. A common occurrence in this mode is the initial formation of a monolayer that accommodates the strain and acquires a critical thickness.

The next stage is the aggregation of three-dimensional islands on the two-dimensional monolayer. Another eventuality is the coverage of the surface with monolayer islands of a preferred size to better accommodate the lattice mismatch strain. This can be followed by adding further layers to these islands. A typical size of such a monolayer island is perhaps 5 nm, and it might contain 12 unit cells.

In addition to the three free energies included in Eq. (4.16), the formation and growth of monolayer islands involves the free energies of the strain due to the lattice mismatch, the free energy associated with the edges of a monolayer island, and the free energy for the growth of a three-dimensional island on a monolayer. A great deal of experimental work has been done studying the adsorption of atoms on substrates for gradually increasing coverage from zero to several monolayers thick. The scanning tunneling microscope images presented in Fig. 4.15 show the successive growth of islands of InAs on a GaAs (001) substrate for several fractional monolayer coverages. We see from the data in Table B.1 (of Appendix B) that the lattice constant $a = 0.604$ nm for InAs, and $a = 0.565$ nm for GaAs, corresponding to a lattice mismatch $f = 6.6\%$ from Eq. (4.17), which is quite large.

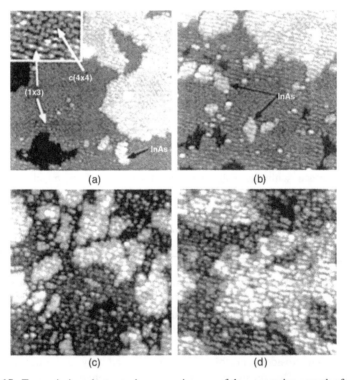

Figure 4.15. Transmission electron microscopy images of the successive growth of islands of InAs on a GaAs (001) substrate for fractional monolayer coverages of (a) 0.1, (b) 0.3, (c) 0.6, and (d) 1.0. The image dimensions are 50×50 nm for (a) and 40×40 nm for (b) to (d). The inset for (a) is an enlarged view of the GaAs substrate. [From J. G. Belk, J. L. Sudijono, M. M. Holmes, C. F. McConville, T. S. Jones, and B. A. Joyce, *Surf. Sci.* **365**, 735 (1996).]

4.4.3. Monolayers

A model system that clearly illustrates the principles and advantages of the self-assembly process is a self-assembled monolayer. The Langmuir–Blodgett technique, which historically preceded the self-assembled approach, had been widely used in the past for the preparation and study of optical coatings, biosensors, ligand-stabilized Au_{55} clusters, antibodies, and enzymes. It involves starting with clusters, forming them into a monolayer at an air–water interface, and then transferring the monolayer to a substrate in the form of a Langmuir–Blodgett film. These films are difficult to prepare, however, and are not sufficiently rugged for most purposes. Self-assembled monolayers, on the other hand, are stronger, easier to make, and utilize a wider variety of available starting materials.

Self-assembled monolayers and multilayers have been prepared on various metallic and inorganic substrates such as Ag, Au, Cu, Ge, Pt, Si, GaAs, SiO_2, and other materials. This has been done with the aid of bonding molecules or ligands such as alkanethiols RSH, sulfides RSR′, disulfides RSSR′, acids RCOOH, and siloxanes $RSiOR_3$, where the symbols R and R′ designate organic molecule groups that bond to, for example, a thiol radical —SH or an acid radical —COOH. The binding to the surface for the thiols, sulfides and disulfides is via the sulfur atom; that is, the entity RS—Au is formed on a gold substrate, and the binding for the acid is RCO_2—$(MO)_n$, where MO denotes a metal oxide substrate ion, and the hydrogen atom H of the acid is released at the formation of the bond. The alkanethiols RSH are the most widely used ligands because of their greater solubility, compatibility with many organic functional groups, and speed of reaction. They spontaneously adsorb on the surface; hence the term *self-assemble* is applicable. In this section we consider the self-assembly of the thiol ligand $X(CH_2)_nSH$, where the terminal group X is methyl (CH_3) and a typical value is $n = 9$ for decanethiol, corresponding to $C_{10}H_{21}$ for R.

To prepare a gold substrate as an extended site for the self-assembly, an electron beam or high-temperature heating element is used to evaporate a polycrystalline layer of gold that is 5–300 nm thick on to a polished support such as a glass slide, a silicon wafer, or mica. The outermost gold layer, although polycrystalline, exposes local regions of a planar hexagonal close-packing atom arrangement, as indicated in Figs. 4.16 and 4.17. Various properties such as the conductivity, opacity, domain size, and surface roughness of the film depend on its thickness. The adsorption sites are in the hollow depressions between triplets of gold atoms on the surface. The number of depressions is the same as the number of gold atoms on the surface. Sometimes Cr or Ti is added to facilitate the adhesion. When molecules in the liquid (or vapor) phase come into contact with the substrate, they spontaneously adsorb on the clean gold surface in an ordered manner; that is, they self-assemble.

The mechanism or process whereby the adsorption takes place involves each particular alkanethiol molecule $CH_3(CH_2)_nSH$ losing the hydrogen of its sulfhydryl group HS—, acquiring a negative charge, and adhering to the surface as a thiolate compound by embedding its terminal S atom in a hollow between a triplet of Au

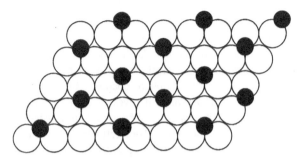

Figure 4.16. Hexagonal close-packed array of adsorbed *n*-alkanethiolate molecules (small solid circles) occupying one-sixth of the threefold sites on the lattice of close-packed gold atoms (large circles).

atoms, as noted above and indicated in Fig. 4.16. The reaction at the surface might be written

$$CH_3(CH_2)_n SH + Au_m \rightarrow CH_3(CH_2)_n S^-(Au_3^+) \cdot Au_{m-3} + \tfrac{1}{2}H_2 \qquad (4.18)$$

where Au_m denotes the outer layer of the gold film that contains m atoms. The quantity (Au_3^+) is the positively charged triplet of gold atoms that forms the hollow depression in the surface where the terminal sulfur ion (S^-) forms a bond with the gold ion Au_3^+. Figure 4.17 gives a schematic view of this adsorption process. The sulfur–gold bond that holds the alkanethiol in place is a fairly strong one ($\sim$44 kcal/mol), so the adhesion is stable. The bonding to the surface takes place at one-sixth of the sites on the (111) close-packed layer, and these sites are occupied in a regular manner so they form a hexagonal close-packed (hcp) layer with the

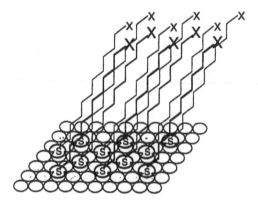

Figure 4.17. Illustration of the self-assembly of a monolayer of *n*-alkanethiolates on gold. The terminal sulfur resides in the hollow between three close-packed gold atoms, as shown in Fig. 4.16. The terminal groups labeled by X represent methyl. (From J. L. Wilber and G. M. Whitesides, in *Nanotechnology*, G. Timp, ed., Springer-Verlag, Berlin, 1999, Chapter 8, p. 336.)

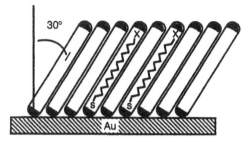

Figure 4.18. Sketch of *n*-alkanethiolate molecules adsorbed on a gold surface at an angle of 30° with respect to the normal. (From J. L. Wilber and G. M. Whitesides, in *Nanotechnology*, G. Timp, ed., Springer-Verlag, Berlin, 1999, Chapter 8, p. 336.)

lattice constant equal to $3^{1/2}a_0 = 0.865$ nm, where $a_0 = 0.4995$ nm is the distance between Au atoms on the surface. Figures 4.17 and 4.18 indicate the location of the alkanethiol sites on the gold close-packed surface.

The alkanethiol molecules RS—, bound together sideways to each other by weak van der Waals forces with a strength of about ~ 1.75 kcal/mol, arrange themselves extending lengthwise up from the gold surface at an angle of about 30° with the normal, as indicated in the sketch of Fig. 4.18. The alkyl chains R extend out ~ 2.2 nm to form the layer of hexadecanethiol $CH_3(CH_2)_{10}S$—. A number of functional groups can be attached at the end of the alkyl chains in place of the methyl group such as acids, alcohols, amines, esters, fluorocarbons, and nitriles. Thin layers of alkanethiols with $n < 6$ tend to exhibit significant disorder, while thicker ones with $n > 6$ are more regular, but polycrystalline, with domains of differing alkane twist angle. Increasing the temperature can cause the domains to alter their size and shape.

To be useful in commercial microstructures, self-assembled monolayers can be arranged in structured regions or patterns on the surface. An alkanethiol "ink" can systematically form or write patterns on a gold surface with alkanethiolate. The monolayer-forming ink can be applied to the surface by a process called *microcontact printing*, which utilizes an elastomer, which is a material with rubberlike properties, as a "stamp" to transfer the pattern. The process can be employed to produce thin radiation-sensitive layers called *resists* for nanoscale lithography. The monolayers themselves can serve for so-called passivation by protecting the underlying surface from corrosion. Alkanediols can assist in colloid preparation by controlling the size and properties of the colloids, and this application can be very helpful in improving the efficacy of catalysts.

PROBLEMS

4.1. Which would be a better catalyst, a nanoparticle with a low porosity or high porosity? Explain your answer.

4.2. Suggest a method to make aluminum nanoparticles water-soluble.

4.3. Why are alkanetiol molecues used for self-assembly on gold? Could they self-assemble on other metals? If so, which metals? Are there other molecules that could self-assemble on gold?

4.4. Develop a method to coat a gram of 50 nm copper nanoparticles with polyurethane.

4.5. What experimental methods would you use to verify that the copper particles of Problem 4.4 are coated? Explain how the methods detect the coating.

4.6. What experimental methods would you use to determine whether flat nano-structured gold substrate has thiol molecules self-assembled on its surface?

4.7. Provide an example of a self-assembly process from a biological system.

4.8. Find the specific surface area of a regular octahedron of side a.

4.9. How much surface area does a regular tetrahedron have relative to a sphere of the same volume?

4.10. Find the fraction of mismatch of (a) a film of GaAs and (b) a film of InSb grown on a [111] surface of germanium.

4.11. In the gold lattice of Fig. 4.16, how far apart are the alkane thiolate molecules from each other?

4.12. What percent of the aluminum nanoparticles in the distribution of Fig. 4.1 have diameters greater than and less than the value at the peak of the distribution?

CHAPTER 5

Polymer and Biological Nanostructures

5.1. POLYMERS

5.1.1. Polymer Structure

Nanoparticles have been prepared from various types of large organic molecules, as well as from polymers formed from organic subunits. In this chapter some of these nanostructures are described, as well as some examples of biological molecules that can be considered nanostructures, and some that resemble polymers. Organic compounds are those compounds that contain the atom carbon (C), with a few exceptions such as carbon monoxide (CO), carbon dioxide (CO_2), and carbonates (e.g., $CaCO_3$) that are classified as inorganic compounds. Almost all organic compounds also contain hydrogen atoms (H), and those that contain only carbon and hydrogen are called *hydrocarbons*. Before proceeding to discuss polymer nanoparticles, it will be helpful to present a little background material on some of the chemical concepts that will be used. Carbon has a valence of 4, and hydrogen has a valence of 1, so they can be written

$$-\overset{\displaystyle |}{\underset{\displaystyle |}{C}}- \qquad H- \tag{5.1}$$

to display their chemical bonds indicated by horizontal (—) and vertical (|) dashes. The compound methane (CH_4), which has a structural formula that may be written in one of two ways

$$
\begin{array}{cc}
\overset{\displaystyle H}{\underset{\displaystyle |}{|}} & \\
H-\overset{\displaystyle |}{\underset{\displaystyle |}{C}}-H \quad \text{or} \quad &
\begin{array}{c} H \\ H\,C\,H \\ H \end{array}
\end{array}
\tag{5.2}
$$

The Physics and Chemistry of Nanosolids. By Frank J. Owens and Charles P. Poole, Jr.
Copyright © 2008 John Wiley & Sons, Inc.

H H H H H
HC—C—C—C—CH
H H H H H

Linear hydrocarbon (pentane)

H H H H
C=C—C=C

π – conjugated butadiene

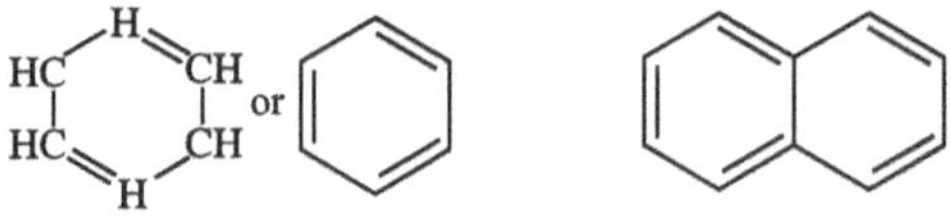

Benzene Naphthalene

Aromatic Compounds

HC≡CH
Acetylene

HC≡C—C≡CH
Diacetylene

Figure 5.1. Examples of hydrocarbon molecules. Acetylene and diacetylene serve as monomers for the formation of polymers.

is one of the simplest organic compounds, and it is a gas at room temperature. Figure 5.1 shows structural formulas of some examples of other hydrocarbons. The linear pentane molecule (C_5H_{12}) has all single bonds with the tetrahedral angle $109°28'$ between them, and the π-conjugated compound butadiene (C_4H_6) has alternating single and double bonds. The simplest aromatic compound, or π-conjugated ring compound (i.e., one with alternating single and double bonds) is benzene (C_6H_6), and the figure shows two ways to represent it. It can also be expressed in the form $H\phi$, where the phenyl group ϕ is a benzene ring that is missing a hydrogen atom, corresponding to —C_6H_5, with the dash (—) denoting an incomplete chemical bond. Naphthalene ($C_{10}H_8$) is the simplest condensed ring (fused ring) aromatic compound. The bottom panel of Fig. 5.1 shows the triple-bonded compounds acetylene and diacetylene. Aromatic compounds can contain other atoms besides carbon and hydrogen, such as chlorine Cl, nitrogen N, oxygen O, and sulfur S, as well as atomic groups or radicals such as amino (—NH_2), nitro (—NO_2), and the acid group (—COOH).

The present chapter emphasizes experimental aspects of polymeric nanoparticles. A polymer is a compound of high molecular weight that is formed from repeating subunits called *monomers*. The monomer generally has a parent compound with a double chemical bond that opens up to form a single bond during the polymerization reaction that forms the polymer. For example, consider the chemical compound styrene ($\phi CH{=}CH_2$), which has the structural formula

$$\begin{matrix} H & & H \\ C & = & C \\ \phi & & H \end{matrix} \tag{5.3}$$

Opening up the double bond forms the monomer

$$
\begin{array}{cc}
\mathrm{H} & \mathrm{H} \\
-\mathrm{C}-\mathrm{C}- \\
\phi & \mathrm{H}
\end{array}
\qquad (5.4)
$$

and a sequence of many of these monomers forms the linear polymer

$$
\begin{array}{cccccccccccccccc}
\mathrm{H} & \mathrm{H} & \mathrm{H} & \mathrm{H} & \mathrm{H} & \mathrm{H} & \mathrm{H} & \mathrm{H} & \mathrm{H} & \mathrm{H} & \mathrm{H} & \mathrm{H} & \mathrm{H} & \mathrm{H} & \mathrm{H} & \mathrm{H} \\
\mathrm{R}-\mathrm{C}-\mathrm{C}-\mathrm{C}-\mathrm{C}-\mathrm{C}-\mathrm{C}-\mathrm{C}-\mathrm{C}-\mathrm{C}-\mathrm{C}-\mathrm{C}-\mathrm{C}-\mathrm{C}-\mathrm{C}-\mathrm{C}-\mathrm{C}-\mathrm{R}' \\
\phi & \mathrm{H} & \phi & \mathrm{H} & \phi & \mathrm{H} & \phi & \mathrm{H} & \phi & \mathrm{H} & \phi & \mathrm{H} & \phi & \mathrm{H} & \phi & \mathrm{H}
\end{array}
\qquad (5.5)
$$

which can be written in the abbreviated form

$$
\mathrm{R}\left[-\mathrm{C}\begin{array}{c}\mathrm{H}\\ \\ \phi\end{array}\!\!-\!\!\mathrm{C}\begin{array}{c}\mathrm{H}\\ \\ \mathrm{H}\end{array}\!\!-\right]_n \mathrm{R}'
\qquad (5.6)
$$

where, in the structural formula (5.5), the index $n = 8$. The groups R and R' are added to the ends to satisfy the chemical bonds of the terminal carbons. This particular compound (5.6) is called *polystyrene* where R and R' are hydrogen and it is referred to as a *linear polymer*, although in reality it is staggered since the chemical bonds C—C—C in the carbon chain subtend the tetrahedral angle $\theta \cong 109°28'$ between them.

The more complex monomer

$$
\begin{array}{cc}
\mathrm{H} & \mathrm{A} \\
-\mathrm{C}-\mathrm{C}- \\
\mathrm{H} & \mathrm{B}
\end{array}
\qquad (5.7)
$$

forms the polymer

$$
\mathrm{R}\left[-\mathrm{C}\begin{array}{c}\mathrm{H}\\ \\ \mathrm{H}\end{array}\!\!-\!\!\mathrm{C}\begin{array}{c}\mathrm{A}\\ \\ \mathrm{B}\end{array}\!\!-\right]_n \mathrm{R}'
\qquad (5.8)
$$

When A is the methyl radical $-\mathrm{CH_3}$ and B is the radical $-\mathrm{COOCH_3}$, then the polymer is polymethylmethacrylate. It generally has a molecular weight between 10^5 and 10^6 daltons (Da, g/mol), and since the molecular weight of the monomer $\mathrm{CH_2}=\mathrm{C(CH_3)CO_2CH_3}$ is 100 Da, there are between 1000 and 10,000 monomers in this polymer. Polymers are also formed from a number of other radicals, such as allyl ($\mathrm{CH_2}=\mathrm{CHCH_2}-$) and vinyl ($\mathrm{CH_2}-\mathrm{CH}-$), where the free bond denoted by ($-$) on the right resides on the carbon atom. Natural rubber is a polymer based on the isoprene molecule

$$
\begin{array}{ccc}
\text{H} & \text{H} & \text{CH}_3 \\
| & | & | \\
\text{H}-\text{C}=\text{C}-\text{C}=\text{CH}_2
\end{array}
\tag{5.9}
$$

which has a pair of double bonds. The second double bond is used to form cross-linkages between the linear polymer chains. The methyl group —CH_3 at the top of the molecule is an example of a sidechain on an otherwise linear molecule.

5.1.2. Sizes of Polymers

Polymers are generally classified by their molecular weight, and to discuss them from the nanoparticle aspect, we need a convenient way to convert molecular weight to a measure of the polymer size d. The volume V in the units cubic nanometers (nm^3) of a substance of molecular weight M_W and density ρ is given by

$$
V = 0.001661 \frac{M_W}{\rho}
\tag{5.10}
$$

where M_W is expressed in daltons or g/mol and ρ, in the conventional units g/cm^3. If the shape of the nanoparticle is fairly uniform, with very little stretching or flattening in any direction, then a rough measure of its size is the cube root of the volume (5.10), which we call the size parameter d:

$$
d = 0.1184 \, (M_W/\rho)^{1/3} \, \text{nm}
\tag{5.11}
$$

This expression is exact for the shape of a cube, but it can be used to estimate average diameters of polymers of various shapes. If the molecule is a sphere of diameter D_0, then we know from solid geometry that its volume is given by $V = \pi D_0^3/6$, and inserting this in Eq. (5.10) provides the expression $d_{sph} = D_0 = 0.1469 \, (M_W/\rho)^{1/3}$ nm for a spherical molecule. For a molecule shaped like a cylinder of diameter D and length (or height) L with the same volume as a sphere of diameter D_0, we have the expression $\pi D_0^3/6 = \pi D^2 L/4$, which gives

$$
D_0 = \left(\frac{3}{2}\right)^{1/3} (D^2 L)^{1/3} = \left(\frac{3}{2}\right)^{1/3} D \left(\frac{L}{D}\right)^{1/3} = \left(\frac{3}{2}\right)^{1/3} L(D/L)^{2/3}
\tag{5.12}
$$

These equivalent relationships permit us to write expressions for the diameter and the length of the cylinder in terms of its length : diameter ratio, and the molecular weight of the molecule

$$D = 0.128 \left(\frac{M_W}{\rho}\right)^{1/3} \left(\frac{D}{L}\right)^{1/3} \tag{5.13}$$

$$L = 0.128 \left(\frac{M_W}{\rho}\right)^{1/3} \left(\frac{L}{D}\right)^{2/3} \tag{5.14}$$

where D and L are measured in nanometers. These expressions are plotted in Figs. 5.2 and 5.3 for $D > L$ and for $L > D$, respectively. The figures can be employed to estimate the size parameter for an axially shaped, flat, or elongated polymer if its molecular weight, density, and length : diameter ratio is known. The curves in these figures were drawn for the density $\rho = 1 \text{ g/cm}^3$, but the correction for density is

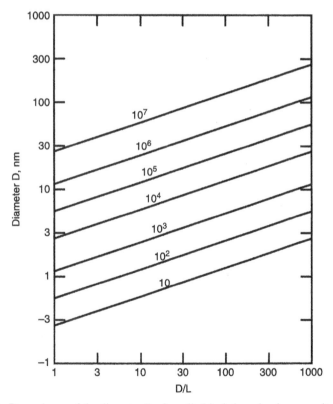

Figure 5.2. Dependence of the diameter D of a cylindrical shaped polymer on its diameter : length ratio D/L for molecular weights ranging from 10 to 10^7 Da, as indicated on the curves. A density $\rho = 1 \text{ g/cm}^3$ was assumed in Eq. (5.13) for plotting these curves.

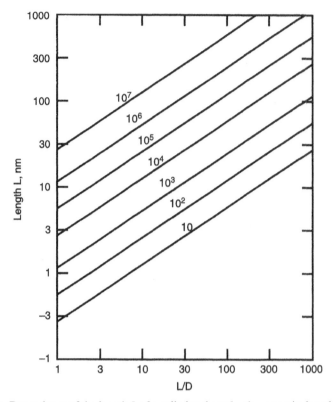

Figure 5.3. Dependence of the length L of a cylinder-shaped polymer on its length : diameter ratio L/D for molecular weights ranging from 10 to 10^7 Da, as indicated on the curves. A density $\rho = 1$ g/cm^3 was assumed in Eq. (5.14) for plotting these curves.

easily made since most polymer densities are close to 1. Typical polymers have molecular weights ranging from 10^4 to 10^7 Da.

5.1.3. Nanocrystals of Polymers

The nanocrystals listed toward the bottom of Table 5.1 were prepared by recrystallization from a polydiacetylene polymer with a backbone formed from the diacetylene molecule, which has the structure sketched at the lower part of Fig. 5.1. The starting compound is RC≡C—C≡CR′; examples of the R and R′ groups are given in Fig. 5.4. The monomer for the polymer is formed by relocating the group R, and converting one triple bond (≡) to a double bond (=) in the following manner

$$\begin{array}{cc} \text{R} & \text{R}'\ \text{h}\nu & \text{R} \quad \text{R}' \\ \text{C}{\equiv}\text{C}{-}\text{C}{\equiv}\text{C} \Rightarrow -\text{C}{\equiv}\text{C}{-}\text{C}{=}\text{C}{-} \end{array} \qquad (5.15)$$

TABLE 5.1. Nanocrystals Prepared by Recrystallization of Compounds Listed in Fig. 5.4 and Recrystallization of Diacetylene Polymers with Sidechains Listed in Fig. 5.4

Material	Type	Crystallite Size
Anthracene $C_{14}H_{10}$	π-Conjugated compound	150 nm–1 μm
PIC	π-Conjugated compound	200–300 nm
Perylene $C_{20}H_{12}$	π-Conjugated compound	50–200 nm
Fullerene C_{60}	π-Conjugated compound	200 nm
4-BCMU	Polydiacetylene	200 nm–1 μm
DCHD	Polydiacetylene	15 nm to 1 μm
DCHD	Polydiacetylene	1-μm-long, 60-nm-wide microfiber
14-8ADA	Polydiacetylene	15–200 nm

Source: Data were obtained from H. Kasai et al., in *Handbook of Nanostructured Materials and Nanotechnology*, H. S. Nalwa, ed., Academic Press, Boston, 2000, Vol. 5, Chapter 8, p. 441.

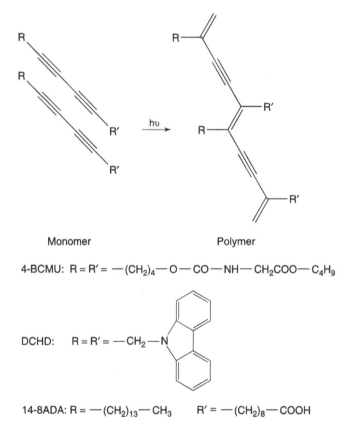

Monomer   Polymer

4-BCMU: R = R′ = —$(CH_2)_4$—O—CO—NH—CH_2COO—C_4H_9

DCHD: R = R′ = —CH_2—N⟨carbazole⟩

14-8ADA: R = —$(CH_2)_{13}$—CH_3 R′ = —$(CH_2)_8$—COOH

Figure 5.4. Chemical structures of diacetylene monomers used to form polyacetylene nanocrystals. (From H. Kasai et al., in *Handbook of Nanostructured Materials and Nanotechnology*, H. S. Nalwa, ed., Academic Press, Boston, 2000, Vol. 5, Chapter 8, p. 441.)

where the resulting open bonds (—) at the ends of the structure are used to attach successive monomers to each other during polymerization via the single bonds. The polymerization process in the solid state is brought about by the application of heat, UV light, or γ-irradiation (via γ rays), as indicated by the photon notation $h\nu$ above the arrow in Eq. (5.15). The polymer backbone is fully conjugated by its system of alternating single and double/triple bonds, and as a result of this conjugation all of the carbon atoms in the backbone lie in the same plane. Both the solid forms and solutions of diacetylene polymers exhibit many bright colors: red, yellow, green, blue, gold.

Diacetylene monomers have the capability of forming perfect crystals in the solid state, with every polymer chain reaching from one end of the crystal to the other. This occurs when the crystal size is less than the usual length of the polymer in a bulk material, and it causes the polymer molecular weight to depend on the nanocrystal size. A typical molecular weight is 10^6 Da. Figure 5.5 shows a 130-nm rectangular microcrystal of the polymer 4-BCMU, which has the chemical structure given in Fig. 5.4. The compound DCHD was found to form both nanocrystals like the one illustrated in Fig. 5.5, as well as nanofibers about 7 μm long, with diameters of about 60 nm.

These materials have a number of important applications, such as in nonlinear optics. Small nanocrystals of the polydiacetylene compound DCHD exhibit quantum size effects, with the excitonic absorption peaks of 70-, 100-, and 150-nm crystals appearing

Figure 5.5. Scanning electron microscope picture of a poly(4-BCMU) single nanocrystal about 130 nm in size. (From H. Kasai et al., in *Handbook of Nanostructured Materials and Nanotechnology*, H. S. Nalwa, ed., Academic Press, Boston, 2000, Vol. 5, Chapter 8, p. 443.)

at wavelengths of 640, 656, and 652 nm, respectively, which is the expected shift toward lower energies (i.e., longer wavelengths) with increasing particle size.

5.1.4. Conductive Polymers

Many nanoparticles are metals, such as the structural magic number particle Au_{55}. The metals under consideration in their bulk form are good conductors of electricity. There are also polymers, called conductive polymers or organic metals, which are good conductors of electricity, and polyacetylene is an example. Many polyaniline-based polymers are close to silver in the galvanic series, which lists metals in the order of their potential or ease of oxidation. Acetylene $HC\equiv CH$ has the monomer

$$\begin{matrix} H & H \\ -C = C- \end{matrix} \qquad (5.16)$$

corresponding to the repeat unit $[-CH=CH-]_n$. Other examples of compounds that produce conducting polymers are the benzene derivative aniline $C_6H_5NH_2$, which may also be written ϕNH_2, and the two 5-membered ring compounds pyrrole C_4H_4NH and thiophene C_4H_4S. The latter two compounds have the same planar structure with the sulfur atom S of thiophene replacing the nitrogen atom N of pyrrole. See Fig. 10.20 of our earlier work.[1] These molecules all have alternating double–single chemical bonds, and hence they form polymers which are π-conjugated. The π-conjugation of the carbon bonds along the oriented polymer chains provides pathways for the flow of conduction electrons, and hence it is responsible for the good electrical conduction along individual polymer nanoparticles. Polarons, or electrons surrounded by clouds of phonons, may also contribute to this intrinsic conductivity. The overall conductivity, however, is less than this intrinsic conductivity, and must take into account the particulate nature of the polymer.

Wessling[2] has proposed an explanation of the high electrical conductivity of conductive polymers such as polyacetylene and polyaniline on the basis of their nanostructure involving primary particles with a metallic core of diameter $\cong 8$ nm surrounded by an amorphous nonconducting layer 0.8 nm thick of the same $[C_2H_2]_n$ composition. Figure 5.6 presents a sketch of the model proposed by Wessling based on scanning electron microscope pictures of conductive polymers. The individual nanoparticles are seen joined together in networks comprising 30–50 particles, with branching every 10 or so particles. Several of the nanoparticles are pictured with their top halves removed to display the inner metallic core and their surrounding amorphous coating. The electrical conductivity mechanism is purely metallic within each particle, and involves thermally activated tunneling of the electrons responsible for the passage of electric current through the outer amorphous layer from one particle to the next. Thus bulk conductive polymers are truly nanomaterials because of their $\cong 10$ nm microstructure. In many cases it is easier to prepare conductive polymers in the nanoparticle range of dimensions than it is to prepare conventional metal particles in this size range.

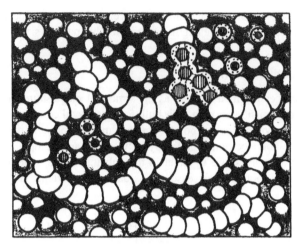

Figure 5.6. Sketch of the nanostructure of a polyacetylene conductive polymer showing the $\cong$ 9.6-nm-diameter nanoparticles. The top halves of several of these nanoparticles have been removed to display the $\cong$ 8-nm-diameter metallic core and the $\cong$ 0.8-nm-thick surrounding amorphous coating. This illustration was reconstructed from scanning electron microscope pictures. (From B. Wessling, in *Handbook of Nanostructured Materials and Nanotechnology*, H. S. Nalwa, ed., Academic Press, Boston, 2000, Vol. 5, Chapter 10, p. 512.)

Polyaniline and its analogs change color with the application of particular voltages and suitable chemicals, that is they are electrochromic and chemochromic. This makes them appropriate candidates for use in light-emitting diodes (LEDs). Other applications are the surface finish of printed-circuit boards, corrosion protection for metal surfaces, semitransparent antistatic coatings for electronic products, polymeric batteries, and electromagnetic shielding.

5.1.5. Block Copolymers

We have seen that a polymer is a very large molecule composed of a chain of individual basic units called *monomers* joined together in sequence. A copolymer is a macromolecule containing two or more types of monomers, and a block copolymer has these basic units or monomer types joined together in long individual sequences called *blocks*[3]. Of particular interest is a diblock polymer $(A)_m(B)_n$, which contains a linear sequence of m monomers of type A joined through a transition section to a linear sequence of n monomers of type B. An example of a diblock polymer is *polyacetylene–transition section–polystyrene* with the following structure:

[Endgroup] – [polyacteylene] – [transition member] – [polystyrene]

$\quad$ – [endgroup] $\hfill$ (5.17)

For a particular case, this may be written in a more detailed manner as

$$C_2H_5-[-CH{=}CH-]_m-CH_2-C\phi_2-[-C\phi H-CH_2-]_n-CH_3 \qquad (5.18)$$

where the endgroups have been chosen as the ethyl $-C_2H_5$ and methyl $-CH_3$ radicals, and the transition member that joins the two polymer sequences is the chemical group $-CH_2-C\phi_2-$. Of greater practical importance are more complex copolymers that contain several or many monomer sequences of the types $(A)_m$ and $(B)_n$.

If the conditions are right, then individual polymers are able to self-assemble to produce copolymers. In many cases one polymer component is water-soluble and the other is not. Some examples of nanostructures fabricated from copolymers are hairy nanospheres, star polymers, and polymer brushes, which are illustrated in Fig. 5.7. The nanosphere can be constructed from one long polymer $(A)_m$ that coils up and develops crosslinks between adjacent lengths of strands to give rigidity to the sphere. The projections from the nanosphere surface are sets of the other copolymer element $(B)_n$ attached to the spherical surface formed from polymer $(A)_m$. If the lengths of the projections formed by $(B)_n$ are short compared, to the sphere diameter, then the nanostructure is called a "hairy nanosphere"; if the sphere is small and the projections are long, the nanostructure is called a "star polymer." If the nanosphere is hollow, then fibers of the polymer $(B)_n$ can also project inward from the inner surface of the spherical shell.

Star polymers are used in industry to improve the melt strength, that is, the mechanical properties of molten plastic materials. Hairy nanospheres have been employed for the removal of organic compounds from water, both in a dispersed form and as solid microparticles. Polymer brushes are effective for dispersing latex and pigment particles in paint. Nanostructures consisting of block copolymers that

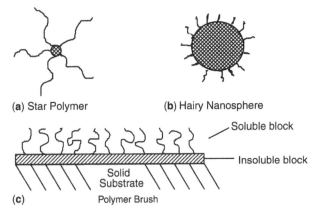

(a) Star Polymer   (b) Hairy Nanosphere

Soluble block

Insoluble block

Solid
Substrate

(c)   Polymer Brush

Figure 5.7. Sketch of a star polymer (a), a hairy nanosphere (b), and a polymer brush (c). (From G. Liu, in *Handbook of Nanostructured Materials and Nanotechnology*, H. S. Nalwa, ed., Academic Press, Boston, 2000, Vol. 5, Chapter 9, pp. 479, 488.)

function as catalysts are utilized in the production of nanosized electronic devices, and find applications for water reclamation. Otsuka et al.[4] pointed out that block co-polymers adsorbed on surfaces in brush or micelle forms, or self-assembled into micelles, provide a powerful tool for manipulating the characteristics of surfaces and interfaces. Block copolymers are expected to have novel applications, especially of the biomedical type.

5.2. BIOLOGICAL NANOSTRUCTURES

5.2.1. Sizes of Biological Nanostructures

It is customary to define nanoparticles or nanostructures as entities in the range of sizes from 1 to 100 nm, so many biological materials are classified as nanoparticles. Bacteria, which range in size between 1 and 10 μm, are in the mesoscopic range, while viruses with dimensions of 10–200 nm are at the upper part of the nanoparticle range. Proteins, which ordinarily come in sizes between 4 and 50 nm, are in the low-nanometer range. The building blocks of proteins are 20 amino acids, each about 0.6 nm in size, which is slightly below the official lower limit of a nanoparticle. More than 100 amino acids occur naturally, but only 20 are involved in protein synthesis. To construct a protein, combinations of these latter amino acids are tied together one after the other by strong peptide chemical bonds and form long chains called *polypeptides* containing hundreds, and in some cases thousands, of amino acids; hence they correspond to nanowires. The polypeptide nanowires undergo twistings and turnings to compact themselves into a relatively small volume corresponding to a polypeptide nanoparticle with a diameter typically ranging from 4 to 50 nm. Thus a protein is a nanoparticle consisting of a compacted polypeptide nanowire. The genetic material deoxyribonucleic acid (DNA) also has the structure of a compacted nanowire. Its building blocks are four nucleotide molecules that bind together in a long double-helix nanowire to form the chromosomes that in humans contain about 140 million nucleotides in sequence. Thus the DNA molecule is a double nanowire, two nucleotide nanowires twisted around each other with a repeat unit every 3.4 nm, and a diameter of 2 nm. This long double-stranded nanowire also undergoes systematic twistings and turnings to become compacted into a chromosome about 6 μm long and 1.4 μm wide. The chromosome itself is not small enough to be a nanoparticle; rather it is in the mesoscopic range of size.

To gain some additional perspective about the overall scope of nanometer range sizes involved in the buildup of biological structures let us consider the tendon as a typical structure.[5] The function of a tendon is to attach a muscle to a bone. From the viewpoint of biology the fundamental building block of a tendon is the assemblage of amino acids (0.6 nm) that form the gelatinlike protein called *collagen* (1 nm) that coils into a triple helix (2 nm). There follows a threefold sequence of fiber-like or fibrillar nanostructures: a microfibril (3.5 nm), a subfibril (10–20 nm), and a fibril itself (50–500 nm). The final two steps in the buildup, namely the cluster of fibers called a *fascicle* (50–300 μm) and the tendon itself (10–50 cm), are far beyond the nanometer range of sizes. The fascicle is considered mesoscopic, and the tendon macroscopic in size. Since the smallest amino acid glycine is about 0.42 nm in size, and some viruses reach 200 nm, it seems appropriate to define a biological nanostructure as being in the nominal range from 0.5 to 200 nm. With this in mind, the remainder of this chapter focuses on some examples of biological structures of nanometer dimension. In addition, we will also comment on some special cases in which artificially constructed nanostructures are of importance in biology.

There are a number of ways to determine or estimate the size parameters d of the fundamental biological building blocks that are amino acids for proteins, and nucleotides for DNA. If the crystal structure is known for the building block molecule, and there are n molecules in the crystallographic unit cell, then one can divide the unit cell volume V_U by n and take the cube root of the result to obtain an average size or average dimension

$$d = \left(\frac{V_U}{n}\right)^{1/3} \tag{5.19}$$

If the crystal structure is orthorhombic, then the unit cell is a rectangular box of length a, width b, and height c with volume $V_U = a \times b \times c$, to give for the average size of the molecule $d = (a \times b \times c/n)^{1/3}$, where n is the number of molecules in the cell. In a typical case n is 2 or 4. For the higher-symmetry tetragonal case, we set $a = b$ in this expression, and the cubic case for $n = 1$ has the special result $a = b = c = d$. One can also deduce the size by reconstructing the molecule from knowledge of its atomic constitution, taking into account the lengths and angles of the chemical bonds between its atoms.

Another common way to determine the size of a biological molecule is to observe it using an electron microscope, which can provide pictures called *electron micrographs* taken from various molecular orientations. This approach is especially useful for larger nanosized objects such as proteins or viruses. Figure 5.8 presents

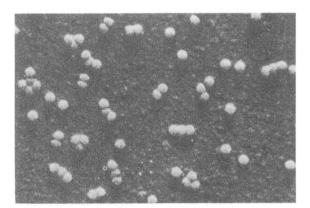

Figure 5.8. Micrograph of poliomyelitis virus, with an amplification of 74,000×. (From R. C. Williams, in *Textbook of Modern Biology*, A. Nason, ed., Wiley, New York, 1965, p. 81.)

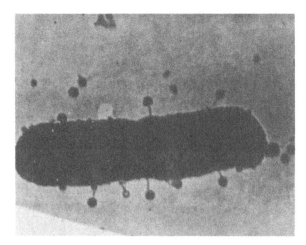

Figure 5.9. Micrograph of bacteriophages attacking a bacterium, with an amplification of 41,580×. (From R. C. Williams, in *Textbook of Modern Biology*, A. Nason, ed., Wiley, New York, 1965, p. 82.)

a micrograph of the poliomyelitis virus enlarged 74,000 times, and Fig. 5.9 shows a bacteriophage attacking a bacterium at an enlargement of 41,600×. A *bacteriophage* is a type of virus that attacks and infects bacteria. The poliomyelitis virus has a diameter of 30 nm, and the bacteriophage shown in Fig. 5.9 has a 40-nm head attached to a tail that is 100 nm long and 13 nm wide. The bacterium in the illustration is 1600 nm (i.e., 1.6 μm) long and 360 nm wide. These figures confirm our earlier observation that viruses are nanoparticles, and that bacteria are mesoscopic, beyond the nanoparticle size range. Figure 5.10 provides sketches of the shapes of four well-known proteins with dimensions ranging from 4 to 76 nm.

Many biological macromolecules such as proteins are characterized by their molecular weight M_W, and it was shown earlier that the size d of a molecule or nanoparticle is related to its molecular weight M_W and its density ρ through the expression

$$d = 0.1184 \left(\frac{M_W}{\rho} \right)^{1/3} \text{nm} \tag{5.20}$$

where M_W is expressed in daltons (Da, g/mol) and ρ, in the conventional units g/cm^3. This formula assumes that the nanoparticle is fairly uniform in shape, with very little stretching or compression in any direction. If the molecule is flat, or perhaps elongated like γ-globulin or fibrinogen, which have the shapes depicted in Fig. 5.10, then either Fig. 5.2 or Fig. 5.3 can be used to deduce the size.

Crystallographic data can be employed to calculate the density of amino acids, and the *Handbook of Chemistry and Physics* reports the densities 1.43, 1.607 and 1.316 g/cm^3 for the amino acids alanine, glycine and valine, respectively, and of course $\rho = 1$ g/cm^3 for water. Proteins have a less compact structure and hence

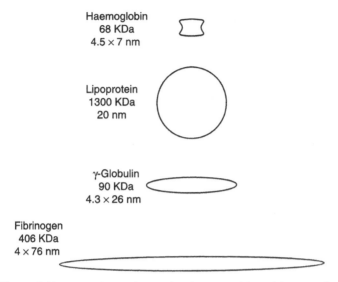

Haemoglobin
68 KDa
4.5 × 7 nm

Lipoprotein
1300 KDa
20 nm

γ-Globulin
90 KDa
4.3 × 26 nm

Fibrinogen
406 KDa
4 × 76 nm

Figure 5.10. Approximate sizes and molecular weights of four proteins.

lower values of ρ than do their constituent amino acids. With these considerations, we arrive at the following approximate expression

$$d = 0.12(M_W)^{1/3} \text{ nm} \tag{5.21}$$

which we will use to estimate the sizes of biological macromolecules. For example, the protein hemoglobin, which has a molecular weight $M_W = 68,000$ Da, has the length parameter $d = 4.8$ nm, well within the nanoparticle range.

The twistings and turnings of polypeptide nanowires to form a compact protein structure held together by weak hydrogen and disulfide (—S—S—) bonds can be somewhat loose, with spaces present between the polypeptide nanowire sections, so the density of the protein is less than that of its constituent amino acids in the crystalline state. This can cause Eq. (5.21) to underestimate the size parameter of a protein. If the molecular weight of a protein is known and the volume is determined from an electromicrograph such as those pictured in Figs. 5.8 and 5.9, then the density ρ in units g/cm^3 can be calculated by an inversion of Eq. (5.10)

$$\rho = 0.001661 \frac{M_W}{V} \tag{5.22}$$

where the molecular weight M_W is in daltons and the volume V is in cubic nanometers.

Table 5.2 lists the molecular weights M_W and various length parameters d for a number of biological nanoparticles, Fig. 5.10 provides estimated molecular weights and dimensions of four proteins, and Table 5.3 lists sizes for biological structures

TABLE 5.2. Typical Sizes of Various Biological Substances in Nanometer Range

Class	Material	M_w, Da	Size d, nm
Amino acids	Glycine (smallest amino acid)	75	0.42
	Tryptophan (largest amino acid)	246	0.67
Other molecules	Steric acid ($C_{17}H_{35}CO_2H$)	284	0.87
	Chlorophyll, in plants	720	1.1
Nucleotides	Cytosine monophosphate (smallest DNA nucleotide)	309	0.81
	Guanine monophosphate (largest DNA nucleotide)	361	0.86
	Adenosine triphosphate (ATP, energy source)	499	0.95
Proteins	Insulin, polypeptide hormone	6,000	2.2
	Hemoglobin, carries oxygen	68,000	4.8
	Albumin, in white of egg (albumen)	69,000	4.8
	Elastin, cell-supporting material	72,000	5.0
	Fibrinogen, for blood clotting	400,000	9.0
	Lipoprotein, carrier of cholesterol (globular shape)	1,300,000	13
	Ribosome (where protein synthesis occurs)		30
	Glycogen granules of liver		150
Virus	Influenza		60
	Tobacco mosaic, length		120
	Bacteriophage T_2		140

TABLE 5.3. Typical Sizes of Various Biological Substances in Mesoscopic Range

Class	Material	Size d, μm
Organelles (structures in cells outside nucleus)	Mitochondrion, where aerobic respiration produces ATP molecules	$0.5 \times 0.6 \times 1.4$
	Chloroplast, site of photosynthesis, length	4.0
	Lysosome (vesicle with enzymes for digesting macromolecules)	0.7
	Vacuole of amoeba	10
Cells	Escherichia coli bacterium, length	8
	Human blood platelet	2
	Leukocytes (white blood cells), globular shape	8–15
	Erythrocytes (red blood cells), disk shape	1.5×8
Miscellaneous	Human chromosome, length	9
	Fascicle in tendon	50–300

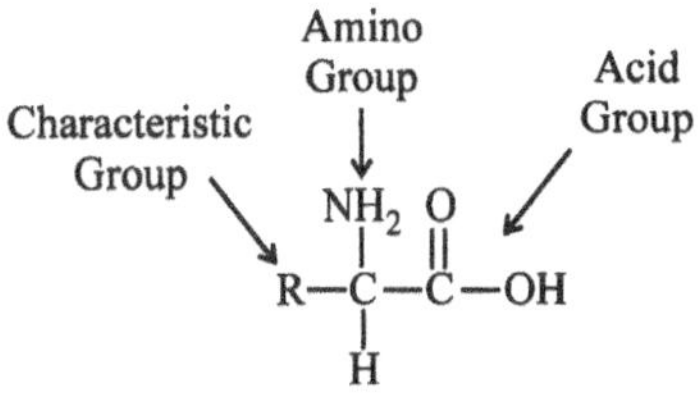

Figure 5.11. Chemical structure of an amino acid.

and quantities that are larger than nanoparticles, in the micrometer region. All of the amino acids have the common structure sketched in Fig. 5.11, with the acid or carboxyl group —COOH at one end; an adjacent carbon atom that is bonded to a hydrogen atom; an amino group NH_2; and a group R, which characterizes the particular amino acid. Figure 5.12 presents the structures of six of the amino acids,

NAME	SYMBOL	Mw, Da size d, nm	RNA WORDS (CODONS)	STRUCTURE
Glycine	Gly	75.07 Da 0.42 nm*	GGU GGA GGC GGG	NH_2-CH_2-COOH
Alanine	Ala	89.09 Da 0.47 nm*	GCU GCA GCC GCG	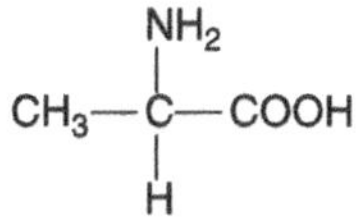
Valine	Val	117.12 Da 0.54 nm	GUU GUA GUC GUG	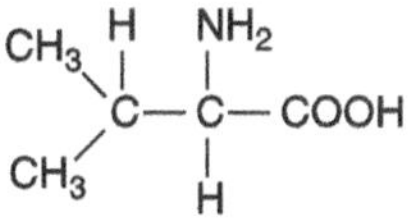
Threonine	Thr	119.12 Da 0.54 nm	ACU ACA ACC ACG	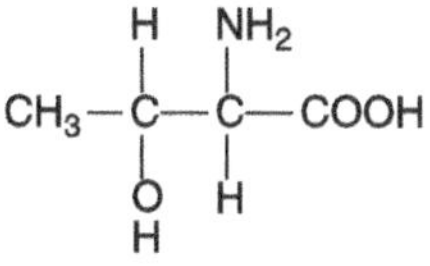
Glutamate (Glutamic acid)	GluN	147.13 Da 0.58 nm	GAA GAG	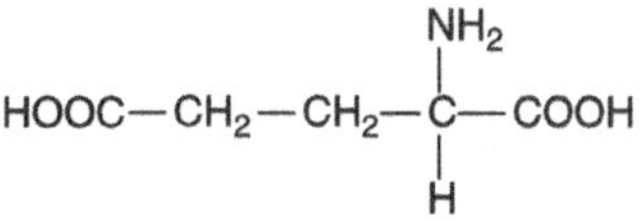
Tryptophan	Try	246.27 Da 0.67 nm	UGG	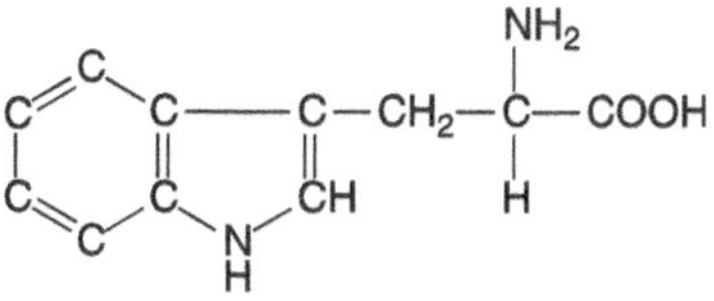

Figure 5.12. Names, symbols, molecular weights M_W, size parameters d, three-letter RNA genetic codewords (codons), and structures of 6 of the 20 amino acids, with the smallest amino acid (glycine) at the top, and the largest (tryptophan) at the bottom.

including the smallest acid glycine for which the R group is simply a hydrogen atom H, and the largest tryptophan in which R is a conjugated double-ring system. The structures of the nucleotide building blocks of DNA and RNA are presented below.

5.2.2. Polypeptide Nanowire and Protein Nanoparticles

Figure 5.13 illustrates the manner in which amino acids combine together in chains through the formation of a peptide bond. To form this bond the hydroxy (—OH) of the carboxyl group of one amino acid combines with the hydrogen atom H of the amino group of the next amino acid, with the establishment of a C—N peptide bond accompanied by the release of water (H_2O), as displayed in the figure. The figure shows the formation of a tripeptide molecule, and a typical protein is composed of one or more very long polypeptide molecules. Small peptides are called *oligopeptides*, and amino acids incorporated into polypeptide chains are often referred to as *amino acid residues* to distinguish them from free or unbound amino acids. The protein hemoglobin, for example, contains four polypeptides, each with about 300 amino acid residues.

The stretched-out polypeptide chain, of the type shown in Fig. 5.14a, is called the primary structure. To become more compact locally, the chains either coil up in a socalled an alpha helix (α-helix), or they combine in so-called beta sheets (β-sheets) held together by hydrogen bonds, as shown in Fig. 5.14b. The sheets might also be called *nanofilms*. These two configurations constitute the *secondary structure*. An overall compactness is achieved by a tertiary structure that consists of a series of twistings and turnings, held in place by disulfide bonds, as shown in Fig. 5.14c. If there is more than one polypeptide present, then they position themselves relative to each other in a quaternary structure, as shown in Fig. 5.14d. This type of quaternary structure packing is a characteristic of globular proteins. Some proteins are globular, and others are elongated, as illustrated in Fig. 5.10. It is clear from the two bottom sketches of Fig. 5.14 that the tertiary and quaternary structures are not

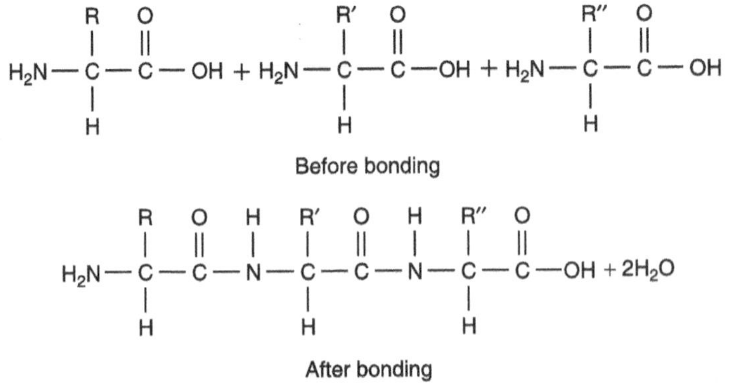

Before bonding

After bonding

Figure 5.13. Formation of a tripeptide chain (bottom) by establishing peptide bonds between three amino acids (top).

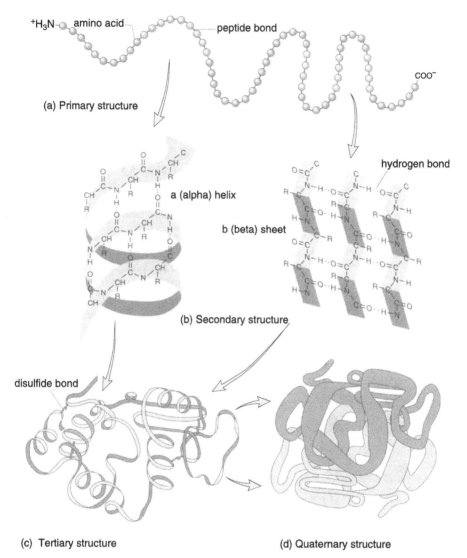

Figure 5.14. The four levels of structure of a protein: (a) primary structure of a stretched-out polypeptide; (b) secondary structures of an α-helix (left) and a ß-sheet (right); (c) tertiary structure of a polypeptide held together by disulfide bonds (—S—S—), (d) quaternary structure formed by two polypeptides. (From S. S. Mader, *Biology*, McGraw-Hill, Boston, 1998, p. 49.)

very closely packed, so the density is lower than that of the amino acids in the crystalline state, as was mentioned above. In practice some of the space within a protein molecule residing in the cytoplasm of a cell will contain water of hydration between the twistings and turnings. We conclude from these considerations that the structure of protein nanoparticles is often complex.

5.2.3. Nucleic Acids

5.2.3.1. DNA Double Nanowire

The basic building block of DNA, which is a nucleotide with the chemical structure sketched in Fig. 5.15, is more complex than an amino acid. It contains a five-membered desoxyribose sugar ring in the center with a phosphate group (PO_4H_2) attached at one end, and a nucleic acid base R attached at the other end. The figure also indicates by arrows on the left side the attachment points to other nucleotides to form the sugar–phosphate backbone of a DNA strand. Figure 5.16 presents the structures of the four-nucleotide bases that can attach to the sugar on the upper right of Fig. 5.15. It is clear from a comparison of Figs. 5.12 and 5.16 that the nucleic acid base molecules are of about the same sizes as the amino acid molecules. The attachment points where the bases bond to the desoxyribose sugar are indicated by vertical arrows in Fig. 5.16, and the bases pair off with each other in the double-stranded DNA in accordance with the horizontal arrows in the figure; that is, cytosine (C) pairs with guanine (G), and thymine (T) pairs with adenine (A), in the manner shown in Fig. 5.17. We see from this figure that the acidphosphate group and the sugar group parts of adjacent nucleotides bond together to form the sugar–phosphate backbone of a DNA strand, resulting in a macroscopically long double-stranded molecule. The complementary base pairs C—G and T—A are held together between the two strands by hydrogen bonds, as shown. Weak hydrogen bonds are used to accomplish this so the double helix can easily unwind for the purposes of transcription (forming RNA) or replication (duplicating itself). The individual strand is 0.34 nm thick, the double helix has a diameter of 2 nm, and the repeat unit containing 10

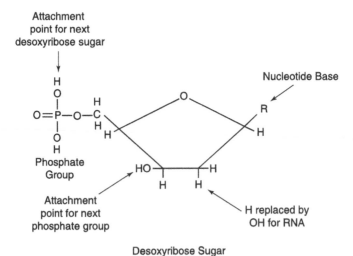

Figure 5.15. Structure of a nucleotide molecule, showing the points of attachment for the next ribose sugar (upper left) and the next phosphate group (lower left), the location of the nucleotide base (upper right), and the hydrogen atom H to be replaced by an hydroxyl group —OH to convert the desoxyribose sugar to ribose.

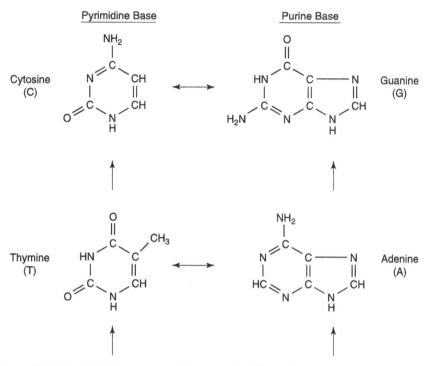

Figure 5.16. Sketch of the structures of the two pyrimidine nucleotide bases cytosine (C) and thymine (T), and the two purine bases guanine (G) and adenine (a). The points of attachment on the desoxyribose sugar of Fig. 5.15, entailing the loss of a hydrogen atom H, are indicated by vertical arrows. The horizontal arrows designate the complementary amino acid pairs.

nucleotide pairs is 3.4 nm long, as indicated in the upper left of Fig. 5.17. The 0.84 nm size of a nucleotide listed in Table 5.2 is greater than the 0.34 distance between base pairs because, in accordance with Fig. 5.15, the distance between the two attachment points on the nucleotide is much less than the overall length of the molecule. It is also clear from Fig. 5.17 that the pairs of nucleotides stretch lengthwise between the sugar–phosphate backbones of the two DNA strands, resulting in a 2-nm separation between them. To accomplish this coupling together of the two nanostrands in an efficient manner a small single-ring pyrimidine base always pairs off with a larger two-ring purine base, namely, cytosine with guanine, and thymine with adenine, as indicated in Fig. 5.17.

The 2-nm-wide strands are many orders of magnitude too long to fit lengthwise in the nucleus of a 6-μm-diameter human cell, so they undergo several stages of coiling, depicted in Fig. 5.18. Figure 5.18a shows the double-stranded DNA that we have been describing. The next coiling stage consists of an ∼ 140-bp (base pair) length of DNA winding around a group of proteins called *histones* to form what is sometimes called a "bead" that has a diameter of 11 nm, as shown in Fig. 5.18b. The histone beads are joined together by lengths of double-stranded DNA in a

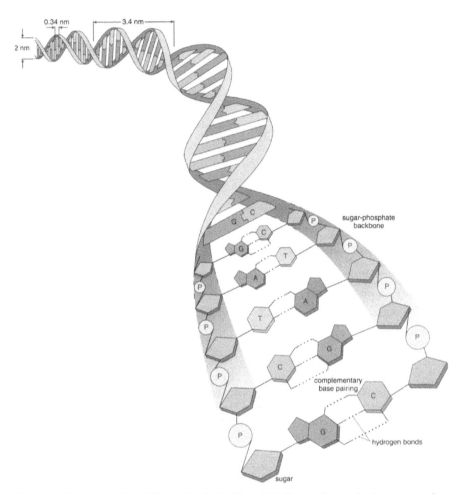

Figure 5.17. Model of the DNA double helix. The width (2 nm), the spacing between nucleotides (0.34 nm), and the length of the repeat unit or pitch (3.4 nm) of the helix are indicated on the upper left. The sugar–phosphate backbones, the arrangements of the nucleotides (C,T,G,A), and the four cases of nucleotide base pairing are sketched in the lower right. (From S. S. Mader, *Biology*, McGraw-Hill, Boston, 1998, p. 227.)

so-called linker region between the beads, shown in Fig. 5.18b. The DNA associated with the histones is called *chromatin*, and the histone bead with encircling DNA strands is called a *nucleosome*. The linker regions provide the nucleosome sequence with the great flexibility that is required for subsequent stages of folding.

In the next stage of compaction the nucleosomes stack one above the other, alternating between two coiled columns and connected by the linker strands, as shown laid out lengthwise in Fig. 5.18c. They now form a structure of 1-mm-long chromatin fibers 30 nm in diameter, a configuration called the "packing" of the nucleosomes. The chromatin fibers then undergo the next-higher order of folding shown in

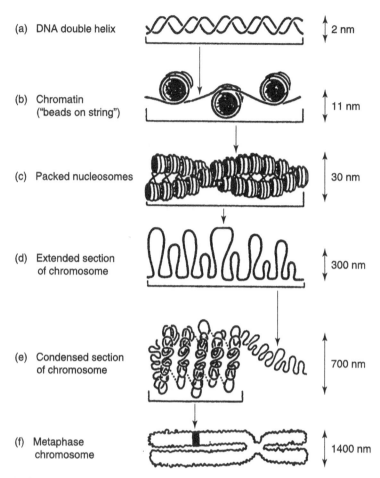

(a) DNA double helix   2 nm

(b) Chromatin
     ("beads on string")   11 nm

(c) Packed nucleosomes   30 nm

(d) Extended section
     of chromosome   300 nm

(e) Condensed section
     of chromosome   700 nm

(f) Metaphase
     chromosome   1400 nm

Figure 5.18. Successive twistings and foldings during the packing of DNA into mammalian chromosomes, with the sizes at successive stages given in nanometers. (From R. J. Nossal and H. Lecar, *Molecular and Cell Biophysics*, Addison-Wesley, Boston, 1991, Fig. 4.9. p. 118.)

Fig. 5.18d, and this becomes condensed into 700-nm-wide hyperfoldings of the 300-nm-wide foldings, in the manner presented in Fig. 5.18e. These various stages of compaction are largely held in place by relatively weak hydrogen bonds, which makes it feasible for the overall structure to unfold, partially or completely, for replication during cell division, or for transcription during the formation of ribonucleic acid (RNA) molecules that bring about or direct the synthesis of proteins. The final condensed structure of what is called the *metaphase chromosome*, sketched in Fig. 5.18f, is small enough to fit inside the nucleus of a cell.

The genome is the complete set of hereditary units called *genes*, each of which is responsible for a particular structure or function in the body, such as determining the color of one's eyes. In human beings the genome consists of 46 chromosomes (two haploid sets of 23 chromosomes), with an average of about 1600 genes arranged lengthwise along each chromosome. The human haploid genome contains $\sim 3.4 \times 10^9$ bp, an average of ~ 150 Mbp (million base pairs) per chromosome.

5.2.3.2. Genetic Code and Protein Synthesis

The DNA molecule that we have just described contains the information for directing the synthesis of proteins by the association of codewords called *codons* with individual amino acids, which bond together to form the proteins. One of the DNA strands, called the *coding strand*, stores this information, while the other strand of the pair, called the *complementary strand*, does not have useful data. The information for the synthesis is contained in a sequence of three-letter words using a four-letter alphabet A, C, G, and T, correlating to the four nucleic acid bases adenine (A), cytosine (C), guanine (G), and thymine (T), whose structures are sketched in Fig. 5.16. Since there are three letters in a codeword and four letters in the alphabet, there are $4^3 = 64$ possible words, and 61 of these serve as codewords for amino acids.

DNA is the carrier of heredity in the human body. This double-stranded molecule has a companion single-stranded molecule called *ribonucleic acid* (RNA), which is involved in the synthesis of proteins using the information transcribed or passed on to it from DNA. To form RNA the DNA uncoils, and sections of the RNA strand are synthesized one nucleotide at a time, in sequence, in a process called *transcription*. The RNA strand structure differs from the DNA strand through the replacement of one hydrogen atom H of the sugar molecule by an hydroxyl group OH, thereby forming the sugar ribose (instead of desoxyribose), as indicated at the lower right of Fig. 5.15. RNA also utilizes the nucleotide base uracil with the structure shown in Fig. 5.19 in place of the base thymine. Both of these nucleic acid macromolecules—DNA and RNA—can be classified as nanowires because their diameters are so small and their stretched-out lengths are so much greater than their diameters.

To carry out the synthesis of a particular protein, a segment of the DNA molecule uncoils, and the region of the double helix that stores the codewords for that particular protein serves as a template for the synthesis of a single-stranded messenger RNA

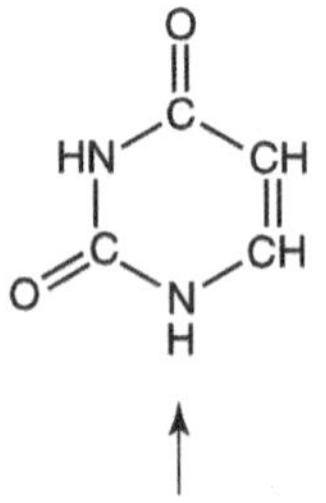

Figure 5.19. Structure of the uracil pyrimidine base nucleic acid, which replaces thymine in the RNA molecule. The point of attachment on the ribose sugar of Fig. 5.15, entailing the loss of a hydrogen atom H, is indicated by a vertical arrow.

molecule (mRNA) containing these codewords. In transcribing the code, each nucleotide base of DNA is replaced by its complementary base on the RNA, with the base uracil substituting for thymine in the RNA. Thus from Fig. 5.17 the transcription takes place by rewriting the codewords in accordance with the scheme

$$A \Rightarrow U$$
$$C \Rightarrow G$$
$$G \Rightarrow C \tag{5.23}$$
$$T \Rightarrow A$$

so each word or codon on the mRNA is a composed of three letters from the set A,C,G,U. Each codon corresponds to an amino acid, and the sequence of codons of the mRNA provides the sequence in which the corresponding amino acids are incorporated into the protein being formed. Since DNA inventories the codewords for thousands of proteins, and mRNA contains the codewords for one protein, the mRNA molecule is very short compared to DNA; it is a short nanowire. The mRNA brings the message of the amino acid sequence from the nucleus of the cell where the transcription from DNA takes place to nanoparticles called *ribosomes* in the cytoplasm region outside the nucleus where the proteins are synthesized. A number of additional protein nanoparticles called *enzymes* function as catalysts for the processes of protein synthesis.

Of the 20 amino acids, 2 have a single word or codon allocated to them, and this word on RNA is UGG for tryptophan, as noted in Fig. 5.12. Other amino acids have between two and six words assigned to them, and some examples are given in Fig. 5.12. In the terminology of linguistics, all the words assigned to a particular amino acid are synonyms, and in the language of science the code is called *degenerate* because there are synonyms. Three of the words, UAA, UAG, and UGA, are stop codons used to indicate the end of a specification of amino acids in a part of a polypeptide. For example, the sequence of codons or words

ACU   GCA   GGC   UAG

corresponds to the triple–amino acid polypeptide *Thr–Ala–Gly*, where the symbols for these amino acids are given in Fig. 5.12, and the final codon UAG indicates the end of the tripeptide.

5.2.3.3. Proteins

Protiens represent another example of biological nanostructures. There are many varieties of proteins, and they play many roles in animals and plants. For example, biological catalysts, called *enzymes*, are ordinarily proteins. In this section we will describe several representative proteins.

The protein hemoglobin, with molecular weight 68,000 Da, consists of four polypeptides, each of which has a sequence of about 300 amino acids. Each of these polypeptides contains a heme molecule ($C_{34}H_{32}O_4N_4Fe$) with an iron (Fe) atom that serves as a site for the attachment of oxygen molecules O_2 to the hemoglobin for transport to the tissues of the body. Every red blood cell or erythrocyte contains about 250 million hemoglobin molecules, so each cell is capable of carrying approximately one billion oxygen molecules.

Collagen constitutes 25–50% of all the proteins in mammals. It is the main component of connective tissue, and hence the major load-bearing constituent of soft tissue. It is found in cartilage, bones, tendons, ligaments, skin, and the cornea of the eye. The fibers of collagen are formed by triple helices of proteins containing repetitions of tripeptide sequences—*GlyProXxx*—where the amino acid denoted by *Xxx* is usually proline (*Pro*) or hydroxyproline (*Hpro*). Collagen differs from other proteins in its high percentage of the amino acids proline (12%) and hydroxyproline (10%), and it also has a high percentage of glycine (34%) and alanine (10%). The proline and hydroxyproline residues have rigid five-membered rings containing nitrogen, so collagen cannot form an α-helix, but these residues do interject bends into the polypeptide chains, which facilitate the establishment of conformations that strengthen the triplet helical structure. The smaller glycine residues make it easy to establish close packing of the strands held together by hydrogen bonds. Artificial genes ranging from 40 to 70 kDa in molecular weight have been prepared that encode tripeptide sequences such as –*GlyProPro*–, which resemble those in natural collagen.

The protein elastin possesses many properties similar to those of collagen. It provides elasticity to skin, lungs, tendons, and arteries in mammals. Elastin has a precursor protein called *tropoelastin* that has a molecular weight of 72 kDa. Its structure involves sequences of *oligopeptides*, short polypeptides that are only four to nine amino acid residues in length. An analog of such a chain has been synthesized that contains the pentapeptide sequence –*ValProGlyValGly*–, which exhibits a reverse turnaround, a *ProGly* dipeptide. Additional amino acid residues can occupy the space between turns, and make use of the high flexibility of glycine as a spacer. Many turns joined together adapt a springlike structure that can be stretched to over 3 times its resting length, and then return to rest without experiencing any residual deformation.

The silk caterpillar *Bombyx mori* produces silk formed from hydrogen-bonded β-sheets of the protein fibroin. The fibers of the sheets are closely packed and highly oriented, which gives them a large tensile strength. The amino acid residues are 46% glycine, 26% alanine, and 12% serine (*Ser*), and the principal repeat sequence is the hexapeptide –*GlyAlaGlyAlaGlySer*–. Artificial fibers with the properties of silk have been prepared in accordance with this hexapeptide sequence, and proteins of this type with molecular weights between 40 and 100 kDa have been expressed (i.e., formed) in the bacterium *Escherichia coli*. The polypeptide sequences –*ArgGlyAspSer*– from the protein fibronectin and –*ValProGlyValGly*– from elastin have been incorporated into artificial silk polymers. The former serves as a substrate for cell culture, and the latter makes the polymer more soluble and easier to process. The larvae of midge spiders (*Chironomus tentans*) produce a silklike fiber with a molecular weight of about 1 MDa.

Deming et al.,[6] have undertaken the de novo design and synthesis of well-defined polypeptides to assess the feasibility of creating novel useful proteins, and to determine the extent to which chain folding and large-scale or supramolecular organization can be controlled at the molecular level. They have incorporated unnatural amino acids into artificial proteins by using intact cellular protein synthesis techniques. "Unnatural" amino acids are those not included in the set of 20, which have DNA codons. These artificial proteins formed folded-chain, layered, or lamellar crystals of controlled surface configuration and thickness. One such crystal was prepared via sequence-controlled crystallization by employing a gene that incorporated the genetic codewords for 36 repeats of the octapeptide $-(GlyAla)_3GlyGlu-$ that could be expressed in the bacterium *E. coli*.

5.2.3.4. Micelles and Vesicles

A *surfactant* (surface-active agent) is an amphiphilic chemical compound, so named because it contains a hydrophilic or water-seeking headgroup at one end and a hydrophobic or water-avoiding (i.e., lipophilic or oil-seeking) tailgroup at the other end, as shown in Fig. 5.20. The hydrophilic part is polar, with a charge that makes it either anionic $(-)$, cationic $(+)$, zwitterionic $(\pm)$, or nonionic in nature, and the lipophilic portion consists of one or perhaps two nonpolar hydrocarbon chains. Surfactants readily adsorb at an oil–water or air–water interface, and decrease the surface tension there. The hydrocarbon chain might consist of a monomer that can participate in a polymerization reaction.

A surfactant molecule is characterized by a dimensionless packing parameter p defined by

$$p = \frac{V_T}{A_H L_T} \tag{5.24}$$

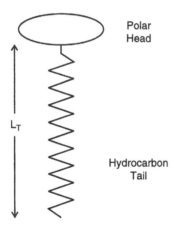

Figure 5.20. Amphiphilic surfactant molecule with a polar hydrophilic head and a nonpolar hydrophobic hydrocarbon tail.

where A_H is the area of the polar head and V_T and L_T are, respectively, the volume and length of the hydrocarbon tail.[7] If the average cross-sectional area of the tail (V_T/L_T) is appreciably less than that of the head, for example, $p < \frac{1}{3}$, then the tails will pack conveniently inside the surface of a sphere enclosing oil with a radius $r > L_T$ suspended in an aqueous medium, as indicated in Fig. 5.21a. Such a structure is called a *micelle*. The elongated or cylindrical micelle of Fig. 5.21b appears over the range $\frac{1}{3} < p < \frac{1}{2}$. Larger fractional values of the packing parameter, $\frac{1}{2} < p < 1$, lead to the formation of vesicles that have a double-layer surface structure, as shown in Fig. 5.21c. For example, sodium di-2-ethylhexyl-phosphate can form nanosized vesicles with $V_T \sim 0.5$ nm^3, $L_T \sim 0.9$ nm, and $A_H \sim 0.7$ nm^2, corresponding to $p \sim 0.8$, which is in the vesicle range. If the packing parameter is unity, $p = 1$, then the average transverse

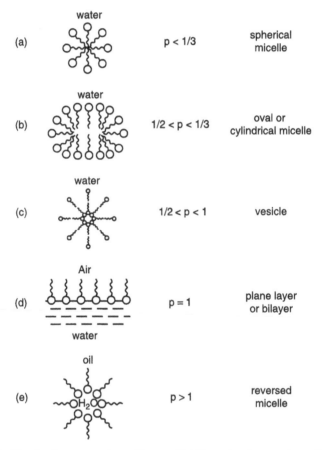

Figure 5.21. Sketch of structures formed by amphiphilic molecules at water–oil or water–air interfaces for various values of the packing parameter p of Eq. (5.24). (Adapted from E. Nakache et al., in *Handbook of Nanostructured Materials and Nanotechnology*, H. S. Nalwa, ed., Academic Press, Boston, 2000, Vol. 5, Chapter 11, p. 580.)

cross-sectional area of the tail V_T/L_T will be the same as the area A_H of the head, and the tails will pack easily at a planar interface, as shown in Fig. 5.21d, or they can form a bilayer. The inverse of a spherical micelle, called an *inverse micelle*, occurs with $p > 1$ for surfactants at the surface of a spherical drop of water in oil, as illustrated in Fig. 5.21e.

An *emulsion* is a cloudy colloidal system of micro meter-range droplets of one immiscible (i.e., nonmixing) liquid dispersed in another, such as oil in water. It is formed by vigorous stirring, and is thermodynamically unstable because the sizes of the droplets tend to grow with time. If surfactants are present, then nanosized particles, ~ 100 nm, can spontaneously form as a thermodynamically stable, transparent microemulsion that persists for a long time. Surfactant molecules in a solvent can organize in various ways, depending on concentration. For low concentrations they can adsorb at an air–water interface. Above a certain surfactant concentration, called the *critical micellar concentration*, distributions of micelles ranging from 2 to 10 nm can form in equilibrium with free surfactant molecules, being continuously constituted and disassembled with lifetimes measured in microseconds or seconds. Synthetic surfactants with more bulky hydrophobic groups, meaning larger packing parameters, produce extended bilayers that can close on themselves to form vesicles that are generally spherical. These structures form above a critical vesicular concentration. Vesicles typically have lifetimes measured in weeks or months, so they are much more stable than micelles.

If the vesicles are formed from natural or synthetic phospholipids, then they are called *unilaminar* or *single-layer liposomes*, that is, liposomes containing only one bilayer. A phospholipid is a lipid (fatty or fatlike) substance containing phosphorus in the form of phosphoric acid, which functions as a structural component of a membrane. The main lipid part is hydrophobic, and the phosphoryl or phosphate part is hydrophilic. The hydration or uptake of water by phospholipids causes them to spontaneously self-assemble into unilaminar liposomes. Mechanical agitation of these unilaminar liposomes can convert them to multilayer liposomes that consist of concentric bilayers. Unilaminar liposomes have diameters ranging from nanometers to micrometers, with bilayers that are 5–10 nm thick. Proteins can be incorporated into unilaminar liposomes to study their function in an environment resembling that of their state in phospholipid bilayers of a living cell.

If polymerizable surfactants are employed, such as those containing acrylate, acrylamido, allyl ($CH_2\!=\!CHCH_2\!-\!$), diallyl, methacrylate, or vinyl ($CH_2\!=\!CH\!-\!$) groups, then polymerization interactions can be carried out. When vesicles are involved then the characteristic time for the polymerization is generally shorter than the vesicle lifetime, so the final polymer is one that would be expected from the monomers or precursors associated with the surfactant. However, when micelles are involved, their lifetimes are generally short compared to the characteristic times of the polymerization processes, and as a result the final product may differ considerably from the starting materials. Surfactants with highly reactive polymerizable groups such as acrylamide or styryl have been found to produce polymers with molecular weights in excess of a million daltons. Those with polymerizable groups of low reactivity such as allyl produce much smaller products, namely, products with degrees of polymerization that can bring them close to the micelle size range prior to the polymerization.

Micelles and bilayers or liposomes have a number of applications in chemistry and biology. In soap solutions, micelles can facilitate dispersal of insoluble organic compounds and cleansing from surfaces. Micelles play a similar role in digestion by permitting components of fat such as fatty acids, phospholipids, cholesterol, and several vitamines (A, D, E, and K) to become soluble in water, and thereby more easily processed by the digestive system. Liposomes can enclose enzymes, and at the appropriate time they can break open and release the enzyme, enabling it to perform its function, such as catalyzing digestive processes.

5.2.3.5. Multilayer Films

Biomimetrics involves the study of synthetic structures that mimic or imitate structures found in biological systems. It makes use of large-scale or supramolecular self-assembly to build up hierarchical structures similar to those found in nature. This approach has been applied to the development of techniques to construct films in a manner that imitates the way nature sequentially adsorbs materials to bring about the biomineralization of surfaces, as discussed below. The process of biomineralization involves the incorporation of inorganic compounds such as those containing calcium into soft living tissue to convert it to a hardened form. Bone contains, for example, many rod-shaped inorganic mineral crystals with typical 5 nm diameters, and lengths ranging from 20 to 200 nm.

The kinetics for the self-assembly of many of these films involved in biomineralization can be approximately modeled as the initial joining together or dimerization of two monomers

$$R + R \Rightarrow R_2 \tag{5.25}$$

with a low equilibrium constant K_D, followed by the step-by-step or sequential addition of more monomers

$$R_n + R \Rightarrow R_{n+1} \tag{5.26}$$

with a much larger equilibrium constant K. These two equilibrium constants exercise control over the rate at which the reaction proceeds. For the case under consideration, $K_D < K$, the concentration of free or unbound monomers C_F always remains below a critical concentration $C_0 = 1/K$, that is, $C_F < C_0$. When the total concentration of free and clustered (i.e., bound) monomers C_T satisfies the condition $C_T < C_0$, then the free monomer concentration C_F increases with increases in C_T. When the reaction proceeds for a long enough time so that C_T exceeds the critical value (i.e., $C_T > C_0$), then the aggregate forms and grows for further increases in C_T. In analogy with this model, self-assembly kinetics often involves a slow dimer formation step followed by faster propagation steps.

There are many cases of multilayer thin films in biology, such as structural colors in insects that change when the films are subjected to pressure, shrinking, or swelling. For example, scale cells from some butterflies can produce iridescent multicoloring affects due to optical interference of thin-film layers or lamellae formed from the secretion of networks of filaments that condense on cell boundaries.

The biomineralization of mollusk cells begins by laying down a sheet of organic material so that calcium carbonate can be deposited on its surface and in its pores and $CaCO_3$ layers can build up. Proteins from the mollusc shell containing high concentrations of particular amino acid residues control the form of the calcium carbonate layering, and these proteins can be altered to vary the layering morphology. Multiple layers either grow in sequences that are organic in nature or contain the rhombohedral calcite form, or the orthorhombic aragonite variety of $CaCO_3$. It is possible to imitate some aspects of these natural biomineralization processes for the preparation of synthetic multilayer thin films, although the resulting films themselves do not closely resemble those in mollusks. For example, consider a positively charged substrate placed in a solution with a negative electrolyte, that is, a solution containing negative ions that can carry electric current. The positive substrate attracts the negative electrolyte, and the latter can adsorb on its surface, forming a *polyion* sheet, as shown in Fig. 5.22. This sheet is rinsed and dried, and then placed into another electrolyte solution from which it adsorbs a second positive layer. The sequential adsorption process can be repeated, as indicated in Fig. 5.22, to form a multilayer of alternating positively and negatively charged polyion sheets. Varying the types of electrolytes in the successive solutions can control the nature of the layering.

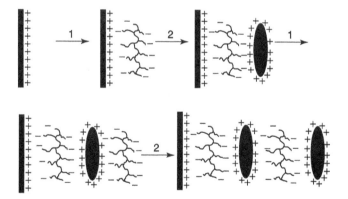

Figure 5.22. Sketch of the sequential adsorption process for the formation of a polyion film. The upper figure shows a positively charged substrate (left) that has adsorbed a negatively charged polyelectrolyte by being dipped into a negative electrolyte solution (center) and then adsorbed a positively charged layer from a positive electrolyte solution (right). The lower figure shows two additional steps in the sequential adsorption process. (From T. M. Cooper, in *Handbook of Nanostructured Materials and Nanotechnology*, H. S. Nalwa, ed., Academic Press, Boston, 2000, Vol. 5, Chapter 13, p. 711.)

PROBLEMS

5.1. Calculate the molecular weight of *trans*-polyacetylence having 1000 monomers.

5.2. Estimate the volume and the length of stretched-out polyacetylene containing 500 monomers.

5.3. Given the average C—C bond length in polyacetylene of 1.4111 Å and the C—C—C bend angle of 125°, calculate the length of the polymer containing 500 monomers from these geometric parameters. How does this calculated length compare with the estimate from Problem 5.2?

5.4. Sketch the structure of a monomer of polyethylene and a monomer of Teflon (polytetrafluroethylene). Given that the molecular weight of the Teflon is 5000 Da, write down the formula for the polymer. Do the same for polyethylene having a molecular weight of 16,800. Are either of these polymers nanostructures?

5.5. Calculate the length of a DNA molecule having 20 repeat units.

5.6. Estimate the size of the amino acids proline and serine.

5.7. Write down the sequence of the complementary DNA strand that pairs with each of the following base sequences: (a) TTAGCC; (b) AGACAT.

5.8. Write down the base sequence of the DNA template from which the following RNA sequence was derived, UGUVACGGA. How many amino acids are coded by the sequence?

5.9. Given the following molecule

$$\Phi—CH_2—\overset{\displaystyle CH_2}{\underset{\displaystyle CH_3 \ Cl^-}{\overset{|}{\underset{|}{N^+}}}}—[CH_2]_{15}—CH_3$$

where the $[CH_2]_{15}$ is a linear chain, would you expect it to form a vesicle or a micelle? The hydrocarbon tail —$[CH_2]_{15}$— CH_3 is soluble in grease, and the ionic part is soluble in water. What is a possible application of this molecule?

REFERENCES

1. C. P. Poole, Jr. and F. J. Owens *Introduction to Nanotechnology*, Wiley-Interscience, New York, 2003.
2. B. Wessling, in *Handbook of Nanostructured materials and Nanotechnology*, H. S. Nalwa, ed., Academic Press, Boston, 2000, Vol. 5, Chapter 10.

3. G. Liu, Polymeric nanostructures, in *Handbook of Nanostructured Materials and Nanotechnology*, H. S. Nalwa, ed., Academic Press, Boston, 2000, Vol. 5, Chapter 9.

4. H. Otsuka, Y. Nagasaki, and K. Kataoka, *Materials today* 3 30 (2001).

5. D. A. Tirrell, ed., *Hierarchical structurs in Biology as a guide for new materials technology*, National Academy Press Washington, DC 1994.

6. T. J. Deming, U. P. Coniticello, and D. A. Tirrell, in Nanotechnology, G. Timp, ed., Springer Verlag, Berlin 1999, Chapter 9.

7. E. Nakache, et al., in *Handbook of Nanostructured Materials and Nanotechnology*, H. S. Nalwa, ed., Academic Press, Boston, 2000 Vol. 5, Chapter 11, p. 580.

CHAPTER 6

Cohesive Energy

In this chapter we discuss the effects of reducing the size of a solid to nanometers on the cohesive energy of the material. The atoms or ions of a solid are held together by interactions between them that can be electrostatic and/or covalent. A crystal is stable if the total energy of the lattice is less than the sum of the energies of the atoms or molecules that make up the crystal when they are free. The energy difference is the cohesive energy of the solid.

6.1. IONIC SOLIDS

Ionic solids are ordered arrays of positive and negative ions such as sodium chloride shown in Fig. 6.1, which is a face-centered cubic (fcc) structure of positive sodium ions and an interpenetration of fcc negative chlorine ions. The interaction potential between the ions of charge Q is electrostatic, $+Q/r$. Ions of opposite sign are attracted, while ions of the same sign repel each other. The total electrostatic energy of any one ion i is U_i given by the sum of all the Coulomb interactions between the ith ion and all positive and negative ions of the lattice

$$U_i = \Sigma_j U_{ij} = \frac{\Sigma_j \pm Q^2}{r_{ij}} \tag{6.1}$$

where r_{ij} is the distance between ions i and j. In the case of the interaction between the nearest neighbor, a term has to be added to Eq. (6.1) to account for the fact that the electron core around the nucleus repels those of the nearest neighbors. The formula for this interaction has been derived experimentally and is given by

$$\lambda \exp\left(\frac{-r_{ij}}{\rho}\right) \tag{6.2}$$

The constants λ and ρ for NaCl are 1.75×10^{-9} ergs and 0.321 Å, respectively.

The Physics and Chemistry of Nanosolids. By Frank J. Owens and Charles P. Poole, Jr.
Copyright © 2008 John Wiley & Sons, Inc.

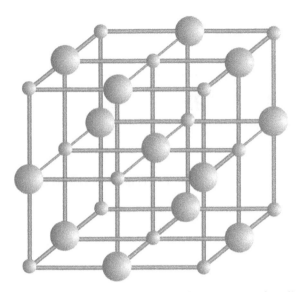

Figure 6.1. Crystal structure of NaCl, which has a FCC arrangement of small sodium positive ions and large chlorine negative ions.

The total energy of a lattice of N ions may be written in the form

$$U_T = NU_i \tag{6.3}$$

where

$$U_i = z\lambda \exp\left(\frac{-R}{\rho}\right) - \frac{\alpha Q^2}{R} \tag{6.4}$$

and z is the number of nearest neighbors. The sums over $1/r_{ij}$ can be represented as multiples p_{ij} of the nearest-neighbor distance R:

$$R_{ij} = p_{ij}R \tag{6.5}$$

The electrostatic interaction (6.1) is then given by

$$\frac{\Sigma_j \pm Q^2}{p_{ij}R} \tag{6.6}$$

where the dimensionless Madelung constant α is defined as

$$\alpha = \Sigma_j \pm \frac{1}{p_{ij}} \tag{6.7}$$

Equation (6.6) represents the Madelung energy, and the total binding energy U_T, the cohesive energy of the ith ion, is given by Eq. (6.4). At the equilibrium separation R_0

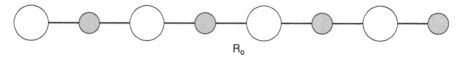

Figure 6.2. One-dimensional hypothetical linear lattice of NaCl.

we have the condition $dU_T/dR = 0$, which from (6.4) yields an equation that allows the determination of R_0:

$$R_0^2 \exp\left(\frac{-R_0}{\rho}\right) = \frac{\rho \alpha Q^2}{z\lambda} \tag{6.8}$$

In order to illustrate how reducing the size of a crystal to nanometer dimensions affects the binding energy of ionic crystals, let us consider a one-dimensional lattice of positive sodium ions and negative chlorine ions having nearest-neighbor distance R_0 as illustrated in Fig. 6.2. The definition of Madelung constant, Eq. (6.7), can be rewritten as

$$\frac{\alpha}{R} = \Sigma_j \frac{(\pm)}{r_{ij}} \tag{6.9}$$

Thus for a single line of ions with alternating positive and negative charges, we obtain

$$\frac{\alpha}{R} = 2\left[\frac{1}{R} - \frac{1}{2R} + \frac{1}{3R} - \frac{1}{4R} + \frac{1}{5R} \cdots\right] \tag{6.10}$$

to give

$$\alpha = 2\left[1 - \frac{1}{2} + \frac{1}{3} - \frac{1}{4} + \frac{1}{5} \cdots\right] \tag{6.11}$$

For an infinite chain it can be shown that,

$$\alpha = 2\log 2 = 1.3863 \tag{6.12}$$

However, we are interested in finite chains of nanometer length. Figure 6.3 is a plot of the Madelung constant versus the number of atoms in the chain calculated by Eq. (6.11). The plot shows that below ~50 atoms, which is approximately 30 nm, there is a marked reduction in the Madelung constant, meaning a reduction of the cohesive energy of the solid. Using Eq. (6.8) it is possible to determine the equilibrium separation of the ions as a function of the length of the chain. As shown in Fig. 6.4, the lattice parameter begins to show a significant increase below ~15 nm. The question arises as to whether these results are typical of three-dimensional (3D) ionic solids. Figure 6.5 shows the results of an X-ray diffraction measurement of the lattice parameter as a function of particle size of CeO_2, which is an ionic crystal having the cubic fluorite structure. The lattice parameter begins to increase below ~25 nm. The Madelung energy and equilibrium lattice parameter of CeO_2

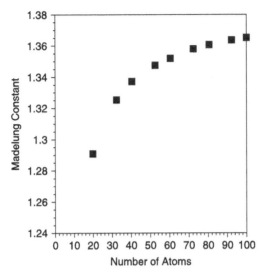

Figure 6.3. Madelung constant versus number of ions in an NaCl linear lattice showing a marked decrease in the constant at 50 atoms that is approximately 30 nm long.

have been calculated as a function of particle size and are in reasonable agreement with the experimental data in Fig. 6.5. As we saw in Chapter 3, the percentage of atoms on the surface of a nanoparticle increases strongly below ~ 10 nm. Ions on the surface will have a lower binding energy than will those inside the particle because they have fewer nearest neighbors, and this contributes to a lowering of

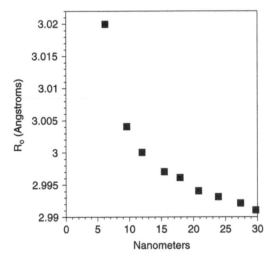

Figure 6.4. Calculated increase in nearest-neighbor separation R_0 in a linear NaCl lattice versus length in nanometers.

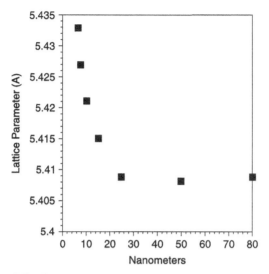

Figure 6.5. X-Ray diffraction measurement of the lattice parameter of the ionic crystal CeO_2 versus particle size showing an increase with reduced particle size. [Adapted from F. Zhang et al., *Appl. Phys. Lett.* **80**, 127 (2002).]

the cohesive energy of the particle. The calculation of the Madelung energy of CeO_2 took this into account. In Fig. 6.6 the percentage of the total cohesive energy per particle in CeO_2 attributable to the surface atoms is plotted versus the diameter of the particle.

The decrease in the cohesive energy of an ionic solid as the particle size is reduced will cause a number of other properties to be altered. One would expect, for example, that a smaller cohesive energy would result in a lower melting temperature. Figure 6.7 is a plot of the melting temperature of a number of alkali halides versus the cohesive energy showing that solids with lower cohesive energies generally have lower melting temperatures. Melting starts at the surface of a material. As discussed in Chapter 3, the percentage of atoms of a nanoparticle that are on the surface increases significantly below ~ 10 nm, and these are more weakly bound than those in the interior of the particle. This will also contribute to a lowering of the melting temperature of the particle. We will return to the effect of nanosizing on melting from another perspective in Chapter 7.

The bulk modulus, which is the inverse of the compressibility, is defined as

$$B = \frac{V d^2 U}{dV^2} \tag{6.13}$$

where U is the total lattice energy given by Eq. (6.3) and V is the volume occupied by N molecules, which is $2NR^3$. The modulus is obtained from Eq. (6.3) as

$$B = \frac{\alpha Q^2}{18 R_0^4} \left(\frac{R_0}{\rho - 2} \right) \tag{6.14}$$

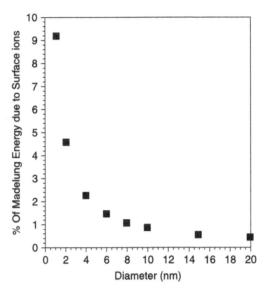

Figure 6.6. Plot of the calculated percentage of the Madelung energy per ion due to surface atoms in CeO_2 versus particle size. [Adapted from V. Perebeinos et al., *Solid State Commun.* **123**, 295 (2002).]

Figure 6.8 is a plot of the bulk modulus for our hypothetical linear NaCl chain showing a strong decrease in the modulus below a length of ~ 15 nm. This means that the compressibility increases, and in effect smaller ionic nanoparticles are softer than larger ones.

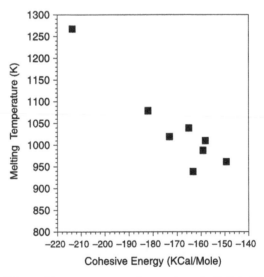

Figure 6.7. Plot of the melting temperature of a number of alkali halides versus cohesive energy.

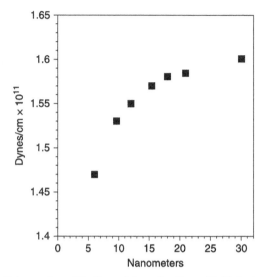

Figure 6.8. Plot of the analog of bulk modulus of a linear NaCl lattice versus the length of the lattice.

6.2. DEFECTS IN IONIC SOLIDS

Real crystals are imperfect. Every lattice site is not occupied. There can be a vacancy at a site called a *Schottky defect* created by moving an atom from its lattice site in the interior to the surface of the crystal. Another kind of defect is a *Frenkel defect*, in which an anion is transferred from its site to an interstitial position. The number of Schottky vacancies present is dependent on the temperature. It can be shown from statistical mechanics that the fraction of sites f vacant at a given temperature is proportional to

$$f = \exp\frac{-E_v}{k_B T} \tag{6.15}$$

where E_v is the defect formation energy, that is, the energy needed to move an atom from its lattice site to the exterior of the crystal. The question we are interested in examining here is how nanosizing affects the fraction of defects in the material, which means that we have to examine the effect of nanosizing on the defect formation energy. For a crystal consisting of nonpolarizable constituents, the defect formation energy is the lattice energy per atom U_i. In the case of ionic solids U_i is an overestimate of the defect formation energy because it neglects the polarization energy produced by the vacancy. When an ion is removed from its site, the ions around the site experience an electric field and are polarized. The potential Φ from this polarization must be included in determining the defect formation energy, which is then

$$E_v = U_i - \frac{e\Phi}{2} \tag{6.16}$$

We consider a simple model for the potential Φ. When a positive ion is removed from its site, the neighboring ions experience and electric field e/r^2, where r is the distance of the ion from the vacancy. The neighboring ions become polarized having a dipole $\mu_+ = \alpha_+ e/r^2$ for positive ions and $\mu_- = \alpha_- e/r^2$ for negative ions, where α is the polarizability of the ion, which will differ between Na^+ ions and Cl^- ions. The potential produced by a dipole is proportional to μ/r^2 and depends on the orientation of the dipole. Thus the induced dipole on the ions produces a potential Φ at the vacancy given by

$$\Phi = \Sigma_j \frac{\alpha_+ e}{r_j^4} + \Sigma_k \frac{\alpha_- e}{r_k^4} \tag{6.17}$$

where r_j is the distance of positive ions from the vacancy and r_k is the distance of negative ions from the vacancy.

To illustrate the effect of nanosizing on the energy to form a vacancy, let us calculate the defect formation energy to remove a positive ion in the middle of our linear NaCl lattice of Fig. 6.2 discussed above. The energy will be $e\Phi/2$. Because the $1/r^4$ term decreases rapidly with r at most, we need only go to next nearest neighbors in the summation to calculate the polarization contribution to the formation energy. In this case the energy will be

$$E_p = \frac{2\alpha_+ e^2}{R_0^4}\left[\frac{1}{2^4} + \frac{1}{4^4}\right] + \frac{2\alpha_- e^2}{R_0^4}\left[1 + \frac{1}{3^4}\right] \tag{6.18}$$

The factor of 2 is included to account for summations in the positive and negative directions from the vacancy. Using the dependence of R_0 on the length of the NaCl chain in Fig. 6.4, we can calculate E_p from Eq. (6.18) as a function of the length of the chain. We can then calculate U_i from Eq. (6.4) using the dependence of R and the Madelung constant on the chain length given in Fig. 6.3 and 6.4. We can then obtain E_v Eq. (6.16). The values of the polarizabilities are those of Pauling, which for Cl^- is 3.66×10^{-24} cm^3 and for Na^+ is 0.21×10^{-24} cm^3. The results plotted in Fig. 6.9 show a reduction in the energy needed to produce a positive ion vacancy with particle size, with a significant decrease occurring below ~ 16 nm. This means that at a given temperature a nanoparticle of NaCl will have a larger fraction of vacancies than will the bulk material. It should be emphasized that the preceding model is oversimplified, but serves the purposes of illustrating that the vacancy formation energy decreases when the materials have dimensions of nanometers. A number of factors have not been taken into account. For example, when the vacancy is produced, there is a change in the positions of the nearest-neighbor ions to the vacancy, referred to as *relaxation*, which has not been factored in the analysis above.

Alkali halides have large bandgaps, and electrical conductivity cannot occur by thermal excitation of electrons from the valence band to the conduction band as in the case of semiconductors. However, at high temperatures many alkali halides can carry current when a potential is applied. It has been observed that increasing the vacancy content increases the conductivity. Doping it with a divalent ion such as

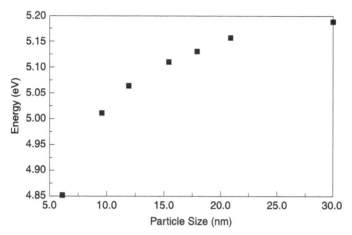

Figure 6.9. Calculated energy needed to produce a positive ion vacancy in a linear NaCl lattice as a function of the length of the lattice.

Ca^{2+} at a Na site can increase the vacancy content of a crystal such as NaCl. In order to maintain charge neutrality, a nearby Na^+ vacancy is produced. The mechanism of conductivity is vacancy movement whereby the positive ions move from one vacancy to another. Thus, if it were possible to measure the conductivity of an individual nanoparticle of an ionic crystal, it would display a higher conductivity because its fraction of vacancies is larger than that in the bulk. Similarly, diffusion of ions, which occurs if there is a concentration gradient of ions, should be enhanced in a nanoparticle of an ionic solid.

6.3. COVALENTLY BONDED SOLIDS

In Chapter 1 we saw that in solids such as germanium, diamond, and silicon the bonding that holds the lattice together is covalent, meaning that there is overlap of the wavefunctions of the outer electrons of the atoms of the crystal analogously to the way atoms are bonded in molecules. To illustrate how covalent bonding is affected by nanosizing, we will consider a one-dimensional chain of acetylene molecues C_2H_2 bonded to each other to form a polymer known as *polyacetylene*, as illustrated in Fig. 6.10. Polyacetylene has important technological applications because when doped with iodine, it becomes a conducting polymer, but it also represents an ideal

Figure 6.10. Structure of polyacetylene polymer, $(C_2H_2)_n$.

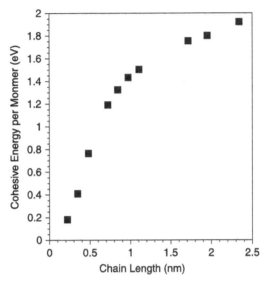

Figure 6.11. Molecular orbital calculation of the binding energy per monomer, (C_2H_2), of polyacetylene versus chain length. [Adapted from F. J. Owens, *Physica* **E25**, 404 (2005).]

one-dimensional system to illustrate how nanosizing can affect properties. The cohesive energy per monomer, which is the acetylene molecule, is given by

$$BE = \frac{BE_p - NBE_m}{N} \qquad (6.19)$$

where BE_p is the total energy of the polymer, BE_m is the total energy of the monomer, and N is the number of monomers in the chain. Figure 6.11 shows the results of a molecular orbital calculation of the binding energy per monomer versus the length of the polymer showing a marked decrease in the cohesive (binding) energy below ~ 1.2 nm. In general, X-ray diffraction studies of three-dimensional (3D) covalently bonded materials show an increase in the lattice spacing as the materials reach nanodimensions. This would cause a reduction of the overlap between neighboring atoms and a lowering of the cohesive energy.

6.4. ORGANIC CRYSTALS

The term *organic crystals* refers to solids in which the individual units are organic molecules such as benzene, naphthalene, or anthracene. The atoms within each molecule are covalently bonded to each other. As an example, consider the crystal structure of solid benzene. The structure of the benzene molecule is shown in Fig. 6.12a and the crystal structure, in Fig. 6.12b. In the latter figure the planes of the flat benzene molecule are perpendicular to the page of the paper. Each circle represents

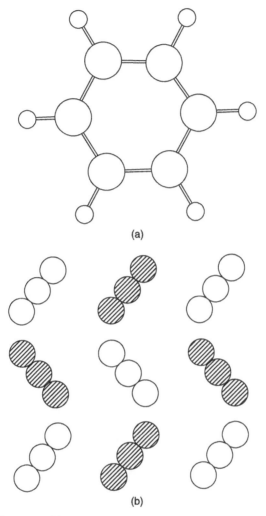

(a)

(b)

Figure 6.12. (a) Structure of benzene molecule and (b) crystal structure of solid benzene, showing the flat benzene molecules oriented perpendicular to the page. The circles represent the top three CH groups of the molecule. The open circles are molecules all in one plane, and the darkened circles are in planes above or below.

the top three CH groups of the molecule. Twelve nearest-neighbor molecules surround each molecule. The molecule at the center has four nearest neighbors in the same plane indicated by the open circles. The dark circles represent benzene molecules, which are above and below the plane of the center molecule by half of the unit cell length. We are interested in the nature of the potential describing the interaction between the molecules that holds them in place in the crystal. Generally it is not covalent involving overlap of the wavefunctions of the neighboring molecules; nor is it ionic bonding as in NaCl because the molecules do not have a net charge.

One approach to characterizing the interaction between organic molecules has been to write potentials between the individual atoms of the molecule and its neighboring molecules. Thus, for benzene, each carbon and hydrogen atom of the benzene molecule would have an interaction with each carbon and hydrogen of the surrounding benzene molecules. The potentials developed are referred to as *nonbonding potentials*. Here we consider one form of the potential called the *Williams potential*, named after the scientist who contributed to its development. These potentials have the form

$$U = \frac{A}{R^6} + B \exp\left[-CR\right] \qquad (6.20)$$

where R is the separation between the atoms of the different molecules. The constants A, B, and C depend on the kind of atoms involved. Thus a carbon–carbon interaction and a carbon–hydrogen interaction will have different sets of parameters A, B, C. The first term in the potential represents interactions between the induced dipoles as explained in Section 3.4.1 for inert-gas clusters. The second term represents the repulsion between the electron cores of the atoms. The parameters have been obtained empirically by fitting them to predicted properties of a large number of organic crystals such as heat of sublimation, compressibility, and crystal structure. A representative set of parameters is presented in Table 6.1. We are interested in understanding how nanosizing affects the cohesive energy of organic crystals. Specifically, at what size does the cohesive energy begin to decrease? Let us again consider a hypothetical one-dimensional lattice to investigate the issue. The lattice shown in Fig. 6.13 consists of C_2 molecules where the bonded carbons are separated from each other by 1.25 Å and the end carbons of each molecule are separated from each other by 3.88 Å. Using the potential of Eq. (6.20) and parameters appropriate to carbon–carbon interactions, it is found that the contribution to the binding energy of the central molecule from the second nearest neighbors is less than 1.2% of that of the nearest neighbors. The third nearest neighbors contribute only 0.09% In other words, the range of effectiveness of the potential in Eq. (6.20) is quite short. This means that for the cohesive energy of the central molecule to decrease, the length of the chain would have to be less than seven molecules or about 3.2 nm. Indeed, X-ray diffraction investigations of very small organic nanocrystals show no increase in lattice parameters, which would be expected if there were a decrease in the cohesive energy.

TABLE 6.1. Some Parameters of Nonbonding Potentials of Eq. (6.20) for Carbon and Hydrogen Atoms

Constants	C—C	C—H	H—H
C, [Å^{-1}]	3.60	3.67	3.74
A, kcal/mol·Å^6	−568	−125	−27.3
B, kcal/mol	83,630	8766	2654

Source: D. E. Williams, *J. Chem. Phys.* **47**, 4680 (1967).

C-C   C-C   C-C   C-C   C-C   C-C   C-C

Figure 6.13. Hypothetical one-dimensional lattice of C—C molecules.

The discussion above applies to symmetric molecules. Figure 6.14 is a contour plot of the electron density of a C—C molecule and a C—H molecule in the plane of the paper calculated by molecular orbital theory. The closer the contour lines are, the higher the electron density. The figure shows that the electron density is symmetrically distributed about the two carbon nuclei but asymmetrically distributed about the C and H nuclei in the C—H molecule. This means the C—H molecule has a dipole moment and the C—C molecule does not. A dipole moment exists when there is a separation between positive and negative charge. If a charge $+q$ is separated by a distance d from a charge $-q$, the dipole moment is defined as qd. The dipole moment of the CH molecule is 1.3839 debyes. The potential of Eq. (6.20) needs to be modified if the molecules of the lattice have a dipole moment.

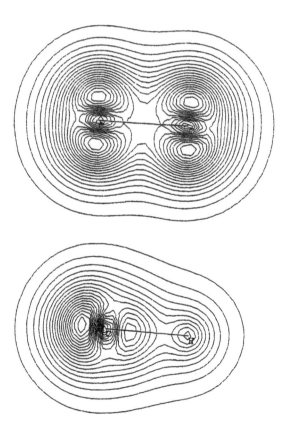

Figure 6.14. Contour plot of electron density in the plane of the paper about the (a) carbon nuclei of the C—C molecule and (b) atoms of the CH molecule.

A term for the dipole–dipole interaction must be added, which is given by

$$U_{dd} = \frac{\mathbf{P} \cdot \mathbf{P}}{R^3} \tag{6.21}$$

where $\mathbf{P}$ is the dipole moment of the molecules. It is a vector pointing from the center of the negative charge to the center of the positive charge, which for the case of the CH molecule would be parallel to the CH axis pointing toward H. To investigate the effect of lattice size on the binding energy, consider a one-dimensional lattice of CH molecules having an arrangement similar to that of the C—C lattice in Fig. 6.13 except that the carbon on the right side of each molecule is replaced by a H atom. Note that this lattice will have a net dipole moment. The CH bond length is 1.13 Å, and the end atoms of each molecule are separated from each other by 3.88 Å. The presence of the $1/R^3$ term increases the effective range of the potential. For example, the third nearest neighbors to the center molecule contribute 0.22% of the binding energy of the molecule compared to 0.099% for a lattice having molecules with no dipole moment. So the length of the chain will still have to be quite short before the binding energy of the central molecule begins to decrease. In another situation that can occur in our hypothetical linear lattice of molecules having dipole moments, the dipoles alternate in direction down the length of the chain. We will leave it as an exercise for the student to determine the effect of chain length on the cohesive energy for this situation.

6.5. INERT-GAS SOLIDS

In Section 3.4.1 we discussed the nature of the interaction between inert-gas atoms and showed that the interaction is described by a Lennard-Jones potential, which has the form

$$U(R) = 4\varepsilon \left[\left(\frac{\sigma}{R}\right)^{12} - \left(\frac{\sigma}{R}\right)^{6} \right] \tag{6.22}$$

For a solid consisting of inert-gas atoms, the cohesive energy is obtained by summing over all pairs of atoms in the crystal

$$U_{tot} = \left(\frac{1}{2}\right) N 4\varepsilon \sum_{j} \left[\left(\frac{\sigma}{p_{ij}R}\right)^{12} - \left(\frac{\sigma}{p_{ij}R}\right)^{6} \right] \tag{6.23}$$

where $p_{ij} R$ is the distance between reference atom i and any other atom j and R is the nearest-neighbor distance. The $\frac{1}{2}$ is needed to avoid counting each pair of atoms twice. For a FCC lattice the summations have been evaluated to be

$$\sum_{j} p_{ij}^{-12} = 12.131, \qquad \sum_{j} p_{ij}^{-6} = 14.454 \tag{6.24}$$

These summations rapidly converge to values very close to the values given above when the summations are carried out to the next-nearest-neighbor atoms.

This means that for inert-gas solids nanosizing will have little effect on the cohesive energy and other properties. For example, in xenon the particle size would have to be less than 0.87 nm for a decrease in the binding energy to occur. Because inert-gas solids have closed electron shells, they are not very polarizable, meaning that the vacancy formation energy is approximately the cohesive energy of the lattice, which, as discussed above, is not affected appreciably by nanosizing. Thus in inert-gas solids we should not expect any significant increase in the fraction of vacancies with nanosizing.

6.6. METALS

Metals conduct electricity because the outer electrons of the atoms of the solid are delocalized and hence free to move about the lattice. This makes the development of a theory of binding energy of metals a bit more complex than for ionic or covalent solids. Because the outer electrons of the atoms are itinerant, the atoms can be considered to be positively charged. The binding energy of a metal can be treated as arising from the Coulomb interaction of a lattice of positive ions embedded in a sea of negative conduction electrons. One relatively simple model is to consider the binding energy to be the interaction of a positive point charge e with a negative charge $-e$ distributed uniformly over a sphere of radius R_0 and volume V_0 equal to the atomic volume. Consider a positive point charge at the center of a sphere of radius R_0 that has a charge $-e$ uniformly distributed over the volume of the sphere. For a sphere of volume V less than the volume V_0, the charge inside the volume V is

$$Q(R) = e - e\left(\frac{V}{V_0}\right) = e - e\left(\frac{R}{R_0}\right)^3 \tag{6.25}$$

The electrostatic potential difference dV between a positive point charge $+e$ at the center of the sphere and a negatively charged shell of thickness dR is

$$dV = 4\pi R^2 \frac{\rho e\left[1 - \left(\dfrac{R}{R_0}\right)\right]^3 dR}{R} \tag{6.26}$$

where $\rho = -e/(4\pi R_0^3/3)$. The total interaction potential for the point charge and the sphere will be

$$V = \int_0^{R_0} dV \tag{6.27}$$

which on substituting the value of ρ gives

$$\frac{-0.9e^2}{R_0} \tag{6.28}$$

It will be shown in a later chapter that the average kinetic energy of a 3D electron gas is $\frac{3}{5}$ $[E_f]$, where E_f is the Fermi energy, which is the energy of the top filled electron level in the solid. The total cohesive energy of the metal is then

$$U_c = \frac{-0.9e^2}{R_0} + \left(\frac{3}{5}\right)[E_f] \qquad (6.29)$$

In Chapter 3 it was shown that the lattice parameters of metals are not strongly dependent on particle size until the particles become very small, less than ~ 1.5 nm, below which there is a slight decrease due to the existence of surface stress, which causes small particles to be in a state of compression. One would not expect the atomic volume to change significantly with nanosizing, and thus the first term in Eq. (6.29) will not change much with nanosizing. In Chapter 8 it will be shown that the magnitude of the Fermi energy E_f increases as the particle size is reduced. From Eq. (6.29) this implies that the cohesive energy of metals decreases as the particle size decreases in the nanometer regime.

In the case of small metal nanoparticles, it is possible to theoretically treat them as molecules and use conventional molecular orbital theory such as density functional theory to calculate the binding energy per atom. Figure 6.15 shows the results of a calculation of the binding energy per atom as a function of the number of atoms in small lithium nanoparticles showing a decrease in the binding energy per atom as the size decreases.

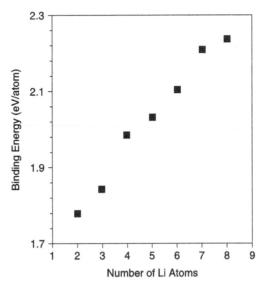

Figure 6.15. Molecular orbital calculation of binding energy per atom in small lithium nano particles versus the number of atoms. (F. J. Owens, unpublished.)

6.7. CONCLUSION

In summary, it is seen that the binding energy of most kinds of solids decreases as the size of the materials is reduced to nanometers, resulting in an expansion of the lattice However, the size at which the decrease is evident depends on the kind of potential that binds the constituents in the solid structure. This change in binding energy affects many properties of the materials such as concentration of defects, compressibility, melting temperature, and intrinsic electrical conductivity.

PROBLEMS

6.1. For a linear NaCl lattice of 10 ions, calculate the binding energy of end atom. How does this compare with the binding energy of the same ion at the center? How does the binding energy of an end ion change as the chain length decreases? What are the implications of this result concerning the effect of chain length on the melting temperature?

6.2. Calculate the energy needed to form a negative ion vacancy at the center of a linear NaCl lattice as a function of the length of the lattice using Eqs. (6.16) and (6.17), and the results plotted in Figs, 6.3 and 6.4 for lattices of 20 atoms and 50 atoms. How is the energy to form vacancy affected by the length of the chain?

6.3. Calculate the polarization potential due to a negative ion vacancy in the 3D sodium chloride lattice taking into account nearest-neighbor and next-nearest-neighbor interactions.

6.4. Using the result from Problem 6.3, calculate the energy needed to form a negative ion vacancy. How does nanosizing affect the formation energy of a negative ion vacancy?

6.5. Consider a hypothetical linear lattice having the following structure

$$
\begin{array}{cccc}
\text{H} & \text{H} & \text{H} & \text{H} \\
\text{C}-\text{C}-\text{C}-\text{C}-\text{C}-\text{C}-\text{C}-\text{C} \\
\text{H} & \text{H} & \text{H} & \text{H}
\end{array}
$$

formed from C^-H^+ molecles. The separation between adjacent CH molecules is 0.388 nm, the CH bond length is 0.113 nm, and one electron has been transferred from the hydrogen atom to the carbon atom. Calculate the dipolar interaction energy between the two centrally located CH molecules. Would the lattice still be stable if all the dipoles were pointed the same direction?

6.6. For the lattice in Problem 6.5, calculate the atom−atom interaction for the nearest neighbors to the central C—H molecule. How does this compare to the strenght of the dipole−dipole interaction?

6.7. The illustration below shows the unit cell of graphite. The dashed lines connect the nearest-neighbor carbon atoms. Assuming only nearest-neighbor interactions, calculate the binding energy per unit cell for interplanar spacing values from 3.5 to 4.2 Å at 0.2-Å intervals. Estimate the separation of the planes at the minimum energy. How does it compare with the experimental value of 3.88 Å?

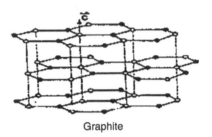

Graphite

6.8. Using (Eq. 6.23), derive an expression for the equilibrium lattice parameter of the rare-gas neon that crystallizes in a FCC structure.

6.9. Derive Eq. (6.28) of the text.

CHAPTER 7

Vibrational Properties

In this chapter we discuss the effect of reducing the size of a crystal to nanometers, and the effect of lowering the dimensionality, on the vibrational behavior of solids.

7.1. THE FINITE ONE-DIMENSIONAL MONATOMIC LATTICE

Some insight into how nanosizing effects the vibrational properties of materials can be obtained by comparing the vibrational dynamics of an infinite and a finite-one dimensional monatomic lattice with atoms of mass M separated from each other by distance a and having a nearest-neighbor simple harmonic interaction of the form Cx^2. Both cases have been analyzed many times. The treatment of the one-dimensional infinite lattice can be found in most basic solid-state physics texts.[1] We will leave it to the student to refer to the texts and just present the results here. The dispersion relationship for the infinite chain of atoms of mass M having only harmonic nearest neighbor interactions is

$$\omega = \left[\frac{4C}{M}\right]^{1/2} \sin\left(\frac{ka}{2}\right) \tag{7.1}$$

Figure 7.1 gives the dispersion relationship, the dependence of the frequency on the wave vector k for this lattice. The important point is that at a given value of k for the infinite lattice, every atom in the lattice has the same frequency. Now let us consider a finite lattice of N particles with free ends and each particle having a harmonic interaction potential with its nearest neighbors. The analysis of this lattice yields vibrational frequencies for the jth atom as[2,3]

$$\omega_j = 2\left[\frac{C}{M}\right]^{1/2} \sin\frac{(j-1)\pi}{2N} \tag{7.2}$$

where j denotes the atoms from left to right in the chain such that $j = 1, 2, \ldots, N$. Now we see that in the finite chain each atom has a different vibrational frequency

The Physics and Chemistry of Nanosolids. By Frank J. Owens and Charles P. Poole, Jr.
Copyright © 2008 John Wiley & Sons, Inc.

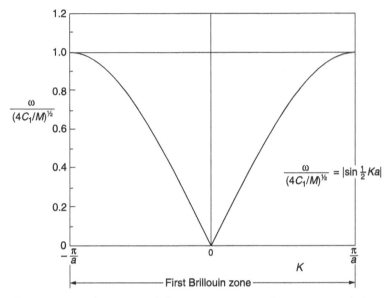

Figure 7.1. The dispersion curve of frequency ω versus the wavevector k for a one-dimensional monatomic infinite lattice.

depending on its position in the chain. Now let us examine how the length of the chain affects the vibrational frequencies. Consider a hypothetical linear lattice of neon atoms where the nearest-neighbor atoms are separated from each other by 1.14 Å and assume that this separation is independent of the length of the lattice, which is a reasonable assumption for an inert-gas lattice as discussed in the previous chapter. Figure 7.2 is a plot of $\omega/[C/M]^{1/2}$ of the center atom of the chain versus the length of the chain in the nanometer regime showing a marked reduction in frequency

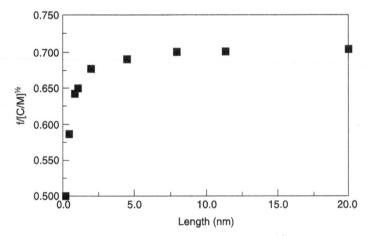

Figure 7.2. Plot of the frequency of vibration divided by $[C/M]^{1/2}$ of the central atom of a finite linear monatomic lattice versus the length.

starting below about 5 nm. As we will see when we examine some experimental results, this decrease in frequency is observed for certain lattice modes as the particle size decreases.

7.2. IONIC SOLIDS

In Chapter 1 we saw that the vibrational frequency of the optical mode at $K = 0$ of an infinitely long line of alternating masses M and m is given by

$$\omega = \left[2C\left(\frac{1}{m} + \frac{1}{M}\right)\right]^{1/2} \tag{7.3}$$

where the force constant C is determined by the condition

$$\frac{d^2U}{dR^2} = 0 \tag{7.4}$$

at the equilibrium separation R_0. In the case of the linear NaCl lattice discussed in the previous chapter the force constant can be obtained by evaluating the second derivative of Eq. (6.4), which is

$$C = \frac{z\lambda}{\rho^2}\exp\frac{-R_0}{\rho} - \frac{2\alpha Q^2}{R_0^3} = 0 \tag{7.5}$$

where the number of nearest neighbors is $z = 2$ for ions in the interior of the chain and $z = 1$ for the end ions. From the increase in the lattice parameter with reduced size, we can calculate C as a function of the length of the chain for the central Cl^- ion. Figure 7.3 is a plot of the calculated force constant versus the length of the chain showing a marked decrease below ~ 15 nm. This means that the vibrational

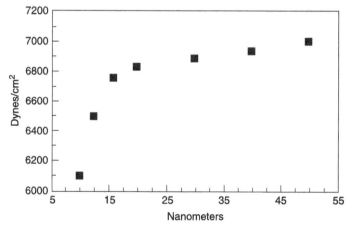

Figure 7.3. Calculated force constant of the optical mode at $K = 0$ for a linear NaCl chain versus the length of the chain.

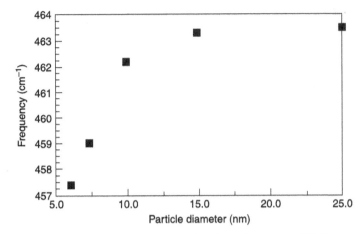

Figure 7.4. Measurement of the Raman active vibrational frequency of CeO_2 versus particle diameter. [Adapted from F. Zhang et al., *Appl. Phys. Lett.* **80**, 127 (2002).]

frequencies will also decrease with reduced length. This is what is experimentally observed in three-dimensional (3D) nanoparticles. Measurements of the frequency versus size in the ionic crystal CeO_2 show a decrease of the frequency. Figure 7.4 is a plot of the measured Raman vibrational frequency versus particle size for this ionic solid showing the onset of a significant decrease in the frequency in the vicinity of 15 nm.

For an infinitely long linear NaCl lattice each atom vibrates with the same frequency given by Eq. (7.3) for $k = 0$. However, this is not true for a short chain. Figure 7.5 shows the results of a calculation of the force constant of the Cl^- ion as

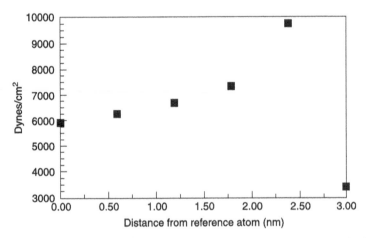

Figure 7.5. Plot of the calculated force constant of the linear NaCl lattice having 20 ions as a function of the distance from center Cl— to the end Cl— showing that the force constant increases for ions closer to the ends of the chain.

a function of its distance from the center of the chain to the end for a 20-ion chain. The results show that except for the end ion, the force constant gradually increases as the ions get closer to the end of the chain. This means the frequencies increase and the amplitudes of vibration decrease as the ions get closer to the end of the linear lattice. Each Cl^- ion in the chain has a different force constant and therefore a different vibrational frequency (as do the Na^+ions), unlike the infinite chain. A consequence of this is that a spectroscopic measurement, such as by Raman or IR will have broader lines in the nanosized materials compared to the bulk material. The broadening is a result of overlap of lines from the slightly different frequencies of the different atoms or molecules in the material having nanometer dimensions. As we will see in the next section, this is what is experimentally observed when materials are reduced to nanometer dimensions. Notice also that ions at the end of the chain have lower force constants than do those in the interior because they have fewer nearby neighbors to bond with. However, there is an important difference between a one-dimensional nanostructure and a two- or three-dimensional structure. In the one-dimensional case, the number of atoms or ions at the end of the chain does not change as the chain gets shorter, whereas in the higher-dimensional nanostructures the number of atoms on the surface increases as the material becomes smaller in the nanometer regime. We saw in Chapter 3 (Fig. 3.17) that the number of surface atoms of gold increases significantly below 10nm. For example, a 2.5-nm lead particle has 76.4% of the atoms on the surface. This means that the vibrational properties of small nanoparticles will be dominated by the dynamical behavior of the surface atoms. Thus a number of mechanisms contribute to the effect of nanosizing on the vibrational frequencies of nanoparticles such as an increase in lattice parameters; an increase in defect concentration, which weakens the binding energy; and an increase in the number of surface atoms or molecules.

7.3. EXPERIMENTAL OBSERVATIONS

7.3.1. Optical and Acoustical Modes

Figure 7.6a is a plot of the frequency the Raman spectrum of the $k = 0$ optical mode of silicon as function of particle size showing a decrease in frequency as the particle size is reduced starting in the vicinity of 10nm. An asymmetric broadening of the Raman line is also observed as the particle size is reduced as shown in Fig. 7.6b, which compares the linewidth in bulk silicon with that in nanosized silicon. These effects have been attributed to phonon confinement, which is discussed below. Subsequently measurements of the effect of nanosizing on optical modes have been made in other materials such as CeO_2, SnO_2, InP, CdSe, and CdS as well as non- stoichiometric $Cd_{0.65}Se_{0.35}$. Brillouin scattering, which is a kind of Raman spectroscopy that measures vibrational frequencies from the acoustic branch of the phonon dispersion curve, has been used to investigate the effect of nanosizing on Si, TiO_2, SnO_2, and CdSe. The frequencies of acoustic modes harden, that is, increase in contrast to the optical modes, as the particle size decreases in the nanometer range.

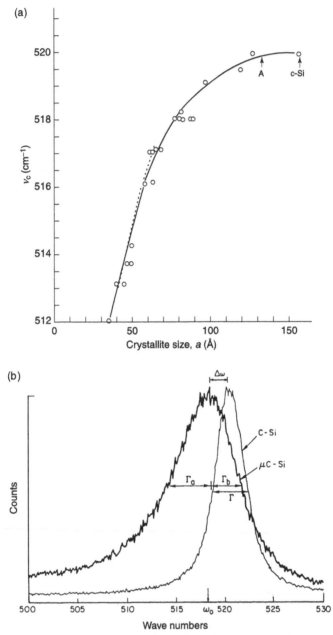

Figure 7.6. (a) Plot of the measured frequency of the optical mode at $k = 0$ of silicon versus particle size showing a reduction in the frequency as the particle becomes smaller; (b) comparison of linewidth of Raman spectra of bulk silicon (C—Si) with nanosized silicon (μC—Si) showing broadening in the nanosized material. [Adapted from Z. Iqbal and S. Veprek, *J. Phys.* **C15**, 377 (1982).]

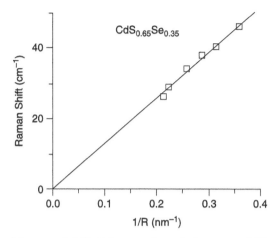

Figure 7.7. Plot of the upward shift of the acoustic mode frequency of $Cd_{0.65}Se_{0.35}$ versus the reciprocal of the radius R, of the particle. [Adapted from P. Verma et al., *J. Appl. Phys.* **88**, 4109 (2000).]

Figure 7.7 illustrates a measurement of the upward shift of the frequencies versus the inverse of the radius of $Cd_{0.65}Se_{0.35}$ nanoparticles. The linear dependence of the shifts on $1/R$ appears to be a general result observed in a number of nanoparticles for the acoustic modes. This increase in frequency of the acoustic modes suggests that phonon confinement is the dominant mechanism causing the shifts. If this lattice expansion of surface atoms were the cause, then the acoustic modes would shift to lower frequencies.

7.3.2. Vibrational Spectroscopy of Surface Layers of Nanoparticles

7.3.2.1. Raman Spectroscopy of Surface Layers
The oxide layers on metal nanoparticles can be detected by Raman spectroscopy. Most metals such as copper, zinc, and aluminum when exposed to air become oxidized on the surface and have a thin layer of metal oxide surrounding the particle. Typically, as shown by the TEM image in Fig. 3.10 for aluminum, the oxide layer is ~3 nanometers thick. Figures 7.8a and 7.8b show respectively a Raman spectroscopy measurement of the vibrational frequency of bulk zinc oxide and 100-nm nanoparticles of zinc. The spectrum from the nanometer zinc oxide layer is very similar to that from zinc oxide except that the lines are asymmetrically broadened and the frequency is at a value lower than that in the bulk material, similar to the effects observed in silicon shown in Fig. 7.6. This means that it arises from ZnO of nanometer dimensions on the surface layer of the particle.

7.3.2.2. Infrared Spectroscopy of Surface Layers
The general principles of infrared (IR) spectroscopy, including Fourier transform infrared spectroscopy

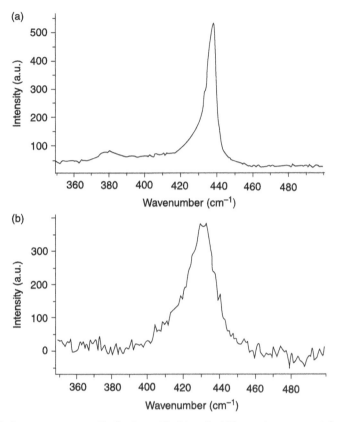

Figure 7.8. Raman spectrum of bulk zinc oxide (a) and a 100-nm zinc nanoparticle (b) arising from the zinc oxide layer on the nanoparticle. [Adapted from F. J. Owens and B. Reddingius, *J. Nanosci. Nanotechnol.* **5**, 836, (2005).]

(FTIR), are explained in Section 2.4.1 These spectroscopic techniques measure the absorption of radiation by high-frequency (i.e., optical branch) phonon vibrations, and they are also sensitive to the presence of particular chemical groups such as hydroxyl (—OH), methyl (—CH_3), imido (—NH), and amino (—NH_2). Each of these groups absorbs IR radiation at a characteristic frequency, and the actual frequency of absorption varies somewhat with the environment.

As an example, Fig. 7.9 shows the FTIR spectrum of titania (TiO_2), which exhibits IR absorption lines from the groups OH, CO, and CO_2. Titania is an important catalyst, and IR studies help elucidate catalytic mechanisms of processes that take place on its surface. This material has the anatase crystal structure at room temperature, and can be prepared with high surface areas for use in catalysis. It is a common practice to activate surfaces of catalysts by cleaning and exposure to particular gases in oxidizing or reducing atmospheres at high temperatures to prepare sites where catalytic reactions can take place. The spectrum of Fig. 7.9 was obtained after adsorbing carbon

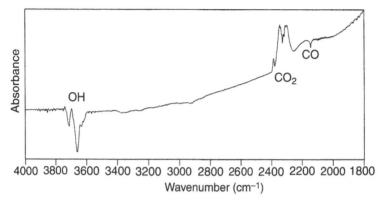

Figure 7.9. Fourier transform infrared (FTIR) spectrum of activated titania nanopowder with carbon monoxide (CO) adsorbed on the surface. The spectrum of the initial activated titania has been subtracted. The negative (downward) adsorption in the OH region indicates the replacement of hydroxyl groups by CO_2 on the surface. [From M.-I. Baraton and L. Merhari, *Nanostruct. Mater.* **10**, 699 (1998).]

monoxide (CO) at $500°C$ on an activated titania nanopowder surface, and subtracting the spectrum of the initial activated surface before adsorption. The strong carbon dioxide (CO_2) IR absorption lines in the spectrum show that the adsorbed carbon monoxide had been oxidized to carbon dioxide on the surface. Note that the OH absorption signal is in the negative (downward) direction. This means that the activated titania surface, whose spectrum had been subtracted, initially had many more OH groups on it did after CO adsorption. Apparently OH groups originally present on the surface have been replaced by CO_2 groups. Also the spectrum exhibits structure in the range from 2100 to $2400 \, cm^{-1}$ due to the vibrational–rotational modes of the CO and CO_2 groups. In addition, the gradually increasing absorption for decreasing wavenumber shown at the right side of the figure corresponds to a broad spectral band that arises from electron transfer between the valence and conduction bands of the N-type titania semiconductor.

To learn more about an IR spectrum, the technique of isotopic substitution can be employed. We know from elementary physics that the frequency of a simple harmonic oscillator ω of mass m and spring constant C is proportional to $(C/m)^{1/2}$, which means that the frequency ω, and the energy E given by $E = \hbar \omega$, both decrease with an increase in mass m. As a result, isotopic substitution, which involves nuclei of different masses, changes the IR absorption frequencies of chemical groups. Thus the replacements of ordinary hydrogen ^1H by the heavier isotope deuterium ^2D (0.015% abundant), ordinary carbon ^{12}C by ^{13}C (1.11% abundant), ordinary ^{14}N by ^{15}N (0.37% abundant), or ordinary ^{16}O by ^{17}O (0.047% abundant) all increase the mass, and hence decrease the IR absorption frequency. The decrease is especially pronounced when deuterium is substituted for ordinary hydrogen since the mass ratio $m_D/m_H = 2$, so the absorption frequency is expected to decrease by the factor $2^{1/2} \cong 1.414$. The FTIR spectrum of boron nitride (BN) nanopowder after

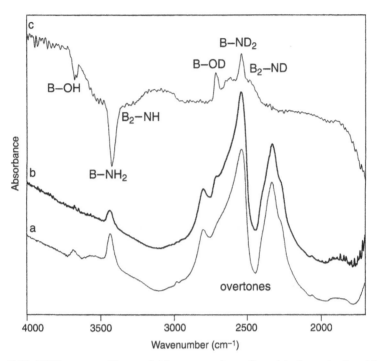

Figure 7.10. FTIR spectra of boron nitride nanopowder surfaces (a) after activation at 875 K, (b) after subsequent deuteration, and (c) difference spectrum of (a) subtracted from (b). [From M.-I. Baraton, L. Merhari, P. Quintard, and V. Lorezenvilli, *Langmuir* **9**, 1486 (1993).]

deuteration (H/D exchange), presented in Fig. 7.10, exhibits this $2^{1/2}$ shift. The figure shows the initial spectrum (a) of the BN nanopowder after activation at 875 K, (b) of the nanopowder after subsequent deuteration, and (c) after subtraction of the two spectra. It is clear that the deuteration converted the initial B—OH, B—NH$_2$, and B$_2$—NH groups on the surface to B—OD, B—ND$_2$, and B$_2$—ND, respectively, and that in each case the shift in wavenumber (i.e., frequency) is close to the expected $2^{1/2}$. The overtone bands that vanish in the subtraction of the spectra are due to harmonics of the fundamental BN lattice vibrations, which are not affected by the H/D exchange at the surface. Boron nitride powder is used commercially for lubrication. Its hexagonal lattice, with planar B$_3$N$_3$ hexagons, resembles that of graphite.

A close comparison of the FTIR spectra from gallium nitride GaN nanoparticles illustrated in Fig. 7.11 with the boron nitride nanoparticle spectra of Fig. 7.10 show how the various chemical groups —OH, —NH$_2$, and —NH and their deuterated analogs have similar vibrational frequencies, but these frequencies are not precisely the same. For example, the frequency of the B—ND$_2$ spectral line of Fig. 7.10 is somewhat lower than that of the Ga—ND$_2$ line of Fig. 7.11, a small shift that results from their somewhat different chemical environments. The H/D exchange results of

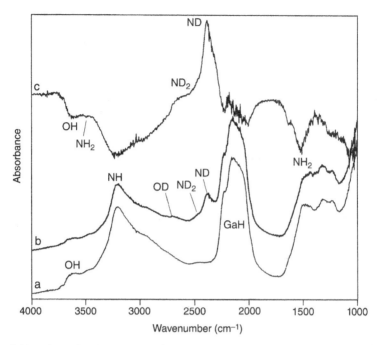

Figure 7.11. FTIR reflection spectra of gallium nitride nanopowder surfaces (a) after activation at 500°C, (b) after subsequent deuteration, and (c) difference spectrum of (a) subtracted from (b). [From M.-I. Baraton, G. Carlson, and K. E. Gonsalves, *Mater. Sci. Eng.* **50**, 42 (1997).]

Fig. 7.11 show that all of the Ga—OH and Ga—NH$_2$ are on the surface, while only some of the NH groups were exchanged. Notice that the strong GaH absorption band near $21,000\,\mathrm{cm}^{-1}$ was not appreciably disturbed by the H/D exchange, suggesting that it arises from hydrogen atoms bound to gallium inside the bulk.

An example that demonstrates the power of infrared spectroscopy to elucidate surface features of nanomaterials is the study of γ-alumina (Al$_2$O$_3$), a catalytic material that can have a large surface area, up to $200-300\,\mathrm{m}^2/\mathrm{g}$, because of its highly porous morphology. It has a defect spinel structure, and its large oxygen atoms form a tetragonally distorted FCC lattice. There are one octahedral (VI) and two tetrahedral (IV) sites per oxygen atom in the lattice, and aluminum ions located at these sites are designated by the notation $_{\mathrm{VI}}\mathrm{Al}^{3+}$ and $_{\mathrm{IV}}\mathrm{Al}^{3+}$, respectively, in Fig. 7.12. There are a total of five configurations assumed by adsorbed hydroxyl groups, which bond to aluminum ions at the surface, and these are sketched in the figure. The first two, types Ia and Ib, involve the simple cases of OH bonded to tetrahedrally and octahedrally coordinated aluminum ions, respectively. The remaining three cases involve the hydroxyl radical bound simultaneously to two or three adjacent trivalent aluminum ions. The frequency shifts assigned to these five surface

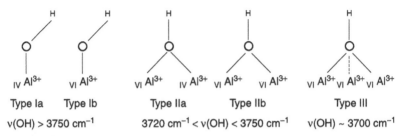

Figure 7.12. Five possible configurations of adsorbed hydroxyl groups bonded at tetrahedral ($_{IV}Al^{3+}$) and octahedral ($_{VI}Al^{3+}$) sites of a γ-alumina surface. (From M.-I. Baraton, in *Handbook of Nanostructured materials and Nanotechnology*, H. S. Nalwa, ed., Academic press, Boston, 2000, Vol. 2, Chapter 2, p. 116.)

species, which are listed in the figure [ν(OH)], are easily distinguished by infrared spectroscopy.

The FTIR spectra from γ-alumina nanopowder before and after deuteration presented in Fig. 7.13 display broad absorption bands with structure arising from the OH and OD groups, respectively, on alumina activated at 600°C, and spectrum (a) of Fig. 7.14 provides an expanded view of the OD region of Fig. 7.11 for γ-alumina activated at 500°C. Analyses of the positions and relative amplitudes of the component lines of these spectra provide information on the distribution of aluminum ions in the octahedral and tetrahedral sites of the atomic layer at the surface. The differences between the γ-alumina and the θ-alumina spectra of Fig. 7.11 provide evidence that the two aluminas differ in their allocations of Al^{3+} to octahedral and tetrahedral sites. In further work, adsorption of the heterocyclic six-membered ring compound pyridine, C_5H_5N, on γ-alumina provided FTIR spectra of the pyridine coordinated to different Al^{3+} Lewis acid sites. These IR spectral results help clarify the coordination state of aluminum ions in the surface layer where Lewis acid sites play an important role in the catalytic activity.

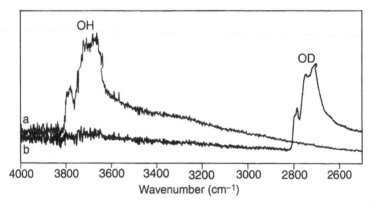

Figure 7.13. FTIR spectra of γ-alumina nanopowder surfaces (a) after activation at 600°C and (b) after subsequent deuteration. (From M.-I. Baraton, p. 115.)

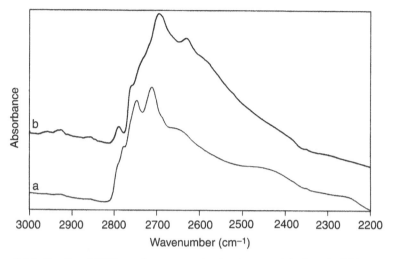

Figure 7.14. Details of FTIR surface spectra in the deuterated hydroxyl (OD) absorption region of (a) γ-Al$_2$O$_3$ and (b) θ-Al$_2$O$_3$ nanopowders after activation at 500°C followed by deuteration. (From M.-I. Baraton, in *Handbook of Nanostructured Materials and Nanotechnology*, H. S. Nalwa, ed., Academic Press, Boston, 2000, Vol. 2, Chapter 2, p. 117.)

7.4. PHONON CONFINEMENT

In the case of our hypothetical linear NaCl lattice, as seen in Fig. 6.3, the force constant of the central Cl$^-$ ion decreases as the length of the chain decreases below 15 nm, which means the frequency also decreases. This reduction is a result of the increase of the lattice parameter as the length of the chain decreases. However, there is another mechanism, called *phonon confinement* which can also cause the frequency to change when materials are reduced to nanosized dimensions. Phonon confinement results from, the fact that the vibrations are now restricted to a finite-sized material and periodic boundary conditions cannot be employed since they assume an infinite lattice. The frequency shifts and broadening shown in Figs. 7.6a and 7.6b for silicon nanoparticles have been attributed entirely to phonon confinement effects. Relatively simple models have been proposed to explain the effect, but first let us look at another way to understand phonon confinement.

The uncertainty principle states that the order of magnitude of the uncertainty in position ΔX times the order of magnitude of the uncertainty in momentum ΔP must at least be h, Planck's constant, divided by 2π:

$$\Delta X \, \Delta P \geq \frac{h}{2\pi} \tag{7.6}$$

Let us assume that ΔX is the diameter of the nanoparticle D and that it can be measured accurately by some technique such as scanning electron microscopy. This means that the uncertainty in momentum P will have a range of values. It can be shown that the momentum of a phonon is $hk/2\pi$. Thus we have

$$D\Delta k \geq 1$$

or the minimum uncertainty in k is

$$\Delta k \sim \frac{1}{D} \qquad (7.7)$$

Raman spectroscopy measures frequencies at the center of the Brillouin zone, $k = 0$. However, that is a precise value that Eq. (7.6) shows becomes uncertain, meaning that there is a spread of values for k for a Raman measurement in small particles. This means that nonzero values of k will contribute to the Raman spectrum. Let us assume that for small values of k the dependence of frequency on k, namely, the dispersion relationship, is linear:

$$\omega = ak \qquad (7.8)$$

From this it follows that

$$\Delta k = C\,\Delta\omega \qquad (7.9)$$

where C is a constant. Thus

$$\Delta\omega = C\left[\frac{1}{D}\right] \qquad (7.10)$$

However, if the dependence of the frequency on k is not linear, then the dependence of the frequency shift on particle diameter will be $1/D^{\gamma}$, which is what is observed with γ ranging from 1 to 1.5 depending on the material.

A phenomalogical model has been developed to explain phonon confinement. The wavefunction of a phonon with wavevector K in an infinite crystal is given by

$$\Psi(k_0, R) = U(k_0, R)\exp(-ik_0 \cdot R) \qquad (7.11)$$

where $U(k_0,R)$ has the periodicity of the lattice. For a spherical particle having a diameter D in the order of nanometers, the phonon, which is restricted to the volume of the crystal, can be treated using a Gaussian weighting function of the form $\exp(-2R^2/D^2)$ to modify Eq. (7.11):

$$\Psi(k_0,R) = A\exp\left(\frac{-4R^2}{D^2}\right)U(k_0, \cdot R)\exp(-ik_0 \cdot R) \qquad (7.12)$$

The idea is that the amplitudes of vibration of the modes will decrease with particle size. There is no physical reason to assume this form except that it does predict the

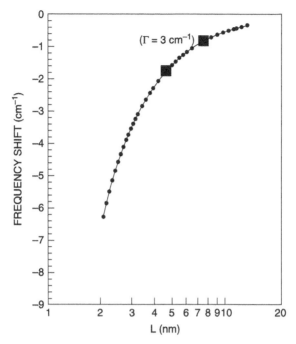

Figure 7.15. Comparison of the measured frequency shift of silicon (■) versus particle size with the prediction from phonon confinement model (•).

observed decrease in frequency of the silicon optical phonon mode with particle size. In Chapter 1 the dispersion relationship of a one-dimensional linear diatomic lattice was discussed. The dispersion relationship is the dependence of the frequency ω on the wavevector k. Raman spectroscopy measures the lattice frequencies at $k = 0$. The theory of phonon confinement predicts a relaxation of this selection rule for nano-sized materials, meaning that there can be contributions to the spectrum for nonzero values of k. Thus, if the frequency decreases with k as does the optical mode of the one-dimensional linear diatomic lattice, there will be a decrease in frequency with nanosizing, which is the most common observation. However, if the frequency increases with k from $k = 0$, the mode frequencies could increase with nanosizing. The results of the analysis lead to a dependence of the vibrational frequency and the linewidth of the spectra on the particle diameter. Figure 7.15 is a semilog plot of the dependence of the frequency shift on the particle diameter from this model compared with experimental results for silicon.

7.5. EFFECT OF DIMENSION ON LATTICE VIBRATIONS

The behavior of lattice vibrations of nanometer-sized materials also depends on the dimensionality. An illustration of the effect of dimensionality is shown by a calculation of the intensity of the two most Raman active vibrations in the one-dimensional polyacetylene chain as a function of chain length in the nanometer range. The results

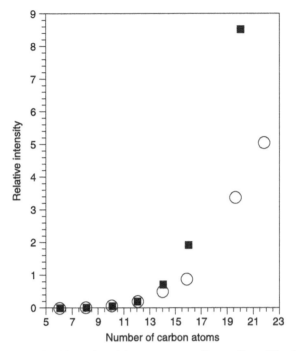

Figure 7.16. The calculated intensity of the two strongest Raman lines (■ and ○) of polyacetylene versus the length of the polymer chain. [Adapted from F. J. Owens, *Physica* **E25**, 404 (2005).]

of the calculation are shown in Fig. 7.16. The calculated frequencies of $1464\,\text{cm}^{-1}$ and $1064\,\text{cm}^{-1}$ are in good agreement with the experimental values of 1450 and $1060\,\text{cm}^{-1}$. The lower-frequency mode is a coupled C—C stretch and C—H bending motion. The higher-frequency mode is a C—C stretch vibration. As shown in the figure, when the length of the chain exceeds 1.5 nm, effectively where the chain begins to become more one-dimensional, there is a marked increase in the intensity of the Raman spectra. Low-dimensional nanostructures tend to have very intense Raman spectra. We will see this again when we discuss carbon nanotubes. The reason for this enhanced intensity is the emergence of singularities called *Van Hove singularities* in the density of electronic states near the bandgap. The electronic density of states $D(E)$, which will be discussed in greater detail in a later chapter, is the number of energy levels per interval of energy dN/dE. Figure 7.17 shows the results of a calculation of the electronic energy levels and density of states for polyacetylene showing the spikes in the density of states. In the Raman experiment the material is electronically excited by laser light, typically in the visible region of the spectrum. If the excitation occurs from an electronic level where the density of states is high, more electrons are involved in the transition and the Raman spectrum becomes more intense.

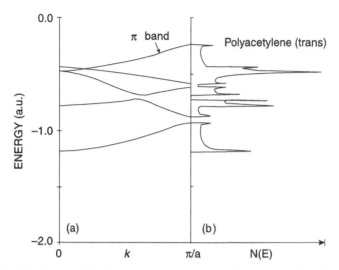

Figure 7.17. Calculation of the electronic energy levels of polyacetylene (a) and the density of states (b) showing peaks in the density of states called *Van Hove singularities*, which occur in low-dimensional materials. (Adapted from J. Bredas, in *Handbook of Conducting Polymers*, B. T. Skotheim, ed., Marcel Dekker, New york, 1986 Vol. 2.)

7.6. EFFECT OF DIMENSION ON VIBRATIONAL DENSITY OF STATES

The vibrational density of states $D(\omega)$ is the number of vibrational modes per interval of frequency $dN/d\omega$. For a one-dimensional line having N atoms, the number of vibrational modes is N. For a three-dimensional lattice it is $3N$. The normal modes may be considered a set of independent oscillators with each oscillator having the energy ε_n given by

$$\varepsilon_n = \left(n + \frac{1}{2}\right)h\omega \tag{7.13}$$

where n is a quantum number having integral values in the range $0,1,2\ldots,n$. The average energy of a harmonic oscillator assuming a Maxwell–Boltzmann distribution function is

$$<\varepsilon> = h\omega\left[\frac{1}{2} + \frac{1}{\exp\{h\omega/k_BT\} - 1}\right] \tag{7.14}$$

The total energy of a collection of oscillators is

$$E = \Sigma_k\langle\varepsilon_k\rangle h\,\omega_k \tag{7.15}$$

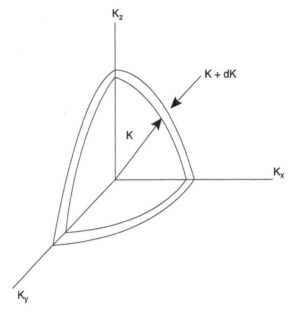

Figure 7.18. A spherical shell in k-space used to obtain the density of states of vibrational frequencies.

When there are a large number of atoms, the allowed frequencies are very close and can be treated as a continuous distribution allowing replacement of the summation by an integral, where $D(\omega)d\omega$ is the number of modes of vibration in the range ω to $\omega + d\omega$. It turns out that it is more convenient to work in k-space, where $k = \omega/c$ and c is the velocity of light, and talk about the number of modes $D(k)dk$ in the interval from k to $k + dk$. As shown in Fig. 7.18, the number of values of K in three dimensions between k and $k + dk$ will be proportional to the volume element on a sphere of radius k which is given by

$$dV = 4\pi k^2 dk = D(k)dk \tag{7.16}$$

The dispersion relationship refers to the dependence of the frequency ω on the K vector. One approximation due to Debye assumes that the relationship is linear, which is valid for low values of k:

$$\omega(k) = uk \tag{7.17}$$

From Eqs. (7.16) and (7.17), we obtain

$$D(\omega)d\omega = B\omega^2 d\omega \tag{7.18}$$

where B is a constant and thus

$$D(\omega) = B\omega^2 \tag{7.19}$$

In three dimensions the total number of modes is $3N$, which means that

$$\int_0^{\omega_d} D(\omega)d\omega = 3N \tag{7.20}$$

where ω_d is the highest frequency that can propagate in the lattice, and is referred to as the *Debye frequency*. Thus, for three dimensions in the Debye approximation, the density of states for $\omega < \omega_d$ is

$$D(\omega) = \frac{9N\omega^2}{\omega_d^3} \tag{7.21}$$

In two dimensions we would carry out an analogous derivation using a circle having area $\pi k,^2$ obtaining

$$D(k)dk = 2\pi k\,dk \tag{7.22}$$

and

$$D(\omega)d\omega = \omega\,d\omega \tag{7.23}$$

$$\int_0^{\omega_d} D(\omega)d\omega = 2N \tag{7.24}$$

giving

$$D(\omega) = 4\omega N/\omega_d^2 \tag{7.25}$$

Following the same procedure, the density of states in one dimension can be obtained to be

$$D(\omega) = \frac{N}{\omega_d} \tag{7.26}$$

Thus the density of phonon states in the Debye approximation depends on the dimensionality of the material. Figure 7.19 shows a plot of phonon density of states versus the frequency in the different dimensions. Notice that the density of states decreases with the number of atoms N, which means that the density of states will decrease in the nanometer range. Figure 7.20 is a plot of the log of the density of states in the Debye approximation at $\omega = \omega_d/2$ versus the diameter of an aluminum nanoparticle

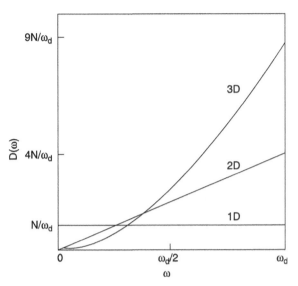

Figure 7.19. A plot of the density of phonon states versus the frequency in the Debye approximation for 3D, 2D, 1D materials.

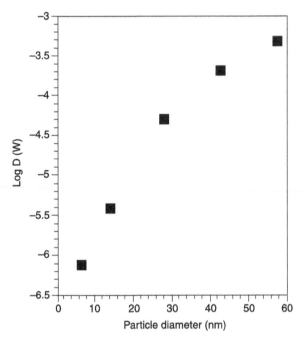

Figure 7.20. Density of states in the Debye approximation at the frequency $\omega_d/2$ versus the diameter of an aluminum nanoparticle.

using the relationship between particle size and the number of atoms in the particle as discussed in Chapter 1. The calculation has not accounted for the dependence of the Debye frequency on particle size, which, as we will see below, decreases with smaller particle size.

7.7. EFFECT OF SIZE ON DEBYE FREQUENCY

Let us examine the effect of reducing the size of a solid on the Debye frequency. A simple assumption is that the fractional change in the vibrational frequency is proportional to the fractional change in the volume of the unit cell of the lattice:

$$\frac{\Delta\omega}{\omega} = \frac{\gamma\Delta V}{V} \tag{7.27}$$

The dimensionless proportionality constant γ is called the *Gruneisen constant*. It is a measure of the anharmonicity of the lattice potential because for a harmonic potential the frequency will not change with the volume of the lattice. Equation (7.27) can be used to estimate the effect of nanosizing on the Debye frequency. Taking the

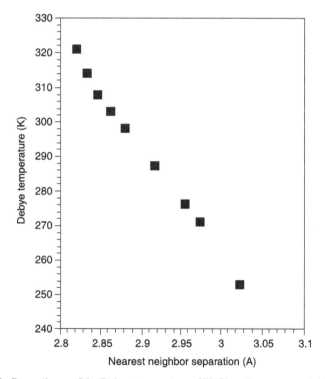

Figure 7.21. Dependence of the Debye temperature of NaCl on the nearest-neighbor distance.

Gruneisen parameter for NaCl as 1.5 at 300 K, it is possible to calculate the change in the Debye temperature as a function of the nearest-neighbor distance. The results are plotted in Fig. 7.21. Because lattice parameters increase as materials reach nano-dimensions we then expect the Debye frequency to decrease.

7.8. MELTING TEMPERATURE

Earlier we argued that the melting temperature of nanosized solids should decrease because the binding energy decreases as the material reaches nanometer dimensions. Here we consider another factor that affects the melting temperature. Figure 7.22 is a plot of the melting temperature of alkali halide structures versus the Debye tempera-ture, $\Theta = \hbar \omega_d / k_B$, of the solids. The plot shows that the lower the Debye tempera-ture, the lower the melting temperature. Many years ago Lindemann proposed that melting occurs when the root mean square (RMS) displacement of an atom from its equilibrium site reaches some fraction f of the nearest-neighbor distance. It had been observed in a number of simple solid structures that at melting the RMS amplitude is approximately 20% of the nearest-neighbor distance r_0. The RMS

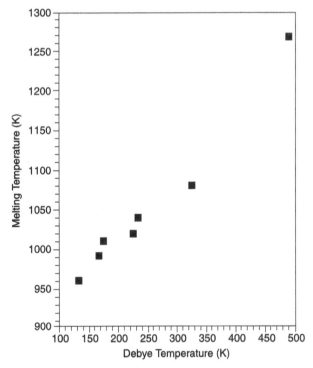

Figure 7.22. Dependence of the melting temperature of alkali halides on the Debye temperature.

amplitude $<A^2>$ of a collection of independent oscillators is given by

$$<A^2>= \int_0^{\omega_d} A^2 D(\omega)d\omega \tag{7.28}$$

where ω_d is the Debye frequency, $D(\omega)$ the density of states, and A the amplitude of vibration, which for a harmonic oscillator is given by

$$A = \left[\left(\frac{h}{M\omega}\right)(N+1)\right]^{1/2} \tag{7.29}$$

At high temperatures the energy of the harmonic oscillator $\hbar\omega(N+1) \sim k_B T$, which allows the elimination of $N+1$ from Eq. (7.29), and then the amplitude of vibration is given by

$$A = \frac{k_B T}{M\omega^2} \tag{7.30}$$

Assuming a Debye density of states given by Eq. (7.21), we get for $<A^2>$

$$<A^2>= \frac{9k_B T}{M\omega_d^2} \tag{7.31}$$

This can be reformulated in terms of the Debye temperature by substituting $\omega_d = \Theta_d k_B/\hbar$, assuming that melting occurs when A is some fraction f of the nearest-neighbor distance r_0 giving an expression for the melting temperature:

$$T_m = \frac{f^2}{9h^2} M k_B \Theta_d^2 r_0^2 \tag{7.32}$$

The lower the frequency of vibration, the greater the amplitude of vibration at a given temperature. Therefore solids having lower vibrational frequencies should have lower melting temperatures. Thus we see that this simple idea predicts that the melting temperature depends on the Debye temperature. Since the Debye temperature decreases with particle size in the nanometer regime, we have a simple explanation for the observed decrease in melting temperature with particle size. The amplitude of vibration of atoms on the surface of a nanoparticle at a given temperature exceeds that in the interior of the particle. This is because they are bonded to fewer atoms of the nanoparticle surface than in the interior of the particle. Perhaps a better index of melting would be a surface Debye frequency, which would be lower than that of the interior. It has been shown theoretically that the Debye temperature for vibrations perpendicular to the surface of a three-dimensional solid are about 71% of the Debye temperature in the center of the lattice.[4] As the particle size decreases and the percentage of atoms on the surface increases, the melting temperature might be expected to decrease. The Lindemann model is very simple, and actually the factors involved in melting are somewhat more complex, but the model did point out the importance of the role of lattice vibrations in the process. Since the mid-1980s a number of more complex models have been developed to treat melting. One model assumes that melting is the result of an instability at high temperature, which occurs when the vibration pressure P_v of the lattice becomes greater than the electrostatic pressure P_e. The vibrational pressure in the Debye approximation is obtained by taking the derivative of the vibrational energy $<E>$ of Eq. (7.15) with respect to the volume $d<E>/dV$. The electrostatic pressure is the derivative of the lattice energy U with respect to volume. The total pressure of the lattice is

$$P_t = P_e - P_v \tag{7.33}$$

Because of the expansion of the lattice parameters with increasing temperature, both P_e and P_v increase with temperature. However, the model shows that at some high temperature P_v becomes greater than P_e, which marks the onset of the instability. There is a good correlation between the predicted temperature of the onset of the instability and the melting temperature of the materials. It would be very interesting to examine how nanosizing affects the temperature of the onset of the instability in this model, but it would take us too far afield to do this here.

7.9. SPECIFIC HEAT

The main contribution to the specific heat is the derivative of the vibrational energy $<E>$ with respect to temperature. In the Debye approximation the vibrational energy $<E>$ is given by

$$<E> = \int_0^{\omega_d} D(\omega)n(\omega)\hbar\omega \, d\omega \tag{7.34}$$

where $D(\omega)$ is the Debye density of states and $n(\omega)$ is given by

$$n(\omega) = \frac{\hbar\omega}{\exp[\hbar\omega/k_BT] - 1} \tag{7.35}$$

For three dimensions, we obtain

$$<E> = (9N/\omega_d) \int_0^{\omega_d} \left[\frac{\hbar\omega^3}{\exp[\hbar\omega/k_BT] - 1} \right] d\omega \tag{7.36}$$

At high temperature where $kT \gg \hbar\omega$, the vibrational energy becomes

$$<E> = 3Nk_B T \qquad (7.37)$$

and the specific heat is

$$C_v = 3Nk_B \qquad (7.38)$$

In two dimensions it is $2Nk_B$ and one dimension it is Nk_B. Since the number of atoms decreases as the size of a particle decreases, we expect the high-temperature specific heat to decrease as a material becomes nanosized. Using the relationship between the number of atoms in a nanoparticle and the diameter of a nanoparticle discussed in Chapter 1, the specific heat can be calculated as a function of the diameter of the nanoparticle. Figure 7.23 is a plot of the log of the high-temperature specific heat of aluminum versus the particle diameter calculated in the Debye model showing a marked decrease as the particle diameter decreases.

Now let us consider the effect of nanosizing on the low-temperature specific heat in the Debye approximation. Letting $x = \hbar\omega/k_B T$ and substituting into

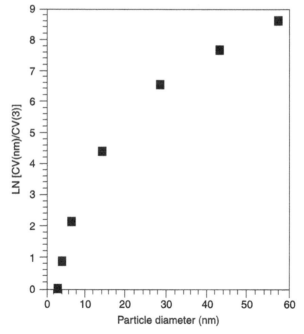

Figure 7.23. Plot of the logarithm of the high-temperature specific heat in the Debye approximation normalized to $3k_B T$ versus the diameter of the nanoparticle.

Eq. (7.36), we get

$$<E> = \frac{9N(k_B T)^4}{\omega_d^3 \hbar^3} \int_0^\infty \frac{x^3 dx}{\exp(x) - 1} \qquad (7.39)$$

The low-temperature behavior can be approximated by integrating from 0 to ∞, which gives

$$<E> = \frac{9N(k_B T)^4}{\omega_d^3 \hbar^3} \frac{\pi^4}{15} \qquad (7.40)$$

C_V is obtained from Eq. (7.40) by differentiation and is found to be,

$$C_v = \frac{2.4 N k_B^4 T^3 \pi^4}{\omega_d^3 \hbar^3} \qquad (7.41)$$

In order to assess how nanosizing affects the low-temperature specific heat, we have to determine how the ratio N/ω_d^3 changes with the particle diameter. A reasonable assumption is that the decrease in ω_d with particles size can be accounted for by the phonon confinement model discussed earlier. Using that model, taking the Debye frequency of aluminum as $308\,\mathrm{cm}^{-1}$, and applying our previously observed relationship between the particle diameter and the number of atoms N, we can estimate how N/ω_d^3 changes with particle size. The results show that this quantity will decrease with particle size, but not as strongly as will the high-temperature specific heat. Thus the low-temperature specific heat also decreases with particle size, but the dependence is weaker than the high-temperature specific heat.

7.10. PLASMONS

In a semiconductor or metal the conduction electrons are not localized on the atoms, and the atoms are thus positively charged. These free electrons can be viewed as an electron "cloud," and this cloud can vibrate, a phenomenon referred to as *plasmon oscillation*. The electron cloud oscillates as a whole with respect to the fixed positive ions. The oscillation can be analyzed by considering a thin slab of the metal shown in Fig. 7.24a, where the gray area is the electron cloud and the + signs are the ionized atoms of the solid. Now let us assume the cloud oscillates in the y direction as shown in Fig. 7.24b, where the cloud has shifted up with a displacement u. The displacement establishes a negative surface charge density $-neu$ on the top surface and a positive charge density $+neu$ on the bottom surface, where n is the number of electrons per unit volume. An electric field E is created inside the slab given by

$$E = 4\pi n e u \qquad (7.42)$$

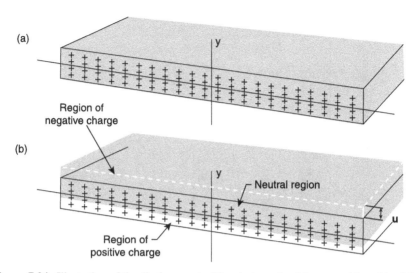

Figure 7.24. Illustration of the displacement of the electron cloud (gray area) in a thin slab of metal that oscillates and is called a *plasmon*.

where n is the number of electrons in the cloud per unit volume. The equation of motion of the oscillating electron cloud is thus

$$\frac{nmd^2u}{dt^2} = -neE = 4\pi n^2 e^2 u \tag{7.43}$$

where m is the mass of the electron. Rewriting this, we have

$$\frac{d^2u}{dt^2} + \omega_p^2 u = 0 \tag{7.44}$$

where ω_p is the plasma frequency given by

$$\omega_p = \left[\frac{4\pi ne^2}{m}\right]^{1/2} \tag{7.45}$$

One might expect this frequency to be affected by nanosizing through the dependence on the number of free electrons per unit volume on particle size. To examine this, let us consider the case of aluminum where we have previously employed the relationship between the number of atoms and the diameter of a nanosized particle. If we assume that each Al atom provides three free electrons, then the number of electrons per unit volume can be calculated. Figure 7.25 is a plot of the relative increase of electron density versus the diameter of a spherical aluminum nanoparticle showing that the electron density increases only at very small particle diameters, less than 5 nm.

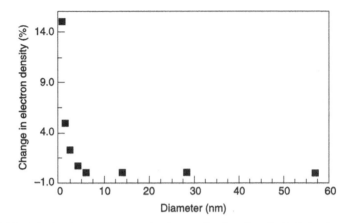

Figure 7.25. Calculated percent change in electron density of a spherical aluminum nanoparticle versus the particle diameter.

From Eq. (7.45) this means that there will be an increase in the plasmon frequency. The calculation does not account for the slight decrease of the lattice parameter of Al discussed in Chapter 3, which will provide a further contribution to increasing the plasmon frequency. However, the experimental measurements of the effect of nanosizing on the frequency of the plasma mode are not clear. Some studies show redshifts and others blueshifts. The increase in electron density with particle size cannot explain a redshift or decrease in frequency. Thus further work needs to be done to resolve this issue.

7.11. SURFACE-ENHANCED RAMAN SPECTROSCOPY

It has been found that when a thin layer of a molecule is adsorbed on an array of gold or silver nanoparticles smaller than the wavelength of the laser light, the intensity of the Raman spectra from the molecule is greatly enhanced. The spectra can be enhanced by a factor of $10^6 - 10^7$. Closely spaced particles provide a further enhancement up to 10^8. Figure 7.26 shows the enhancement of the intensity of the Raman spectra of the ring vibrations of pyridine deposited on a film of 8-nm nanoparticles of silver compared to a film of 30-nm silver nanoparticles, showing an enhancement of the intensity of the spectra on the smaller particles. The effect is a result of the interaction between the laser used in Raman spectroscopy and the surface plasmon vibrations discussed above. The time-dependent E field of the laser excites the plasma mode. The induced oscillation of the plasma mode in turn produces an intense electromagnetic field at the surface of the particle having the vibrational frequency of the plasmon. The vibration of the charge cloud with respect to the fixed ions of the lattice is analogous to the vibration of a dipole, which will emit radiation having the same frequency as that of the vibrating dipole. The intense electromagnetic

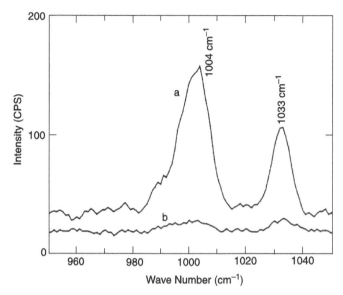

Figure 7.26. Raman spectrum of ring vibrations of pyridine deposited on silver films having grain sizes of (a) 8 nm and (b) 30 nm showing enhancement of the intensity for the smaller particles. [Adapted from H. Seki, *J. Chem. Phys.* **76**, 4418 (1982).]

field at the surface of the particle produces further excitation from the ground vibrational states of the molecule on the surface to a virtual level, further enhancing the intensity of the Raman spectrum. This effect can also be observed in infrared spectroscopy.

7.12. PHASE TRANSITIONS

A *phase transition* is a change in the structure of a crystal that occurs at some temperature and pressure. Phase transitions can be classified as first-order or second-order. In a first-order phase transition the structural change occurs abruptly at a specific temperature or pressure. A second-order phase transition occurs more gradually as a function of temperature or pressure. We will confine ourselves here to a discussion of second-order phase transitions because the mechanism is better understood.

The Lyddane–Sachs–Teller formula relates the transverse optical phonon frequency ω_t and the longitudinal optical phonon frequency ω_l to the ratio of the low-frequency ε_0 and high-frequency dielectric constant $\varepsilon(\infty)$ for a diatomic cubic crystal.

$$\frac{\omega_l^2}{\omega_t^2} = \frac{\varepsilon_0}{\varepsilon(\infty)} \tag{7.46}$$

Some phase transitions, besides involving a change in structure, also involve a change in polarization. When the structure changes, the dipoles in the lattice become ordered, and the crystal acquires a net polarization. These materials are referred to as *ferroelectric*, and $BaTiO_3$ is an example of such a material. In such transitions ε_0 is observed to diverge near the transition temperature having a temperature dependence proportional to $1/(T - T_c)$, suggesting from the Lyddane–Sachs–Teller formula that ω_t is approaching zero with a temperature dependence of $(T - T_c)^{1/2}$. This process is referred to as *mode softening*, and there have been many verifications of the $(T - T_c)^{1/2}$ temperature dependence of some low-frequency mode as the temperature of a phase transition is approached. This softening is also observed in nonferroelectric phase transitions. It was first observed by Raman spectroscopy in the α–β transition in quartz. It is now generally recognized that a second-order displacive phase transition in a crystal is often the result of a softening of a vibrational mode in a crystal. The first understanding of why a mode softens in a lattice was given independently by Cochran in 1959.[5] Here we will present a very simple version of the theory, which illustrates the salient physics of the process. Then, in the context of this model, we will examine how nano sizing affects the mechanism.

Let us consider a diatomic cubic lattice of rigid ions, which means that we are ignoring the charge cloud polarizability. The displacement of a charge Q from its equilibrium site due to vibration is equivalent to the creation of a dipole, $Q(U_1 - U_2)$. The transverse optical mode at $k = 0$ involves a displacement of all positive and negative ions by amounts U_1 and U_2 in the same direction. This results in a polarization of the lattice

$$P = \frac{Q(U_1 - U_2)}{V} \tag{7.47}$$

where V is the volume of the unit cell. In an applied electric field E, the equations of motion of the positive and negative ions are respectively

$$\frac{M_1 d^2 U_1}{dt^2} = Q\left(E + \frac{P}{3\varepsilon_0}\right) - \beta(U_1 - U_2) \tag{7.48}$$

where β is the force constant. The equation of motion for M_2 is

$$\frac{M_2 d^2 U_2}{dt^2} = -Q\left(E + \frac{P}{3\varepsilon_0}\right) - \beta(U_2 - U_1) \tag{7.49}$$

Subtracting Eq. (7.49) from (7.48) gives

$$\frac{M_1 d^2 U_1}{dt^2} - \frac{M_2 d^2 U_2}{dt^2} = 2Q\left(E + \frac{P}{3\varepsilon_0}\right) - 2\beta(U_1 - U_2) \tag{7.50}$$

Now we transform to the center of mass system:

$$U_1 = \frac{M_2 U}{(M_1 + M_2)} \tag{7.51}$$

$$U_2 = \frac{-M_1 U}{M_1 + M_2} \tag{7.52}$$

For $E = 0$, we get

$$\frac{M d^2 U}{dt^2} = 2\left(\beta - \frac{Q^2}{3\varepsilon_0 V}\right) U \tag{7.53}$$

so the force constant of the transverse optical mode is

$$k = 2\left(\beta - \frac{Q^2}{3\varepsilon_0 V}\right) \tag{7.54}$$

Similarly, it can be shown that the force constant of the longitudinal mode is

$$k = 2\left(\beta + \frac{Q^2}{3\varepsilon_0 V}\right) \tag{7.55}$$

Equation (7.54) shows that the transverse optical mode could decrease as a function of temperature and pressure because of a cancellation of the short-range interaction term by the long-range Coulombic term. Thus in certain solids there can be mode softening, which, when k approaches zero, will induce a phase transition. It was mentioned earlier that the smaller the frequency, the larger the amplitude of vibration. As the frequency decreases because of the softening of k the amplitude of vibration increases, and the atoms make larger excursions from the equilibrium positions and eventually find a new potential minimum to reside in, which means a different crystal structure.

The question now is how does nanosizing affect the mode-softening process and the temperature at which phase transitions occur. Reducing an ionic crystal to nano-dimensions will effect the force constant k in Eq. (7.54) through the effect of size on β, V, and ε_0. We saw for the case of our linear NaCl ionic lattice that β decreases below ~ 15 nm and the lattice parameter increases and thus V increases. This indicates that k for the optical mode decreases with particle size, which would mean that the temperature of the phase transition would lower with particle size. Indeed, experimental measurements confirm this. Figure 7.27 is a plot of the measurement of the transition temperature of lead titanate versus the particle diameter showing the expected reduction of the transition temperature with size in the nanometer region.

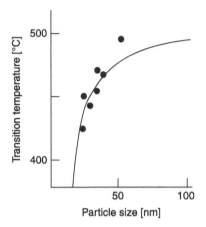

Figure 7.27. Plot of a measurement of the ferroelectric transition temperature of lead titanate versus the diameter of the particles. [Adapted from K. Ishikawa, *Phys. Rev.* **B37**, 5852 (1988).]

PROBLEMS

7.1. For the finite linear xenon lattice having an interatomic separation 4.35 Å $\varepsilon = 320 \times 10^{-16}$ erg, and $\sigma = 3.98$ Å, determine how small the length of the chain would have to be for a reduction in vibrational frequency to occur.

7.2. For a finite xenon chain of 10 atoms, calculate the relative change in frequency of the end atom on the right of the chain as a function of the chain length.

7.3. For a linear monatomic xenon lattice having 10 atoms, calculate the frequency of each atom in the chain. How does it change from the left end to the right end?

7.4. Derive the expression for the phonon density of states in one dimension [Eq. (7.26) of the text] in the Debye approximation.

7.5. Derive the phonon density of states for a three-dimensional crystal where the dispersion relationship has the form $\omega = uk^2$.

7.6. In Fig. 1.13 (of Chapter 1), the dependence of the frequencies of the optical and acoustical modes of a one-dimensional diatomic chain are plotted versus the wavevector k. For the linear NaCl lattice, how are these plots affected as the length of the chain shortens?

7.7. Obtain an approximate expression for the melting temperature of a two-dimensional monoatomic lattice using the Debye approximation for the density of vibrational states and the Lindemann model.

7.8. Equation (7.45) assumes that there is no resistance to the movement of the charge cloud. Is this assumption valid? If not, what would be the source of resistance? Assuming that the resistance to movement of the cloud is proportional to the velocity of the electron cloud, supply a differential equation that describes this situation. What do you think the solution of this equarion would look like?

7.9. Calculate the plasma frequency of copper versus the particle diameter in the nanometer region assuming that the only reason for such a change arises from a change in the electron density. How is the frequency affected by a reduced particle size?

7.10. The dispersion relationship of the plasma frequency is horizontal up to $\sim \frac{1}{2}k$. Would you expect phonon confinement effects to occur for plasmon frequencies in nanometer-sized particles? Explain the reason for your answer.

REFERENCES

1. See, for example C. Kittel *Introduction to Solid State Physics*, Wiley, (2004).
2. J. D. Louck *Am. J. Phys.* 30, 585 (1962).
3. J. R. Clem and R. P. Goodwin *Am. J. Phys.* 34, 460 (1966).
4. A. A. Maradudin and J. Melngailis, *Phys. Rev.* 133, A1188 (1964).
5. W. Cochran *Phys. Rev. Lett.* 3, 412 (1959).

CHAPTER 8

Electronic Properties

As discussed in Chapter 1, the distinguishing characteristic of the electronic structure of solids is the existence of bands of energy rather than discrete energy levels that exist in atoms or molecules. It was seen that in the case of insulators there is a large energy gap between the top filled band and the first empty band above it. In the insulating alkali halides such as NaCl, the gap is 8.6 eV. Semiconductors have much smaller gaps. For example, the semiconductor germanium has a bandgap of only 0.67 eV. Metals are distinguished from semiconductors and insulators because the highest occupied band is unfilled, which means that the energy of the outer electrons can easily be increased by the application of an electric field, and thus metals can conduct electricity. In this chapter we will examine how the electronic structure of ionic, covalent organic, and metal solids is affected when the dimensions of the solid are nanometers. Initially we will look at theories of the electronic structure of bulk materials and examine what they predict when the materials have nanometer dimensions. This will be followed by representative experimental studies of the effect of nanosizing on the electronic structure.

8.1. IONIC SOLIDS

In an ionic solid such as KCl the top filled band is derived from the Cl^- $3p$ orbital. The first unfilled band is derived from the K^+ $4s$ orbital. The bandgap is the energy needed to transfer an electron from the Cl^- $3p$ level to the K^+ $4s$ level. Von Hipple developed the following method to estimate the bandgap of an ionic solid such as KCl:

1. A negative ion in the lattice is moved to outside the crystal requiring an amount of energy $\alpha e^2/R_0$, where α is the Madelung constant and R_0 the equilibrium separation between the nearest-neighbor ions.

2. An electron is then removed from this ion requiring an energy equal to the electron affinity of the neutral atom, designated A, which is a measure of an atom's ability to attract an electron.

The Physics and Chemistry of Nanosolids. By Frank J. Owens and Charles P. Poole, Jr.
Copyright © 2008 John Wiley & Sons, Inc.

3. This neutral atom is then put back into the lattice, which requires no work because there is no electrostatic interaction with the ions of the lattice.
4. A positive ion is then removed from the lattice requiring an amount of energy $\alpha e^2/R_0$.
5. An electron is now added to the positive ion requiring energy $-I$ where I is the ionization energy, which is the energy needed to remove an electron from the neutral atom.
6. This neutral atom is put back in the lattice requiring no expenditure of energy.

Thus the total energy expended in transferring the electron from the negative ion to the positive ion to produce two neutral atoms is

$$\frac{2\alpha e^2}{R_0} + A - I \tag{8.1}$$

This equation provides a rough estimate of the bandgap, but gives an overestimate of the magnitude of the gap. The reason is that the energy to remove the ions is not simply $\alpha e^2/R_0$. It must include the polarization energy as discussed in Chapter 6. A better estimate of the bandgap would be

$$E_g = E_{v+} + E_{v-} + A - I \tag{8.2}$$

where E_{v+} and E_{v-} are the energies needed to produce a positive and negative ion vacancy, respectively, given by Eq. (6.16). These energies have been calculated for KCl and NaCl, and are given in Table 8.1 along with the values of I and A and the calculated and experimental values of the bandgap. The bandgap calculated using Eq. (8.2) is in reasonably good agreement with the experimental values.

Now let us use Eq. (8.2) to see how the bandgap of the linear NaCl lattice discussed in Chapter 6 changes with the length of the chain in the nanometer regime. In Chapter 6 we have calculated the dependence of α, R_0, and E_{v+} on the chain length. We need the dependence of E_{v-} on chain length, which is an exercise for the student in Chapter 6. With these results we can calculate the dependence of the

TABLE 8.1. Cohesive Energy U, E_v^+ Formation Energy for a Positive ion Vacancy, E_v^- Formation Energy for a Negative Ion Vacancy, Electron Affinity A of Cl, and Ionization Energy I of Na and KCl, the Calculated Bandgap and the Experimental Gap Eg

Crystal	U, eV	E_v^+, eV	E_v^-, eV	A of Cl, eV	I of + ion, eV	Calculated E_g, eV	Experimental E_g, eV
KCl	6.997	4.47	4.79	3.61	4.34	8.53	8.5
NaCl	7.733	4.62	5.18	3.61	5.14	8.27	8.6

Source: From N. F. Mott and R. W. Gurney, *Electronic Processes in Ionic Solids*, Dover Publications, 1964, p. 61.

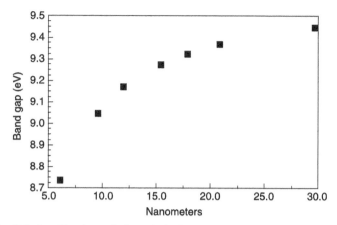

Figure 8.1. Calculated bandgap of a hypothetical linear NaCl lattice using Eq. (8.2) versus the length of the chain.

bandgap of our hypothetical linear NaCl lattice on chain length using Eq. (8.2). The results plotted in Fig. 8.1 show that the bandgap decreases as the chain becomes shorter in the nanometer region. Figure 8.2 is the result of a calculation of the band structure of the filled bands of three-dimensional KCl showing that as the lattice parameter is increased, which occurs on nanosizing, the separation between the bands also increases, which is opposite to the result obtained above for the

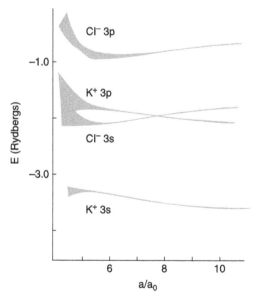

Figure 8.2. Calculation of occupied bands in a three-dimensional KCl lattice versus lattice parameter. [Adapted from L. P. Howard, *Phys. Rev.* **109**, 1927 (1958).]

one-dimensional lattice. However, the calculations did not include the first unfilled band, so we cannot be sure whether there is a difference in the behavior of the bandgap between the one-dimensional case and the three-dimensional case. Notice, however, the narrowing of the bands to discrete energy levels as the lattice parameter is increased.

8.2. COVALENTLY BONDED SOLIDS

As discussed previously, bonding in covalent solids involves overlap of wavefunctions of the valence electrons of nearest neighbors analogous to the bonding in molecules, which was discussed in Chapter 3. Let us consider our previously discussed polyacetylene chain representing a one-dimensional covalently bonded structure. Molecular orbital theory such as density functional theory can be used to calculate the properties of this structure as a function of length in nanometers. The bandgap, the energy required to raise an electron from the top filled level to the first unfilled energy level, can be calculated. Figure 8.3 shows the calculated bandgap versus the length of the chain showing that the bandgap increases as the chain shortens. The energy of this absorption determines the color of the material. This illustrates how controlling size in the nanometer range can be used to engineer optical properties. The calculated bandgap at the longest length of 4.1 nm is in reasonable agreement with the bandgap of 1.5 eV measured in longer chains of polyacetylene. Notice that the increase in the bandgap in the covalent structure with reduced length contrasts to the linear ionic system, where the bandgap decreases with

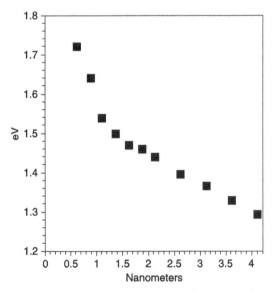

Figure 8.3. Molecular orbital calculation of the bandgap of a polyacetylene chain versus chain length. [Adapted from F. J. Owens, *Physica* **E25**, 404 (2005).]

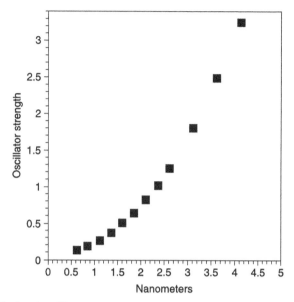

Figure 8.4. Calculated oscillator strength of bandgap transition versus length of polyacetylene chain. [Adapted from F. J. Owens, *Physica* **E25**, 404(2005).]

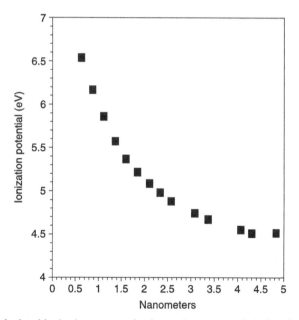

Figure 8.5. Calculated ionization energy of polyacetylene versus chain length. [Adapted from F. J. Owens, *Physica* **E25**, 404 (2005).]

reduced size. The transition probability, sometimes referred to as the *oscillator strength*, determines the intensity of the color of the material. Figure 8.4 presents a plot of the calculated oscillator strength versus chain length showing a marked increase as the chain lengthens. This increase is a result of enhancement of the density of states near the top of the valence band, which occurs when the chain lengthens. As seen in Fig. 7.17 of the previous chapter, there are large spikes in the density of states near the top of the valence band of polyacetylene, which means that there are more electrons available to make the transition. Other electronic properties are also affected by nanosizing. The ionization energy of the chain is the energy needed to remove an electron from the chain. Figure 8.5 is a plot of the calculated ionization energy versus chain length showing a strong increase in the ionization energy as the length of the chain decreases. The electron affinity, which is a measure of the chains ability to attract an electron, also decreases by nanosizing. Figure 8.6 shows the results of a calculation of the electron affinity of the chain as a function of the length.

8.3. METALS

In this section we describe some of the models used to understand the electronic structure of metals and examine how the electronic structure is affected when the dimensions of the solid become nanometers.

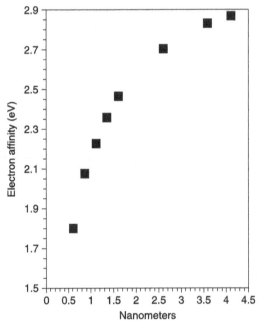

Figure 8.6. Calculated electron affinity of polyacetylene versus chain length. [Adapted from F. J. Owens, *Physica* **E25**, 404 (2005).]

8.3.1. Effect of Lattice Parameter on Electronic Structure

Before we consider various models for the electronic structure of metals and how nanosizing affects it, let us qualitatively look at how the electronic structure of a metal depends on the separation between the atoms. Consider the metal sodium. Sodium has two electrons in the $1s$ level, two in the $2s$ level, six at $2p$, and one at $3s$ which is half-filled. The first unoccupied level is the $3p$. Figure 8.7 is a schematic diagram of how the top half-filled $3s$ level and the first unoccupied $3p$ level depend on the separation between the sodium atoms in the lattice where R_0 is the equilibrium separation. As the lattice parameter is reduced, the $3s$ and $3p$ levels broaden and eventually overlap, creating the top partly filled band.

Most of the experimental observations of the effect of size of metal nanoparticles on the lattice parameters show that it decreases as the particle diameter decreases. The decrease is attributed to the effect of surface stress. The surface stress causes small particles to be in a state of compression where the internal pressure is inversely proportional to the radius of the particle. Figure 8.8 is a plot of the measured decrease in the lattice parameter of copper versus the diameter in angstroms. Notice that the changes don't occur until the diameter reaches a very small value of 0.9 nm. In the case of gold, measurements show that at 3.5 nm the lattice parameter has decreased to 0.36% of the bulk value. In aluminum significant changes are not observed until the particle size is below 1.8 nm. In our discussion of models of the electronic properties of metals in the following section, we will assume that the lattice parameter is not significantly dependent on particle size for values greater than 4 nm. As we will see, it turns out that the number of atoms in the metal nanoparticle has a much more significant influence on the electronic structure.

8.3.2. Free-Electron Model

One of the simplest models of the electronic structure of metals treats the conduction electrons as though they see no potential at all, but are confined to the volume of the solid. The model is best applicable to monovalent metals such as lithium, sodium, or

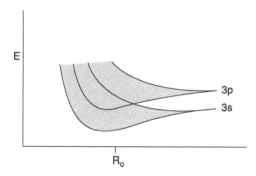

Figure 8.7. Schematic of the top half-occupied ($3s$) and first empty ($3p$) energy level of sodium as a function of sodium atom separation in sodium metal showing narrowing of bands as separation increases.

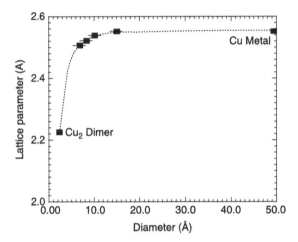

Figure 8.8. Measurement of lattice parameter of copper nanoparticles versus particle diameter. [Adapted from P. A. Montano, *Phys. Rev. Lett.* **56**, 2076 (1986).]

potassium, where the ion cores occupy only ~15% of the volume of the solid. The energies are obtained by solving the Schrödinger wave equation for $V[r] = 0$, with boundary conditions. For the case of a one-dimensional system the wave equation has the form

$$-\left[\frac{h^2}{2m}\right]\frac{d^2\Psi_n}{dx^2} = E_n\psi_n \tag{8.3}$$

where Ψ_n is the wavefunction of the electron in the nth state and E_n is the energy. The boundary condition for a one-dimensional lattice of length L are

$$\Psi_n(0) = 0 \quad \text{and} \quad \Psi_n(L) = 0 \tag{8.4}$$

The eigenvalues obtained by solving Eq. (8.3) are

$$E_n = \frac{h^2}{2m}\left[\frac{n}{2L}\right]^2 \tag{8.5}$$

where n is a quantum number having integer values 0, 1, 2, 3, ... (This treatment can be found in many basic books on solid-state physics, such as Kittel's book, *Introduction to Solid State Physics*.) Equation (8.5) is useful in understanding how the electronic structure of metals is affected when the dimensions are nanometers.

The separation between the energy levels of state n and $n + 1$ is

$$E_{n+1} - E_n = \frac{h^2}{8mL^2}[1 + 2n] \tag{8.6}$$

We see from Eq. (8.6) that as the length of the chain, L decreases, the separation between energy levels increases, and eventually the band structure opens up into a set of discrete levels as was qualitatively discussed in Chapter 3. This also means that the density of states will decrease with size. When the energy levels are filled with electrons, only two electrons are allowed in each level because of the Pauli exclusion principle. These two electrons must have different spin quantum numbers m_s of $+\frac{1}{2}$ and $-\frac{1}{2}$, meaning that the two electron spins in each level n are antiparallel, and there is no net spin value in the level. The Fermi energy is the energy of the top filled level, which, for a monovalent metal, will have the quantum number $n_f = N/2$, where N is the number of atoms in the solid. Thus, for the one-dimensional solid the Fermi energy is obtained from Eq. (8.5) as

$$E_f = \frac{h^2}{2m}\left[\frac{N}{4L}\right]^2 \tag{8.7}$$

In order to examine the effect of nanosizing on the Fermi energy, consider a lattice of N atoms separated from each other by a distance 3.5Å, which is the interatomic separation in the lithium lattice. Using Eq. (8.7), the relative change in the Fermi energy can be calculated as a function of the length of the chain. Figure 8.9 is a plot of the Fermi energy versus the chain length normalized to a 30 nm length. Below ~ 10 nm there is a significant increase in the Fermi energy, which is not surprising since the separation between the levels is increasing. In a manner analogous to the derivation of Eq. (8.7) for the Fermi level in one dimension, the Fermi level in the free-electron model of metals in three dimensions is obtained as

$$E_f = \frac{h^2}{2m}\left[\frac{3\pi^2 N}{V}\right]^{2/3} \tag{8.8}$$

Figure 8.10 shows a calculation of the relative change in Fermi level for the face-centered copper lattice versus size, applying the relationship between the number of atoms in a particle and diameter of the nanoparticle discussed in Chapter 1. Interestingly, the increase in Fermi level with reduced size is considerably weaker than in the one-dimensional case shown in Fig. 8.9, providing an example of the importance of dimensionality in determining the effects of nanosizing on the electronic structure of metals.

In the discussion above, we have seen that the energy bands begin to open up when the dimensions of the solid are nanometers, which implies that the density of states decreases. Here we further explore the effect of reducing the solid to nanometer dimensions on the density of states in the context of the free-electron model.

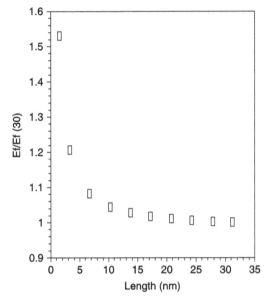

Figure 8.9. Relative change of Fermi level of a linear metallic chain versus length as predicted by the free electron model.

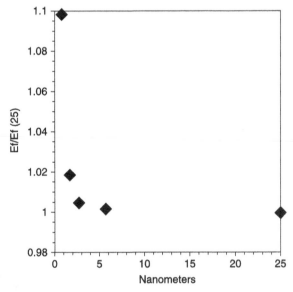

Figure 8.10. Relative change of Fermi level of copper nanoparticles versus particle size calculated using the free-electron model.

In Chapter 9, on quantum dots, the effect of the dimension of the material on the density of states will be discussed. Consider the one-dimensional case discussed above. The density of states is defined as the number of electronic states per unit energy dn/dE. From Eq. (8.5) we have

$$\frac{dn}{dE} = 4L\left(\frac{m}{2E}\right)^{1/2}\frac{1}{h} \qquad (8.9)$$

To account for the fact that there are two values of m_s for each n, Eq. (8.8) is multiplied by a factor of 2. In the case of the linear lattice the density of states is linearly dependent on the length of the chain, which means that the density of states decreases as inferred above as the chain length shortens. It can be shown that the density of electronic states in the free-electron model of metals has the following form in three dimensions:

$$\frac{dn}{dE} = \frac{V}{2\pi^2}(2\,mh^2)^{3/2}E^{1/2} \qquad (8.10)$$

The density of states in three dimensions is proportional to the volume of the crystal V, meaning that in three dimensions the density of electronic states also decreases as the dimensions reach nanometers.

8.3.3. The Tight-Binding Model

The tight-binding model takes a different approach to treating the electronic structure of metals. It assumes that when the electron is close to the atom of the solid, it will have a wavefunction $\Phi_a(r - l)$ corresponding to the wavefunction of the free atom. When the electron is far from the atom, the wavefunction can be describes as that of a free electron, i.e. $\exp(ik \cdot l)$. The wavefunction is written in this model as

$$\Psi_k[r] = \Sigma_l \exp(ik \cdot l)\Phi_a(r - l) \qquad (8.11)$$

The function looks like a series of strongly localized atomic orbitals whose amplitude is modulated by a phase factor as illustrated in Fig. 8.11. The model is appropriate to materials having atoms with d orbitals such as transition metals, which are compact and form narrow, well-defined bands. The energy levels are determined by

$$E = E_0 + <\Psi_k^*[r]H\Psi_k[r]> \qquad (8.12)$$

which is

$$E = E_0 + \Sigma_l \exp(ik \cdot l) <\Phi_a(r - l)H\Phi_a(r)> \qquad (8.13)$$

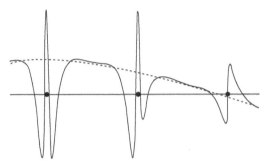

Figure 8.11. Illustration of the wavefunction in a solid in the tight-binding model approximation. (Adapted from J. M. Ziman, *Principles of Theory of Solids*, Cambridge Univ. Press, 1964, p. 81.)

where H is the difference between the potential of the free atom and the potential the atom sees in the crystal, $H = V[r] - V_a[r]$. For the case of a monatomic cubic crystal having lattice parameter a, and assuming only nearest-neighbor interactions, Eq. (8.13) becomes

$$E = E_0 + 2t(\cos K_x a + \cos K_y a + \cos K_z a) \qquad (8.14)$$

where t is given by

$$\langle \Phi_a(r - l)H\Phi_a(r)\rangle \qquad (8.15)$$

It turns out that t is negative. The $\Phi_a(r)$ for the top occupied band is the highest occupied atomic orbital of the free atom, and E_0 is the energy of the top occupied orbital of the free atom. Figure 8.12 shows the dependence of E on k along a cube axis. The lowest energy of the band is $E_0 - 2t$ occurring at $k = 0$ and the highest energy occurs at $E_0 + 2t$, which means that the bandwidth is $4t$. The bandwidth depends on t, which involves integrals of the atomic orbitals over the potential in the crystal given by Eq. (8.15). If, as in the case of gold, the lattice parameter decreases with nanosizing, then the bands widen because t increases and the separation between the bands decreases. Figure 8.13 shows how the bands are affected in this model as the lattice parameter changes.

Consider a finite one dimensional lattice of length L having N atoms separated from each other by a. The number of states in an energy band will be N corresponding to the allowed wavelengths of the wavefunction that fit into L such that $\psi_N(0) = 0$ and $\psi_N(L) = 0$. In the tight-binding model the dependence of E on k in a band will be

$$E = E_0 - 2t \cos ka \qquad (8.16)$$

where k can have values from 0 to $N\pi/L$. The width of the band is $2t$, and the density of states in the band is approximately $N/2t$. Thus, as in the free-electron model, the

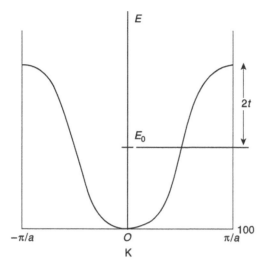

Figure 8.12. Dependence of the energy on the k vector for a cubic lattice along a cube axis. (Adapted from J. M. Ziman, *Principles of Theory of Solids*, Cambridge Univ. Press, 1964, p. 81.)

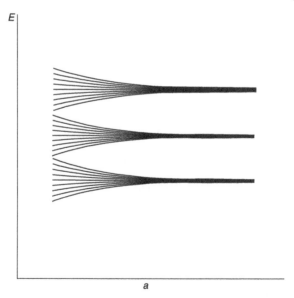

Figure 8.13. Schematic of narrowing of energy bands with increasing lattice parameter for a cubic lattice in the tight-binding model. (Adapted from J. M. Ziman, *Principles of Theory of Solids*, Cambridge Univ. Press, 1964, p. 81.)

density of states decreases with the number of atoms in the particle and therefore the size of the particle. The tight-binding models allows us to obtain other parameters of the electronic structure such as the group velocity v. A wavepacket is a collection of waves close to a particular k. The velocity of the packet is given by

$$v = \frac{1}{\hbar}\frac{dE}{dk} \qquad (8.17)$$

For the one-dimensional lattice v is obtained by differentiating Eq. (8.16)

$$v = \left(\frac{1}{\hbar}\right)2ta \sin ka \qquad (8.18)$$

One can also calculate the effective mass, which we will leave as an exercise for the student.

There are many other theoretical treatments of the electronic structure of metals that we could examine to see how reducing the dimensions to nanometers affects the electronic structure. We have chosen two quite different models, the free-electron model and the tight-binding model, to examine this issue. In general these various theories all show that when the dimensions approach nanometers, the bandwidth changes, the separation between the bands changes, the Fermi level increases, the density of states decreases, and the bandgap changes, as well as the velocity and the reduced mass.

8.4. MEASUREMENTS OF ELECTRONIC STRUCTURE OF NANOPARTICLES

In the previous sections we examined some of the models of bulk solids to see how nanosizing affects the electronic structure. In this section we present some representative experimental results on the effect of nanosizing on the electronic structure of various materials.

8.4.1. Semiconducting Nanoparticles

One way to measure the effect of nanosizing on the bandgap of a nanosized semiconductor is to measure the absorption of light as a function of the wavelength for different particle sizes. An absorption will occur when the energy of the light photon is equal to or greater than the bandgap. When this happens, an electron is excited from the valence band to the conduction band and light energy is absorbed. Nanosized cadmium selenide is a direct-bandgap semiconductor. The nanocrystals are single crystals having the wurtzite structure. Figure 8.14 is a plot of the optical absorption versus wavelength for CdSe particles of different sizes. Figure 8.15 is a plot of the measured bandgap versus the inverse square of the particle radius

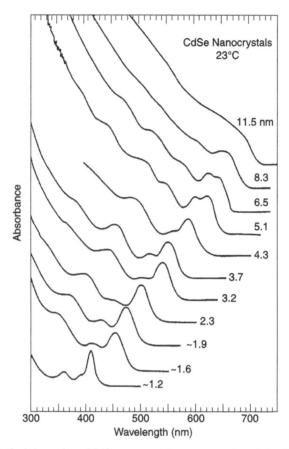

Figure 8.14. Optical absorption of CdSe nanocrystals versus wavelength for different indicated particle sizes. [Adapted from M. L. Steigerwald et al., *J. Am. Chem. Soc.* **110**, 3046 (1988).]

showing a nearly straight-line dependence. The measurements are on nanocrystals in which all three dimensions are nanometers in length. But what if only one or two dimensions are reduced? A *nanowire* or *quantum wire* is a nanostructure in which two dimensions have nanometer size, and the other is large, generally greater than micrometers. Figure 8.16 is the result of the measurement of the bandgap of a silicon nanowire versus its diameter showing the bandgap increasing as the diameter of the wire is reduced.

Another way to study the effect of nanosizing on electronic structure is to measure the wavelength dependence of the photoinduced luminescence of particles of different sizes. In this experiment light is used to excite an electron from the valence band to the conduction band. When the electron returns to the valence band, light of a different wavelength is emitted. Figure 8.17 presents a measurement of the photoluminescence of the semiconductor InP for different particle sizes, showing how the emission peak shifts to higher energies as the particle size decreases. The concept

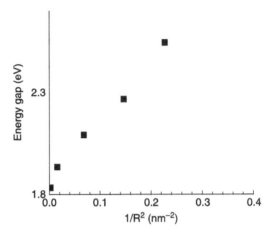

Figure 8.15. Plot of the bandgap of CdSe versus the the square of the reciprocal of the radius of the particle. [Adapted from A. I. Ekimov et al., *Solid State Commun.* **56**, 921 (1985).]

of the exciton, a bound hole–electron pair, was discussed in Section 1.4.3. In Section 3.3.1 there was a brief discussion of the hydrogen-atom-like energy levels of the exciton, and the effect of weak confinement on them. *Weak confinement* refers to the situation where the diameter of the particle is somewhat larger than the radius of the exciton, thereby limiting its range of motion and producing the blueshift. Figure 8.18 shows the optical absorption spectra of copper chloride (CuCl). The lowest energy absorption at 3.20 eV is due to the bandgap. The higher-energy absorption at 3.275 eV is due to a transition between the hydrogenic levels of the exciton.

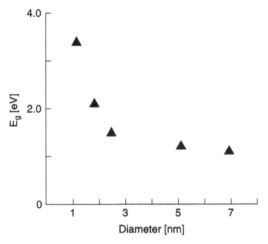

Figure 8.16. Plot of the bandgap of a silicon quantum wire as a function of the diameter of the wire. [Adapted from D. D. D. Ma et al., *Science* **299**, 1874 (2003).]

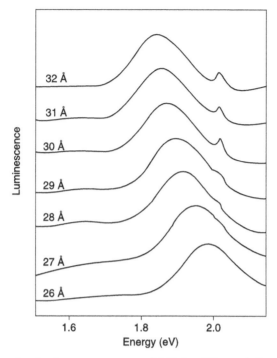

Figure 8.17. The photoluminescence spectra of InP for different indicated particle sizes. [Adapted from A. A. Guzwuan et al., *J. Phys. Chem.* **100**, 7212 (1996).]

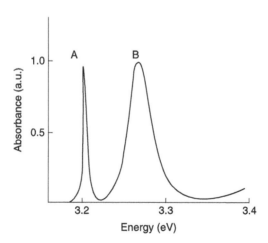

Figure 8.18. Optical absorption spectrum of CuCl. Line *A* arises from the bandgap; line *B*, from excitons. [Adapted from A. I. Ekimov, *JETP* **61**, 893 (1985).]

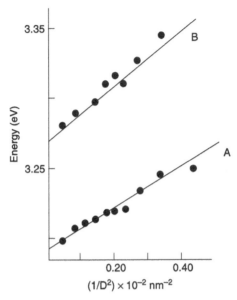

Figure 8.19. Dependence of the optical absorption of the A and B lines from CuCl from Fig. 8.18 on the reciprocal of the square of the radius of nanoparticles of the material. [Adapted from A. I. Ekimov, *JETP* **61**, 893 (1985).]

Figure 8.19 is a plot of the energy of the absorption versus the reciprocal of the square of the radius of the particles. The linear dependence means that the bandgap and the excitonic transition both depend on the radius of the particle as $1/R^2$. Theoretical treatments have predicted this dependence by expressing the bandgap as a function of particle radius as follows

$$E_g(R) = E_{g0} + \frac{\pi^2 h^2}{2\mu R^2} \tag{8.19}$$

where R is the radius of the particle, E_{g0} is the bandgap of the bulk material, and μ is the reduced mass of the hole–electron pair, where $1/\mu = (1/m_e^* + 1/m_h^*)$, m_e^* is the effective mass of the electron, and m_h^* is the effective mass of the hole.

X-Ray photoelectron spectroscopy (XPS), which is discussed in Section 2.4.2, measures the number of emitted electrons and their energies when a material is subjected to X-ray irradiation. Because the electrons come from discrete inner energy levels, the data yield a series of peaks whose separation corresponds to the separation of the energy levels of the material. XPS has been used to investigate how nanosizing effects the electronic structure of materials. Nanoparticles of zinc sulfide, which has a bandgap of 3.75 eV in the bulk and is a semiconductor, have been studied by XPS. In this study the particles were passivied with thioglycerol. Figures 8.20a–8.20c show the XPS spectra of the sulfur $2p$ level for particles of size 3.5, 2.5, and 1.8 nm, respectively. The broad spectrum has been resolved into three peaks labeled

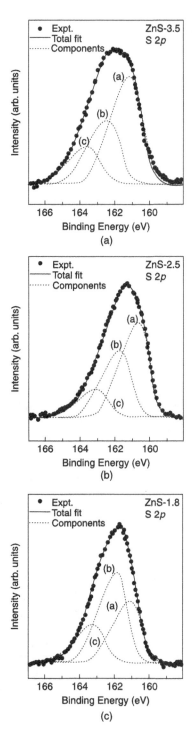

Figure 8.20. X-Ray photoelectron spectra of ZnS for electrons originating from the sulfur $2p$ level for particles of sizes (a) 3.5 nm, (b) 2.5 nm, and (c) 1.8 nm.

(a), (b), and (c) in the figure. Notice that the peak labeled (a) decreases in intensity as the particle size is decreased and the peaks labeled (b) increase as the particle size decreases. Since the volume–surface ratio of atoms in the particle decreases with particle size, these data suggest that the component (a) is from sulfur in the core of the nanoparticle. Component (b) is from sulfur atoms bonded to zinc on the surface of the particles. Component (c) is likely from the sulfurs in the thioglycerol passivating layer. This result underlines the importance of the role of the surface atoms of small nanoparticles in its effect on the electronic structure of the particle.

8.4.2. Organic Solids

Figure 8.21 is a scanning electron microscope picture of perylene nanocrystals made by the rapid precipitation method discussed in Section 3.5.5. The average size of the particles is about 200 nm. Perylene is a condensed ring compound having chemical formula $C_{20}H_{12}$. Excitons can exist in organic solids such as anthracene and perylene. In copper oxide the separation between the hole and the electron of the exciton is large relative to the lattice parameter. Such an exciton is referred to as a *Mott–Wannier exciton*, and its energy levels are accounted for by a hydrogen-atom-like model. In organic solids the exciton is localized on or near a molecule of the lattice, and its energy levels are largely determined by the molecular electronic structure. This highly localized tightly bound exciton is known as a *Frenkel exciton*. It is

Figure 8.21. Scanning electron microscope image of perylene nanocrystals made by the rapid prercipation method. (Adapted from H. Kasai et al., in *Handbook of Nanostructured Materials and Nanotechnology*, H. S. Nalwa, ed., Academic Press, Boston, 2000, Vol. 5.)

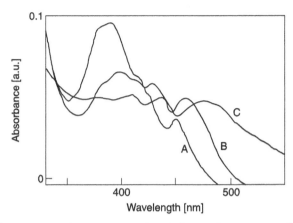

Figure 8.22. Optical absorption versus wavelength for pyrelene nanocrystals of (A) 50 nm, (B) 200 nm, and (C) >1 μm. (Adapted from H. Kasai et al., in *Handbook of Nanostructured Materials and Nanotechnology*, H. S. Nalwa, ed., Academic Press, Boston, 2000, Vol. 5.)

essentially an excited state of the molecule that can hop from one molecule of the lattice to another. Figure 8.22 shows the optical absorption versus wavelength of perylene nanocrystals of (A) 50 nm, (B) 200 nm, and (C) >1 μm. The peak at 480 nm in the large particle, which is due to the exciton, is shifted to shorter wavelengths as the particle size decreases. Figure 8.23 shows the dependence of the wavelength of the exciton absorption on the crystal size in perylene. It should be noted that because shifts are observed in particles ranging from 50 to 200 nm, the effects cannot be due to confinement, which manifests itself at much lower sizes. For example, in semiconductors confinement effects become evident below ∼10 nm. The reasons for this effect are not totally understood. One proposal is that as the crystals get smaller the

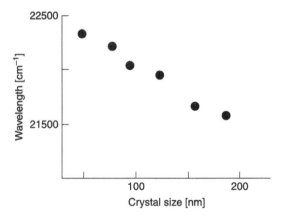

Figure 8.23. Dependence of the peak wavelength of the optical absorption of perylene on the crystal size. (Adapted from H. Kasai et al., in *Handbook of Nanostructured Materials and Nanotechnology*, H. S. Nalwa, ed., Academic Press, Boston, 2000, Vol. 5, p. 438.)

interaction between the molecules of the crystal weakens, which causes changes in the wavelength of the optical absorption.

8.4.3. Metals

Ultraviolet photoelectron spectroscopy (UPS), which has been used to measure the evolution of the electronic structure of copper nanoparticles with a charge of -1,

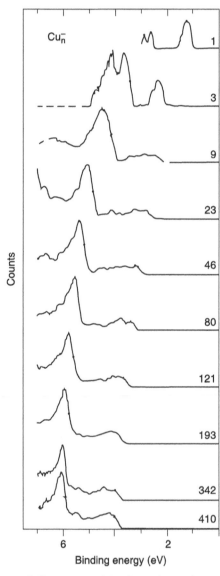

Figure 8.24. UPS spectra of Cu_n^- nanoparticles for an increasing number of atoms n in the particle. [Adapted from O. Cheshnovsky et al., *Phys. Rev. Lett.* **64**, 1785 (1990).]

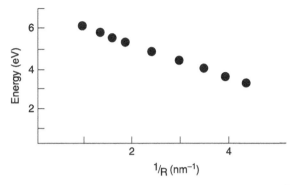

Figure 8.25. Electron affinity of Cu_n^- nanoparticles versus the reciprocal of the radius of the particle determined from UPS measurements. [Adapted from O. Cheshnovsky et al., *Phys. Rev. Lett.* **64**, 1785 (1990).]

is experimentally analogous to XPS discussed in Section 2.4.2, except that ultraviolet light is used to excite electrons from the energy levels of the material. Because the energy of UV light is less than that of X-rays, the electrons that are emitted come from the valence band region. Hence UPS provides a way to measure the effects of nanosizing on the valence band of a material. Figure 8.24 shows the UPS spectra of Cu_n^- nanoparticles having increasing numbers of Cu atoms ranging from a single Cu^- ion to a nanoparticle of 410 atoms with a charge of −1. This measurement represents an elegant display of how the energy levels in the valence band region of an isolated ion change as the number of atoms in the particle increases, eventually forming the band structure of a bulk material at 410 atoms. The measurement shows the evolution of discrete levels in the single ion to broad lines resulting from the crowding of the energy levels that occurs in a bulk material. Notice that the threshold for emission, which corresponds to electrons emitted from the top $3d$ level, shifts to higher energies as the particle size increases. This threshold energy provides a measure of the electron affinity of neutral copper nanoparticles. The electron affinity is a measure of the nanoparticles ability to attract an electron. Figure 8.25 gives a plot of the measured electron affinity versus $1/R$, where R is the radius of the nanoparticle showing that the electron affinity of a neutral Cu nanoparticle decreases as the particle size decreases.

PROBLEMS

8.1. Calculate the bandgap of RbF using the von Hippel cycle. Take the polarizability of F as $1 \times 10^{-24}\,cm^3$ and that of Rb^+ as $1.4 \times 10^{-24}\,cm^3$. The electron affinity A of fluorine is 4.154 eV, and the ionization potential I of rubidium is 4.177 eV. How does the result compare with the experimental value of 10.4 eV? What are the reasons for the difference?

8.2. Calculate the wave velocity for a one-dimensional monatomic lattice in the free-electron approximation. How is the velocity affected when the length of the chain is reduced?

8.3. Calculate the effective mass of the electron for the one-dimensional monatomic lattice in the free-electron model. How is it affected by a reduction in the length of the chain?

8.4. It can be shown that the heat capacity of electrons in the tree-dimensional free-electron model is given by

$$C_{el} = \left(\frac{1}{3}\pi^2\right)D(E_F)k_B^2 T$$

where $D(E_F)$ is the density of states at the Fermi level. How is C_{el} affected by nanosizing at constant temperaure? Plot the relative changes in C_{el} versus particle size for copper at constant temperature in the nanometer regime.

8.5. Use the tight-binding model for a linear C—C lattice having a C—C separation of a to obtain the dispersion relationship along the chain axis taking into acount only nearest-neighbor interactions. How is the energy gap affected by nanosizing in this model?

8.6. Calculate the velocity and reduced mass for the linear C—C lattice from the results of Problem 8.5. How are they affected by nanosizing in the tight-binding model?

8.7. Cadium sulfide has a bandgap in the bulk of 2.482 eV. The effective mass of the electron is 0.19, and that of the hole is 0.8. Calculate the effect of nanosizing on the bandgap. Plot the magnitude of the gap from 1 to 10 nm.

CHAPTER 9

Quantum Wells, Wires, and Dots

9.1. INTRODUCTION

When the size or dimension of a material is continuously reduced from a large or macroscopic value, such as a meter or a centimeter, to a very small size, the properties remain the same at first, then small changes begin to occur, until finally when the size drops below 100 nm dramatic changes in properties can take place. If one dimension is reduced to the nanorange while the other two dimensions remain large, we obtain a structure known as a *quantum well*. If two dimensions are thus reduced and one remains large, the resulting structure is referred to as a *quantum wire*. The limiting case of this process of size reduction in which all three dimensions reach the low-nanometer range is called a *quantum dot*. The word *quantum* is associated with these three types of nanostructures because some of their changes in properties arise from the quantum-mechanical nature of physics in the domain of the ultra small. Figure 9.1 illustrates these processes of diminishing the size for the case of rectilinear geometry, and Fig. 9.2 presents the corresponding reductions in curvilinear geometry. In this chapter we will be interested in all three of these methods for decreasing the size. In other words, we will probe the dimensionality effects that occur when one, two, or all three dimensions become small. Of particular interest will be how the electronic properties are altered by these changes. Henini[1] and Jacak et al.[2] have surveyed the field of quantum dot nanostructures.

9.2. FABRICATING QUANTUM NANOSTRUCTURES

One approach to the preparation of a nanostructure, called the *bottom–up approach*, is to collect, consolidate, and fashion individual atoms and molecules into the structure. It sometimes begins with precipitation from a supersaturated solution, and is often carried out by a sequence of chemical reactions controlled by catalysts. It is a process that is widespread in biology, where, for example, catalysts called *enzymes* assemble amino acids to construct proteins and much of the living tissue that forms and supports the organs of the body. In the present chapter we will

The Physics and Chemistry of Nanosolids. By Frank J. Owens and Charles P. Poole, Jr.
Copyright © 2008 John Wiley & Sons, Inc.

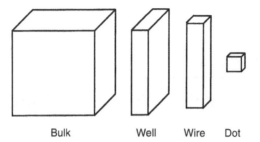

Figure 9.1. Progressive generation of rectangular nanostructures.

confine ourselves to discussing the initial steps in the preparation of nanoparticles from solution.

The opposite approach to the preparation of nanostructures, called the *top–down method*, starts with a large-scale object or pattern and gradually reduces its dimension(s). This can be accomplished by a technique called *lithography*, which shines radiation through a template onto a surface coated with a radiation-sensitive material called a *resist*; the resist is then removed, and the surface is chemically treated to produce the nanostructure.

9.2.1. Solution Fabrication

The bottom–up approach to nanoparticle fabrication can begin with the precipitation or nucleation of a nonoparticle from a supersaturated solution, so we will examine this procedure. When a material or solute dissolves in a solvent, the solute breaks up into its parts, which enter the solution and establish an equilibrium with solvent molecules. The process is endothermic, with the energy E_{diss} gained by going into solution exceeding the cohesive energy E_{coh} that must be supplied to break up the solid, so there is a net gain in energy with the value $\Delta E = E_{diss} - E_{coh}$. It is more

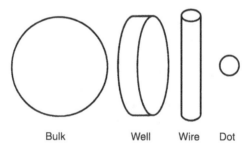

Figure 9.2. Progressive generation of curvilinear nanostructures.

convenient to express this gain in energy ΔE in terms of the change in energy per unit volume β

$$\beta = \frac{E_{\text{diss}} - E_{\text{coh}}}{V} \tag{9.1}$$

of the initial solute material, which has the volume V before going into solution.

It is also necessary to account for the energetics associated with establishing the state of supersaturation. Consider a liquid with a concentration of solute C that exceeds the value C_0 that exists at saturation when the maximum amount of solute at equilibrium has dissolved. The extent of supersaturation σ is defined by the expression

$$\sigma = \frac{C - C_0}{C_0} \tag{9.2}$$

The energy per unit volume V expended to bring about this state of supersturation, which we denote by α, is given by

$$\alpha = \frac{k_B T}{V} \ln \frac{C}{C_0} = k_B \frac{T}{V} \ln(1 + \sigma) \tag{9.3}$$

When a nanoparticle forms by precipitation from the solution, we are interested in the change in the Gibbs free energy ΔG associated with the volume V occupied by the particle. There will be a loss of supersaturation energy α, a decrease in the volume energy β, and an acquisition of surface energy γ as follows

$$\Delta G = -\frac{4\pi}{3} r^3 (\alpha + \beta) + 4\pi r^2 \gamma \tag{9.4}$$

where a spherical shape is assumed for the precipitate and, of course, $\alpha(C)$ depends on the concentration C. We see from Fig. 9.3 that the Gibbs free energy increases with the particle radius for small r, passes through a maximum at the critical radius r_c, and then decreases in value with a further growth in particle size. At the maximum of the curve the derivative $d(\Delta G)/dt = 0$, and the critical free energy ΔG_c at this critical radius $r = r_c$ constitutes an energy barrier that must be overcome to form a stable precipitate. This is the case because smaller precipitates with $r < r_c$ lower their free energy by decreasing their radii and dissolving back into the solvent. In contrast to this, larger particles, with $r > r_c$, lower their free energy by growing in size. This behavior follows from the general principle that physical and chemical systems evolve in the direction of lower free energy.

Evaluating the derivative $d(\Delta G)/dt$ and setting the result equal to zero provides the following expressions for the critical radius r_c and the critical energy ΔG_c

$$r_c = \frac{2\gamma}{\alpha + \beta} \tag{9.5}$$

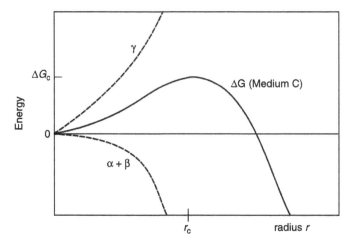

Figure 9.3. Change in Gibbs free energy ΔG (solid curve) of a nanoparticle that precipitates out of a supersaturated solution as a function of the particle radius r. The dashed curves give the contributions from the surface tension energy (γ) and from the sum of the supersaturation (α) and condensation (β) energies. The curves are drawn for the condition $\alpha + \beta = \gamma$. This ΔG curve is the medium-concentration one presented in Fig. 9.4. The critical radius r_c and critical free energy ΔG_c are indicated.

$$\Delta G_c = \frac{16\pi}{3} \frac{\gamma^3}{(\alpha + \beta)^2} \tag{9.6}$$

We know from Eq. (9.3) that the energy density α associated with supersaturation is proportional to the temperature T and the logarithm of the concentration C, so the low concentration curve in Fig. 9.4 corresponds to a higher temperature, and to a larger value of α, than do the other two curves, Since α is in the denominators of Eqs. (9.5) and (9.6), it follows that lower concentrations C provide larger values of critical radii r_c, and critical energy ΔG_c, as is clear from Fig. 9.4.

As more and more nanoparticles form in a solution, the concentration decreases and the critical radius r_c increases. As a result, some already formed nanoparticles can find themselves with radii less than r_c. When this happens, the smaller ones begin to dissolve while their larger counterparts continue to grow. This is an example of *Ostwald ripening*, or the growth of larger structures at the expense of smaller ones. Cao[3] shows that when diffusion processes are the dominant mechanism for particle growth, there is a tendency for the difference between the radii of two particles moderately close together in size to decrease as the growth proceeds, thereby promoting the uniformity of particle sizes. This applies, of course, only to particles with radii greater than r_c. For the attainment of uniform sizes, it is preferable for all the particles to precipitate from the solution at the same time, so their growth can proceed in parallel. The diffusion growth mechanism also favors the formation of relatively large nanoparticles.

As the precipitated nanoparticles increase in number and grow in size, the concentration C of the surrounding solution decreases until eventually dropping below the

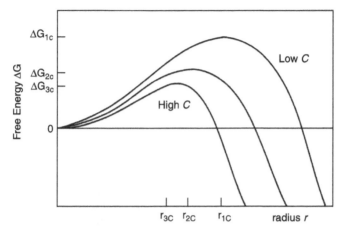

Figure 9.4. Radius dependence of the change in Gibbs free energy ΔG of a nanoparticle that precipitates out of a supersaturated solution for supersaturation energy densities $\alpha(C)$ corresponding to three values of the concentration C as follows: $\alpha = 0.8\gamma - \beta$ (low C), $\alpha = \gamma - \beta$ (medium C), and $\alpha = 1.2\gamma - \beta$ (high C). The medium concentration curve is the one displayed in Fig. 9.3. The three critical radii r_{ic} and the three critical Gibbs free energies ΔG_{ic} are indicated.

equilibrium value C_0. Further nucleation will stop, but growth can continue. Solute molecules diffuse toward the nanoparticles and adhere to them. If there is a growth surface or substrate present in the solution, the nanoparticles will diffuse to it, become adsorbed on it, and continue to grow there. When the adsorbed species are more strongly attracted to the substrate than to each other, the subsequent growth process is layer by layer, starting with laying down a monolayer, then the addition of a second layer, and so on. In thin-film technology this is termed *Frank–van der Merwe growth*. The other extreme, wherein the absorbed species bind to each other more strongly than to the substrate, produces island growth or *Volmer–Weber growth*, in which aggregates of nanoparticles form on the surface and then grow in size. The end result is a collection of quantum dots. Figure 9.5 illustrates these two cases. The figure also presents an intermediate case called *layer–island* or *Stranski–Krastanov growth*, in which aggregates form on a fully covered surface.

9.2.2. Lithography

The reverse approach to the preparation of nanostructures is called the *top–down method*, which begins with a large–scale object or pattern that gradually reduces in dimension(s). This can be accomplished by a technique called *lithography*, which projects radiation through a template onto a surface coated with a radiation-sensitive material termed a *resist*, which is then removed, and the surface is chemically treated to produce the nanostructure. A typical resist consists of the polymer polymethylmethacrylate $[C_5O_2H_8]_n$, which has a molecular weight ranging from 10^5 to 10^6 Da [a dalton (Da, g/mol) is a unit of atomic mass]. The lithographic

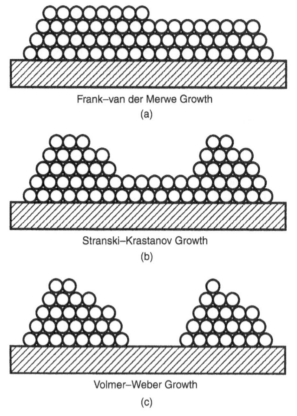

Figure 9.5. Formation of a thin-film layer by layer (a), aggregated or island growth (c), and the intermediate-type island–layer growth (b).

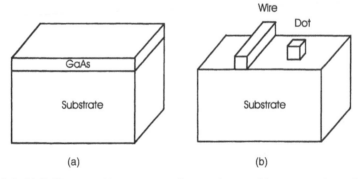

Figure 9.6. (a) Gallium arsenide quantum well on a substrate; (b) quantum wire and quantum dot formed from the well by lithography.

process will be illustrated by starting with a square quantum well fabricated from a material such as GaAs and located on a substrate, as shown in Fig. 9.6a. The final product that results is either a quantum wire or a quantum dot, as shown in Fig. 9.6b. The steps to be followed in this process are outlined in Fig. 9.7.

The first step of the lithographic procedure is to place a radiation-sensitive resist on the surface of the sample substrate, as shown in Fig. 9.7a. The sample is then irradiated by an electron beam in the region where the nanostructure will be located, as shown in Fig. 9.7b. This can be done by using either a radiation mask that contains the nanostructure pattern, as shown, or a scanning electron beam that strikes the surface only in the desired region. The radiation chemically modifies the exposed area of the resist so that it becomes soluble in a developer. The third step (c) in the process is the application of the developer to remove the irradiated portions of the resist. The fourth step (d) is the insertion of an etching mask into the hole in the resist, and the fifth step (e) consists in lifting off the remaining parts of the resist. In the step sixth (f) the areas of the quantum well that are not covered by the etching mask are chemically etched away to produce the quantum structure shown in Fig. 9.7f, covered by the etching mask. Finally the etching mask is

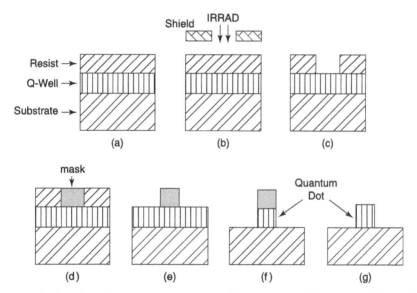

Figure 9.7. Steps in the formation of a quantum wire or quantum dot by electron-beam lithography: (a) initial quantum well on a substrate covered by a resist; (b) radiation with sample shielded by a template; (c) configuration after dissolving the irradiated portion of the resist by a developer; (d) disposition after the addition of an etching mask; (e) arrangement after removal of the remainder of the resist; (f) configuration after etching away the unwanted quantum well material; (g) final nanostructure on the substrate after removal of the etching mask. (See also T. P. Sidiki and C. M. S. Torres, in *Handbook of Nanostructured Materials and Nanotechnology*, H. S. Nalwa, ed., Academic Press, Boston, 2000, vol. 3, Chapter 5, p. 250.)

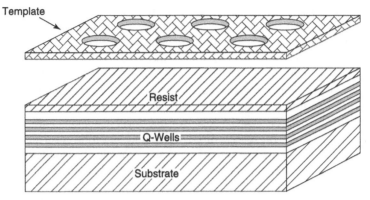

Figure 9.8. Four-cycle multiple quantum well arrangement mounted on a substrate and covered by a resist. A radiation-shielding template for lithography is shown at the top.

removed, if necessary, to provide the desired quantum structure (g), which might be the quantum wire or quantum dot shown in Fig. 9.6b.

We have described the most common process called *electron-beam lithography*, which makes use of an electron beam for the radiation. There are other types of lithography that employ neutral atom beams (e.g., Li, Na, K, Rb, Cs); charged ion beams (e.g., Ga^+); or electromagnetic radiation such as visible light, ultraviolet light, or X rays. When laser beams are utilized, frequency doublers and quadruplers can bring the wavelength into a range (e.g., $\lambda \sim 150\,nm$) that is convenient for quantum dot fabrication. Photochemical etching can be applied to a surface activated by laser light.

The lithographic technique can be used to make quantum structures that are more complex than the quantum wire and quantum dot shown in Fig. 9.6b. For example, one might start with a multiple-quantum-well structure of the type illustrated in Fig. 9.8, place the resist on top of it, and make use of a mask film or template with six circles cut out of it, as portrayed at the top of Fig. 9.8. Following the lithographic procedure outlined in Fig. 9.7, one can produce the 24-quantum-dot array consisting of six columns, each containing four stacked quantum dots, as sketched in Fig. 9.9. As an example of the advantages of fabricating quantum dot arrays, it

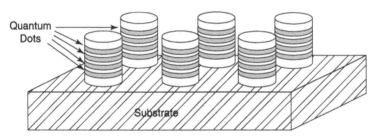

Figure 9.9. Quantum dot array formed by lithography from the initial configuration of Fig 9.8. This 24-fold quantum dot array consists of six columns of four stacked quantum dots.

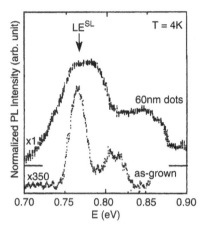

Figure 9.10. Photoluminescence spectrum of an array of 60-nm-diameter quantum dots formed by lithography, compared with the spectrum of the initial as-grown multiple quantum well. The intense peak at 0.7654 eV is attributed to localized excitons (LE) in the superlattice (SL). The spectra were taken at temperature 4 K. (From T. P. Sidiki and C. M. S. Torres, in *Handbook of Nanostructured Materials and Nanotechnology*, H. S. Nalwa, ed., Academic Press, Boston, 2000, Vol. 3, Chapter 5, p. 251.)

has been found experimentally that the arrays produce a greatly enhanced photoluminescent output of light.[4] Figure 9.10 shows a photoluminescence (PL) spectrum from a quantum dot array that is much more than 100 times stronger than the spectrum obtained from the initial multiple quantum wells. The principles behind the photoluminescence technique are described in Section 8.4.1 The main peak of the spectrum of Fig. 9.10 was attributed to a localized exciton (LE), as will be explained in Section 9.4. This magnitude of enhancement shown in the figure has been obtained from initial samples containing, typically, a 15-period superlattice (SL) of alternating 3 nm thick layers of Si and $Si_{0.7}Ge_{0.3}$ patterned into quantum dot arrays consisting of 300-nm-high columns with 60 nm diameters and 200 nm separations.

9.3. SIZE AND DIMENSIONALITY EFFECTS

9.3.1. Size Effects

Now that we have seen how to make nanostructures, it is appropriate to say something about their sizes relative to various parameters of the system. If as a typical material we select the type III–V semiconductor GaAs, the lattice constant from Table B.1 (of Appendix B) is $a = 0.565$ nm, and the volume of the unit cell is $(0.565)^3 = 0.180$ nm^3. The unit cell contains four Ga and four As atoms. Each of these atoms lies on a face-centered cubic (fcc) lattice, shown sketched in Fig. 1.3 (of Chapter 1) and the two lattices are displaced with respect to each other by the amount $\frac{1}{4}\frac{1}{4}\frac{1}{4}$ along the unit cell body diagonal, as shown in Fig. 1.8. This puts each Ga atom in the center of a tetrahedron of As atoms corresponding to the grouping $GaAs_4$, and each arsenic atom has a corresponding configuration $AsGa_4$. There are about 22 of each atom type per cubic nanometer, and a cube-shaped quantum dot 10 nm on a side contains 5.56×10^3 unit cells.

The question arises as to how many of the atoms are on the surface, and it will be helpful to have a mathematical expression for this in terms of the size of a particle with the zinc blende structure of GaAs that has the shape of a cube. If the initial cube is taken in the form of Fig. 1.6 and nanostructures containing n^3 of these unit cells are built up, then it can be shown that the number of atoms N_S on the surface, the total number of atoms N_T, and the size or dimension d of the cube are given by

$$N_S = 12n^2 \qquad (9.7)$$

$$N_T = 8n^3 + 6n^2 + 3n \qquad (9.8)$$

$$d = na = 0.565\,n \qquad (9.9)$$

Here $a = 0.565$ nm is the lattice constant of GaAs, and most of the lattice constants of other zinc blende semiconductors given in Table B.1 are close to this value. These equations (9.7)–(9.9) are for a cubic GaAs nanoparticle with its faces in the xy, yz, and zx planes, respectively. Table 9.1 tabulates N_S, N_T, d, and the fraction of atoms on the surface N_S/N_T, for various values of n. The large percentage of atoms on the surface for small n is one of the principal factors that differentiates nanostructure properties from bulk material properties. An analogous table could easily be

TABLE 9.1. Number of Atoms on Surface N_S, number in Volume N_V, and Fraction of Atoms N_S/N_V on Surface of a Nanoparticle[a]

n	Size na, nm	Total Number of Atoms	Number of Surface Atoms	% of Atoms on Surface
2	1.13	94	48	51.1
3	1.70	279	108	38.7
4	2.26	620	192	31.0
5	2.33	1165	300	25.8
6	3.39	1962	432	22.0
10	5.65	8630	1200	13.9
15	8.48	2.84×10^4	2700	9.5
25	14.1	1.29×10^5	7500	5.8
50	28.3	1.02×10^6	3.0×10^4	2.9
100	56.5	8.06×10^6	1.2×10^5	1.5

[a] The particle has a diamond lattice structure in the shape of a cube n unit cells on a side, having a width na, where a is the unit cell dimension. Column 2 gives sizes for GaAs, in which $a = 0.565$ nm.

constructed for cylindrical shaped quantum structures of the types illustrated in Figs. 9.2 and 9.9.

Comparing Table 9.1, which is for a diamond structure nanoparticle in the shape of a cube, with Table 1.1, which is for a FCC nanoparticle with an approximately spherical shape, it is clear that the results are qualitatively the same. We see from the comparison that the fcc nanoparticle of Chapter 1 has a greater percentage of its atoms on the surface for the same total number of atoms in the particle. This is expected because the calculation was carried out in such a way the only one of the two types of atoms in the GaAs structure contributes to the surface.

9.3.2. Size Effects on Conduction Electrons

The most technologically important properties of conductors and semiconductors are their electrical properties, and these depend on the dynamics of their charge carrier transport mechanisms. A charge carrier such as an electron experiences forward motion in an applied electric field periodically interrupted by scattering off phonons and defects. An electron or hole moving with a drift velocity v will, on average, experience a scattering event every τ seconds, and travel a distance l called the *mean free path* between collisions, where

$$l = v\tau \tag{9.10}$$

This is called *intraband scattering* because the charge carrier remains in the same band after scattering, such as the valence band in the case of holes. Mean free paths in metals depend strongly on the impurity content, and in ordinary metals typical values might be in the low-nanometer range, perhaps from 2 to 50 nm. In very pure samples they will, of course, be much longer. The resistivity of a polycrystalline conductor or semiconductor composed of microcrystallites with diameters significantly greater than the mean free path resembles that of a network of interconnected resistors, but when the microcrystallite dimensions approach or become less than l, the resistivity depends mainly on scattering off boundaries between crystallites. Both varieties of metallic nanostructures are common.

Various types of defects in a lattice can interrupt the forward motion of conduction electrons, and limit the mean free path. Examples of zero-dimensional defects are missing atoms called *vacancies* or *Schottky defects*, and extra atoms called *interstitial atoms* located between standard lattice sites. A vacancy–interstitial pair is called a *Frenkel defect*. Examples of one-dimensional dislocations are a lattice defect at an edge, or a partial line of missing atoms. Common two-dimensional defects are a boundary between grains, and a stacking fault arising from a sudden change in the stacking arrangement of close-packed planes. A vacant space called a *pore*, a cluster of vacancies, and a precipitate of another phase are three-dimensional defects. All of these can bring about the scattering of electrons, and thereby limit the electrical conductivity. Some nanostructures are so small that they are unlikely to have any internal defects.

TABLE 9.2. Conduction Electron Content of Smaller (on Left) and Larger (on Right) Quantum Structures Made from Bulk Material Containing Donor Concentrations of $10^{14}-10^{18}/cm^3$

Quantum Structure	Size	Electron Content	Size	Electron Content
Bulk material	—	$10^{14}-10^{18}/cm^3$	—	$10^{14}-10^{18}/cm^3$
Quantum well	10 nm thick	$1-10^4/\mu m^2$	100 nm thick	$10-10^5/\mu m^2$
Quantum wire	10 × 10-nm cross section	$10^{-2}-10^2/\mu m$	100 × 100-nm cross section	$1-10^4/\mu m$
Quantum dot	10 nm on a side	$10^{-4}-1$	100 nm on a side	$10^{-1}-10^3$

Another size effect arises from the level of doping of a semiconductor. For typical doping levels of $10^{14}-10^{18}$ donors/cm^3, a quantum dot cube 100 nm on a side would have, on average, $10^{-1}-10^3$ conduction electrons. The former figure of 10^{-1} electrons/cm^3 means that on the average only one quantum dot in 10 will have one of these electrons. A smaller quantum dot cube only 10 nm on a side would have, on average, one electron for the 10^{18} doping level, and be very unlikely to have any conduction electrons for the 10^{14} doping level. A similar analysis can be conducted for quantum wires and quantum wells, and the results shown in Table 9.2 demonstrate that these quantum structures are typically characterized by very small numbers or concentrations of electrons that can carry current. This results in the phenomena of single-electron tunneling[5, 6] and the Coulomb blockade to be discussed below.

9.3.3. Conduction Electrons and Dimensionality

We are accustomed to studying electronic systems that exist in three dimensions, and are large or macroscopic in size. In this case the conduction electrons are delocalized, and move freely throughout the entire conducting medium such as a copper wire. It is clear that in practice all of the wire dimensions are very large compared to the distances between atoms. The situation changes when one or more dimensions of the copper shrinks to such an extent as to approach several times the spacings between the atoms in the lattice. When this occurs the delocalization becomes impeded, and the electrons experience confinement. For example, consider a flat plate of copper that is 10 cm long, 10 cm wide, and only 3.6 nm thick. This thickness corresponds to the length of only 10 unit cells, which means that 20% of the atoms are in unit cells at the surface of the copper. The conduction electrons would be delocalized in the plane of the plate, but confined in the narrow dimension, a configuration referred to as a *quantum well*. A *quantum wire* is a structure such as a very thin copper wire that is long in one dimension but has a nanometer size as its diameter. The electrons are delocalized and move freely along the wire, but are confined in the transverse directions. Finally, a quantum dot, which for example might have the shape of a tiny cube, a short cylinder, or a sphere with low-nanometer dimensions,

TABLE 9.3. Delocalization and Confinement Dimensionalities of Quantum Nanostructures

Quantum Structure	Delocalization Dimensions	Confinement Dimensions
Bulk conductor	3 (x,y,z)	0
Quantum well	2 (x,y)	1 (z)
Quantum wire	1 (z)	2 (x,y)
Quantum dot	0	3 (x,y,z)

exhibits confinement in all three spatial dimensions, so there is no delocalization. Figures 9.1 and 9.2, as well as Table 9.3, summarize these scenarios.

9.3.4. Fermi Gas and Density of States

Many of the properties of good conductors of electricity are explained by the assumption that the valence electrons of a metal dissociate themselves from their atoms and become delocalized conduction electrons that move freely through the background of positive ions such as Na^+ or Ag^+. On average, they travel a mean free path distance l between collisions, as mentioned in Section 9.3.2. These electrons act like a gas called a *Fermi gas* in their ability to move with very little hindrance throughout the metal. They have an energy of motion called *kinetic energy*, $E = \frac{1}{2} mv^2 = p^2/2m$, where m is the mass of the electron, v is its speed or velocity, and $\mathbf{p} = m\mathbf{v}$ is its momentum. This model provides a good explanation of Ohm's law, whereby the voltage V and current I are proportional to each other through the resistance R, that is, $V = IR$.

In a quantum-mechanical description the component of the electron's momentum along the x direction p_x has the value $p_x = \hbar k_x$, where $\hbar = h/2\pi$, h is Planck's universal constant of nature, and the quantity k_x is the x component of the wavevector $\mathbf{k}$. Each particular electron has unique k_x, k_y, and k_z values, and we saw in Section 1.3.2 that the k_x,k_y,k_z values of the various electrons form a lattice in k-space, which is called *reciprocal space*. At the temperature of absolute zero, the electrons of the Fermi gas occupy all the lattice points in reciprocal space out to a distance k_F from the origin $k = 0$, corresponding to a value of the energy called the *Fermi energy* E_F, which is given by

$$E_F = \frac{\hbar^2 k_F^2}{2m} \tag{9.11}$$

We assume that the sample is a cube of side L, so its volume V in ordinary coordinate space is $V = L^3$. The distance between two adjacent electrons in k-space is $2\pi/L$, and at the temperature of absolute zero all the conduction electrons are equally spread out inside a sphere of radius k_F, and of volume $4\pi k_F^3/3$ in k-space, as was explained in Section 1.3. This equal density is plotted in Fig. 9.11a for the temperature absolute

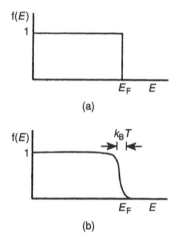

Figure 9.11. Fermi–Dirac distribution function $f(E)$, indicating equal density in k-space, plotted for the temperatures (a) $T = 0$ and (b) $0 < T << T_F$. (From C. P. Poole, *Handbook of Physics*, Wiley, New York, 1998, p. 138.)

zero or 0 K, and in Fig. 9.11b we see that deviations from equal density occur only near the Fermi energy level E_F at higher temperatures.

The number of conduction electrons with a particular energy depends on the value of the energy and also on the dimensionality of the space. This is because in one dimension the size of the Fermi region containing electrons has the length $2\,k_F$, in two dimensions it has the area of the Fermi circle πk_F^2, and in three dimensions it has the volume of the Fermi sphere $4\pi k_F^3/3$. These expressions are listed in column 3 of Table A.1 of Appendix A. If we divide each of these Fermi regions by the size of the corresponding k-space unit cell listed in column 2 of this table, and make use of Eq. (9.11) to eliminate k_F, we obtain the dependence of the number of electrons N on the energy E given on the left side of Table 9.4, and shown plotted in Fig. 9.12. The slopes of the lines $N(E)$ shown in Fig. 9.12 provide the density of states $D(E)$, which is defined more precisely by the mathematical derivative $D(E) = dN/dE$, corresponding to the expression $dN = D(E)\,dE$. This means that the number of electrons dN with an energy E within the narrow range of

TABLE 9.4. Number of Electrons $N(E)$ and Density of States $D(E) = dN(E)/dE$ as Function of Energy E for Conduction Electrons Delocalized in One, Two and Three Spatial Dimensions[a]

Number of Electrons $N(E)$	Density of States $D(E)$	Delocalization Dimensions
$N(E) = K_1 E^{1/2}$	$D(E) = \frac{1}{2}K_1 E^{-1/2}$	1
$N(E) = K_2 E$	$D(E) = K_2$	2
$N(E) = K_3 E^{3/2}$	$D(E) = \frac{3}{2}K_3 \, E^{1/2}$	3

[a]The values of the constants K_1, K_2, and K_3 are given in Table A.2 (of Appendix A).

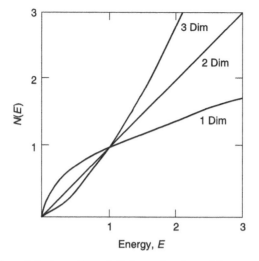

Figure 9.12. Number of electrons $N(E)$ plotted as a function of the energy E for conduction electrons delocalized in one (quantum wire), two (quantum well), and three (bulk material) dimensions (Dim).

energy $dE = E_2 - E_1$ is proportional to the density of states at that value of energy. The resulting formulaes for $D(E)$ for the various dimensions are listed in the middle column of Table 9.4, and are shown plotted in Fig. 9.13. We see that the density of states decreases with increasing energy for one dimension, is constant for two

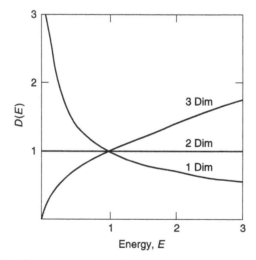

Figure 9.13. Density of states $D(E) = dN(E)/dE$ plotted as a function of the energy E for conduction electrons delocalized in one (quantum wire), two (quantum well), and three (bulk) dimensions.

dimensions, and increases with increasing energy for three dimensions. Thus the density of states has quite a different behavior for the three cases. These equations and plots of the density of states are very important in determining the electrical, thermal, and other properties of metals and semiconductors, and clarify why these features can be so dependent on the dimensionally. Examples of how various properties of materials depend on the density of states are given in Section 9.3.7.

9.3.5. Potential Wells

In the previous section we discussed the delocalization aspects of conduction electrons in a bulk metal. These electrons were referred to as *free electrons*, but perhaps *unconfined electrons* would be a more appropriate term for them. This is because when the size of a conductor diminishes to the nanoregion, these electrons begin to experience the effects of confinement, meaning that their motion becomes limited by the physical size of the region or domain in which they move. The influence of electrostatic forces becomes more pronounced, and the electrons become restricted by a potential barrier that must be overcome before they can move more freely. More explicitly, the electrons become sequestered in a *potential well*, an enclosed region of negative energies. A simple model that exhibits the principal characteristics of such a potential well is a *square well*, in which the boundary is very sharp or abrupt. *Square wells*, can exist in one, two, three, and higher dimensions, and for simplicity we will describe a one-dimensional case.

Standard quantum-mechanical texts show that for an infinitely deep square potential well of width a in one dimension the coordinate x has the range of values $-\frac{1}{2}a \leq x \leq \frac{1}{2}a$ inside the well, and the energies there are given by the expressions

$$E_n = \frac{\pi^2 \hbar^2}{2\,ma^2} n^2 \qquad (9.12a)$$

$$= E_0 n^2 \qquad (9.12b)$$

which are plotted in Fig. 9.14, where $E_0 = \pi^2 \hbar^2/2\,ma^2$ is the ground-state energy and the quantum number n assumes the values $n = 1,2,3,\ldots$. The electrons that are present fill up the energy levels starting from the bottom, until all available electrons are in place. An infinite square well has an infinite number of energy levels, with ever-widening spacings as the quantum number n increases. If the well is finite, then its quantized energies E_n all lie below the corresponding infinite well energies, and there are only a limited number of them. Figure 9.15 illustrates the case for a finite well of potential depth $V_0 = 7E_0$ that has only three allowed energies. No matter how shallow the well, there is always at least one bound state E_1.

The electrons confined to the potential well move back and forth along the direction x, and the probability of finding an electron at a particular value of x is given by the square of the wavefunction $|\psi_n(x)|^2$ for the particular level n where the electron is located. There are even and odd wavefunctions $\psi_n(x)$, which alternate for the levels in

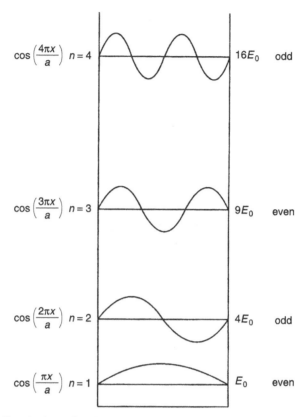

Figure 9.14. Sketch of wavefunctions for the four lowest energy levels, E_n for $n = 1$–4, of the one-dimensional infinite square well. For each level the form of the wavefunction is given on the left, and its parity (even or odd) is indicated on the right. (From C. P. Poole, Jr., *Handbook of Physics*, Wiley, New York, 1998, p. 289.)

the one-dimensional square well, and for the infinite square well we have the unnormalized expressions

$$\psi_n = \cos\frac{n\pi x}{a} \qquad n = 1,3,5,\ldots \qquad \text{even parity} \qquad (9.13)$$

$$\psi_n = \sin\frac{n\pi x}{a} \qquad n = 2,4,6,\ldots \qquad \text{odd parity} \qquad (9.14)$$

These wavefunctions are sketched in Fig. 9.14 for the infinite well. The property called *parity* is defined as even when $\psi_n(-x) = \psi_n(x)$, and it is odd when $\psi_n(-x) = -\psi_n(x)$.

There is another important variety of potential well, corresponding to a well with a curved cross section. For a circular cross section of radius a in two dimensions, the

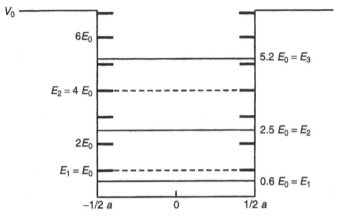

Figure 9.15. Sketch of a one-dimensional square well showing how the energy levels E_n of a finite well (right side, solid horizontal lines) lie below their infinite well counterparts (left side, dashed lines). (From C. P. Poole, Jr., *Handbook of Physics*, Wiley, New York, 1998, p. 285.)

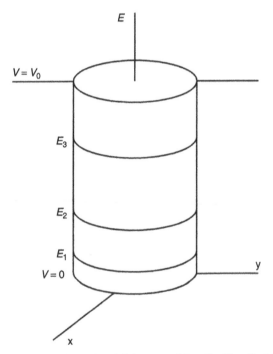

Figure 9.16. Sketch of a two-dimensional finite potential well with cylindrical geometry and three energy levels, E_1, E_2, and E_3.

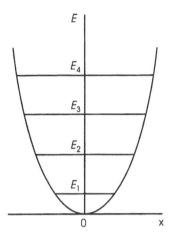

Figure 9.17. Sketch of a one-dimensional parabolic potential well showing the positions of the four lowest energy levels.

potential is given by $V = 0$ in the range $0 \leqslant \rho \leqslant \# a$, and has the value V_0 at the top and outside, where $\rho = (x^2 + y^2)^{1/2}$, and $\tan \varphi = y/x$ in polar coordinates. The finite well sketched in Fig. 9.16 has only three allowed energy levels with the values E_1, E_2, and E_3.

Another type of commonly used potential well is the *parabolic well*, which is characterized by the potentials $V(x) = \frac{1}{2}kx^2$, $V(\rho) = \frac{1}{2}k\rho^2$ and $V(r) = \frac{1}{2}kr^2$ in one, two, and three dimensions, respectively, and Fig. 9.17 provides a sketch of the potential in the one-dimensional case.

Another characteristic of a particular energy state E_n is the number of electrons that can occupy it, and this depends on the number of different combinations of quantum numbers that correspond to this state. From Eq. (9.12) we see that the one-dimensional square well has only one allowed value of the quantum number n for each energy state. An electron also has a spin quantum number m_s that can take on two values, $m_s = +\frac{1}{2}$ and $m_s = -\frac{1}{2}$, for spin states up and down, respectively, and for the square well both spin states $m_s = \pm\frac{1}{2}$ have the same energy. According to the Pauli exclusion principle of quantum mechanics, no two electrons can have the same set of quantum numbers, so each square well energy state E_n can be occupied by two electrons, one with spinup, and one with spindown. The number of combinations of quantum numbers corresponding to each spin state is called its *degeneracy*, and so the degeneracy of all the one-dimensional square well energy levels is 2.

The energy of a two-dimensional square well

$$E_n = \frac{\pi^2/h^2}{2ma^2}(n_x^2 + n_y^2) = E_0n^2 \qquad (9.15)$$

depends on two quantum numbers $n_x = 0,1,2,3, \ldots$, and $n_y = 0,1,2,3, \ldots$, where $n^2 = n_x^2 + n_y^2$. This means that the lowest energy state $E_1 = E_0$ has two possibilities, namely $n_x = 0, n_y = 1$ and $n_x = 1, n_y = 0$, so the total degeneracy (including spin direction) is 4. The energy state $E_5 = 25E_0$ has more possibilities since it can have, for example, $n_x = 0$, $n_y = 5$, or $n_x = 3$ and $n_y = 4$, and so on, so its degeneracy is 8.

9.3.6. Partial Confinement

In the previous section we examined the confinement of electrons in various dimensions and found that it always leads to a qualitatively similar spectrum of discrete energies. This is true for a broad class of potential wells, irrespective of their dimensionality and shape. We also examined, in Section 9.3.4, the Fermi gas model for delocalized electrons in these same dimensions and found that the model leads to energies and densities of states that differ quite significantly from each other. This means that many electronic and other properties of metals and semiconductors change dramatically when the dimensionality changes. Some nanostructures of technological interest exhibit both potential well confinement and Fermi gas delocalization, confinement in one or two dimensions, and delocalization in two or one dimensions, so it will be instructive to show how these two strikingly different behaviors coexist.

In a three-dimensional Fermi sphere the energy varies from $E = 0$ at the origin to $E = E_F$ at the Fermi surface, and similarly for the one- and two-dimensional analogs. When there is confinement in one or two directions the conduction electrons will distribute themselves among the corresponding potential well levels that lie below the Fermi level along confinement coordinate directions, in accordance with their respective degeneracies d_i, and for each case the electrons will delocalize in the remaining dimensions by populating Fermi gas levels in the delocalization direction of the reciprocal lattice. Table 9.5 lists formulas for the energy dependence of the number of electrons $N(E)$ for quantum dots that exhibit total confinement, quantum wires and quantum wells that involve partial confinement, and bulk material where there is no confinement. The density of states formulas $D(E)$ for these four cases are also

TABLE 9.5. Number of Electrons $N(E)$ and Density of States $D(E) = dN(E)/dE$ as function of Energy E for Electrons Delocalized/Confined in Quantum Dots, Wires, and Wells and Bulk material[a]

			Dimensions	
Type	Number of Electrons $N(E)$	Density of States $D(E)$	Delocalized	Confined
Dot	$N(E) = K_0 3d_i \, \Theta(E - E_{iW})$	$D(E) = K_0 \, 3d_i \, \delta(E - E_{iW})^2$	0	3
Wire	$N(E) = K_1 3d_i (E - E_{iW})^{1/2}$	$D(E) = \frac{1}{2} K_1 3d_i (E - E_{iW})^{-1/2}$	1	2
Well	$N(E) = K_2 3d_i (E - E_{iW})$	$D(E) = K_2 \, 3d_i$	2	1
Bulk	$N(E) = K_3 (E)^{3/2}$	$D(E) = \frac{3}{2} K_3 (E)^{1/2}$	3	0

[a]The degeneracies d_i of the confined (square or parabolic well) energy levels depend on the particular level. The Heaviside step function $\Theta(x)$ is zero for $x < 0$ and one for $x > 0$; the delta function $\delta(x)$ is zero for $x \neq 0$, is infinity for $x = 0$, and integrates to a unit area. The values of the constants K_1, K_2, and K_3 are given in Table A.3.

listed in the table. The summations in these expressions are over the various confinement well levels i.

Figure 9.18 shows plots of the energy dependence $N(E)$ and the density of states $D(E)$ for the four types of nanostructures listed in Table 9.5. We see that the number of electrons $N(E)$ increases with energy E, so the four nanostructure types vary only qualitatively from each other. However, it is the density of states $D(E)$ that determines the various electronic and other properties, and these differ dramatically for each of the four nanostructure types. This means that the nature of the dimensionality and the confinement associated with a particular nanostructure have a pronounced effect on its properties. These considerations can be used to predict properties of nanostructures, and one can also identify types of nanostructures from their properties.

9.3.7. Properties Dependent on Density of States

We have discussed the density of states $D(E)$ of conduction electrons, and have shown how it is strongly affected by the dimensionality of a material. Phonons or quantized lattice vibrations also have a density of states $D_{ph}(E)$ that depends on the dimensionality, and, like its electronic counterpart, it influences some properties of solids, but our principal interest is in the density of states $D(E)$ of the electrons. The effect of dimensionality on the density of states is discussed in Appendix A. In this section we will mention some of the properties of solids that depend on $D(E)$ and describe some experiments for measuring it.

The specific heat of a solid C is the amount of heat that must be added to it to raise its temperature by one degree celsius (centigrade). The main contribution to this heat is the amount that excites lattice vibrations, and this depends on the phonon density of states $D_{ph}(E)$. At low temperatures there is also a contribution to the specific heat C_{el} of a conductor arising from the conduction electrons, and this depends on the electronic density of states at the Fermi level: $C_{el} = \pi^2 D(E_F) k_B^2 T/3$, where k_B is the Boltzmann constant.

The susceptibility $\chi = M/H$ of a magnetic material is a measure of the magnetization M or magnetic moment per unit volume that is induced in the material by the application of an applied magnetic field H. The component of the susceptibility arising from the conduction electrons, called the *Pauli susceptibility*, is given by the expression $\chi_{el} = \mu_B^2 D(E_F)$, where μ_B is the unit magnetic moment called the *Bohr magneton*, and hence χ_{el} is characterized by its proportionality to the electronic density of states $D(E)$ at the Fermi level, and its lack of dependence on temperature.

When a good conductor such as aluminum is bombarded by fast electrons with just enough energy to remove an electron from a particular Al inner-core energy level, the vacant level left behind constitutes a hole in the inner-core band. An electron from the conduction band of the aluminum can fall into the vacant inner-core level to occupy it, with the simultaneous emission of an X ray in the process. The intensity of the emitted X radiation is proportional to the density of states of the conduction electrons because the number of electrons with each particular energy that jumps down to fill the hole is proportional to $D(E)$. Therefore a plot of the emitted

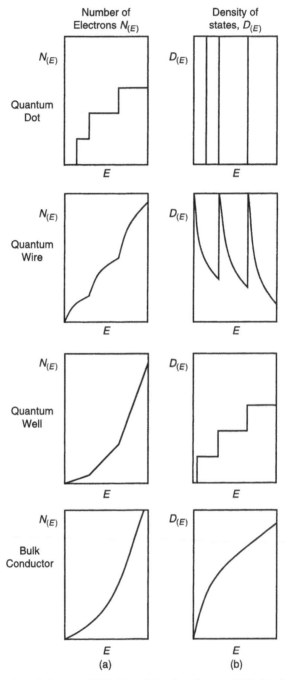

Figure 9.18. Number of electrons $N(E)$ (a) and density of states $D(E)$ (b) plotted against the energy for four types of quantum structures in the square well–Fermi gas approximations.

X ray intensity versus the X ray energy E has a shape very similar to a plot of $D(E)$ versus E. These emitted X rays for aluminum are in the energy range 56–77 eV.

Some other properties and experiments that depend on the density of states and can provide information on it are photoemission spectroscopy, Seebeck effect (thermopower) measurements, concentrations of electrons and holes in semiconductors, optical absorption determinations of the dielectric constant, the Fermi contact term in nuclear magnetic resonance (NMR), the de Haas–van Alphen effect, the superconducting energy gap, and Josephson junction tunneling in superconductors. It would take us too far afield to discuss any of these topics. Experimental measurements of these various properties permit us to determine the form of the density of states $D(E)$, both at the Fermi level E_F and over a broad range of temperature.

9.4. EXCITONS

Excitons, which were introduced in Section 1.4.3, are a common occurrence in semiconductors. When an atom at a lattice site loses an electron the atom acquires a positive charge that is called a *hole*. If the hole remains localized at the lattice site, and the detached negative electron remains in its neighborhood, it will be attracted to the positively charged hole through the Coulomb interaction, and can become bound to form a hydrogen type atom. Technically speaking, this is called a Mott–Wannier type of exciton. The Coulomb force of attraction between two charges $Q_e = -e$ and $Q_h = +e$ separated by a distance r is given by $F = -ke^2/\varepsilon r^2$, where e is the electronic charge, k is a universal constant, and ε is the dielectric constant of the medium. The exciton has a Rydberg series of energies E sketched in Fig. 1.32, and a radius given by Eq. (1.22) of that chapter: $a_{\text{eff}} = 0.0529(\varepsilon/\varepsilon_0)/ (m^*/m_0)$, where $\varepsilon/\varepsilon_0$ is the ratio of the dielectric constant of the medium to that of free space, and m^*/m_0 is the ratio of the effective mass of the exciton to that of a free electron. Using the dielectric constant and electron effective mass values from Tables B.11 and B.8, respectively, we obtain for GaAs

$$E = 4.5\,\text{meV}, \qquad a_{\text{eff}} = 12.0\,\text{nm} \qquad (9.16)$$

showing that the exciton has a radius comparable to the dimensions of a typical nanostructure.

The exciton radius can be taken as an index of the extent of confinement experienced by a nanoparticle. Two limiting regions of confinement can be identified according to the ratio of the dimension d of the nanoparticle to the exciton radius a_{eff} namely, the weak confinement regime with $d > a_{\text{eff}}$ (but not $d \gg a_{\text{eff}}$) and the strong confinement regime $d < a_{\text{eff}}$. The more extended limit $d \gg a_{\text{eff}}$ corresponds to no confinement. Under weak confinement conditions the exciton can undergo unrestricted translational motion, just as in the bulk material, but for strong confinement this translation motion becomes restricted. There is an increase in the spatial overlap of electron and hole wavefunctions with decreasing particle size, and this has the effect of enhancing the electron–hole interaction. As a result, the energy splitting becomes greater between the radiative and nonradiative exciton states. An optical index of the confinement is the blue shift (shift to higher energies) in the optical absorption edge and the increased exciton energy with decreasing nanoparticle size. Another result of the confinement is the appearance at room temperature of excitonic features in the absorption spectra, which are observed only at low temperatures in the bulk material. Further details on exciton spectra are provided in Section 1.4.3.

9.5. SINGLE-ELECTRON TUNNELING

We have been discussing quantum dots, wires, and wells in isolation, such as the ones depicted in Figs. 9.1 and 9.2. To make them useful, they need coupling to their surroundings, to each other, or to electrodes that can add or subtract electrons from them. Figure 9.19 shows an isolated quantum dot or island coupled through tunneling to two leads, a source lead that supplies electrons, and a drain lead that removes electrons for use in the external circuit. The applied voltage V_{sd} causes direct current I to flow, with electrons tunneling into and out of the quantum dot. In accordance with Ohm's law $V = IR$, the current flow I through the circuit of Fig. 9.19 equals the applied source–drain voltage V_{sd} divided by the resistance R, and the main contribution to the value of R arises from the process of electron tunneling from source to quantum dot, and from quantum dot to drain. Figure 9.20 shows the addition to the circuit of a capacitor-coupled gate terminal. The applied gate voltage V_g provides a controlling electrode or gate that regulates the resistance R of the active region of the quantum dot, and consequently regulates the current flow I between the source and drain terminals. This device, as described, functions as a voltage-controlled (or field-effect-controlled) transistor commonly referred to as an FET. For large or macroscopic dimensions the current flow is continuous, and the discreteness of the individual electrons passing through the device manifests itself by the presence

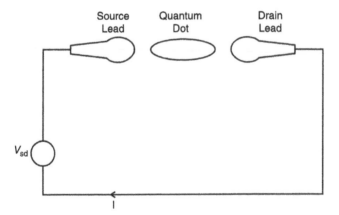

Figure 9.19. Quantum dot coupled to an external circuit through source and drain leads.

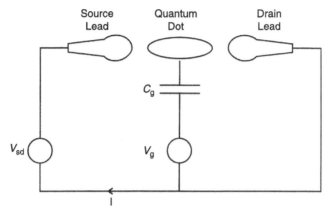

Figure 9.20. Quantum dot coupled through source and drain leads to an external circuit containing an applied bias voltage V_{sd}, with an additional capacitor-coupled terminal through which the gate voltage V_g controls the resistance of the electrically active region.

of current fluctuations or shot noise. Our present interest is in the passage of electrons, one by one, through nanostructures, based on circuitry of the type sketched in Fig. 9.20.

For an FET-type nanostructure the dimensions of the quantum dot are in the low-nanometer range, and the attached electrodes can have cross sections comparable in size. For disk and spherical shaped dots of radius r, the capacitance is given by

$$C = 8\varepsilon_0 \left(\frac{\varepsilon}{\varepsilon_0}\right) r \qquad \text{disk} \qquad (9.17)$$

$$C = 4\pi\varepsilon_0 \left(\frac{\varepsilon}{\varepsilon_0}\right) r \qquad \text{sphere} \qquad (9.18)$$

where $\varepsilon/\varepsilon_0$ is the dimensionless dielectric constant of the semiconducting material that forms the dot, and the dielectric constant of free space has the value $\varepsilon_0 = 8.8542 \times 10^{-12}$ F/m. For the typical quantum dot material GaAs we have $\varepsilon/\varepsilon_0 = 13.2$, which gives the very small value $C = 1.47 \times 10^{-18} r$ farad (F) for a spherical shape, where the radius r is in nanometers. The electrostatic energy E of a capacitor of charge Q is changed by the amount $\Delta E - eQ/C$ when a single electron is added or subtracted, corresponding to the change in potential $\Delta V = \Delta E/Q$

$$\Delta V = \frac{e}{C} \approx 0.109/r \quad \text{volts (V)} \qquad (9.19)$$

where r is in nanometers. For a nanostructure of radius $r = 10$ nm this gives a change in potential of 11 mV, which is easily measurable. It is large enough to impede the tunneling of the next electron.

Two quantum conditions[6] must be satisfied before we can observe the discrete nature of the single-electron charge transfer to a quantum dot: (1) the capacitor charging energy $e^2/2C$ must exceed the thermal energy $k_B T$ arising from the random vibrations of the atoms in the solid, and (2) the Heisenberg uncertainty principle must be satisfied by the product of the capacitor energy $e^2/2C$ and the time $\tau = R_T C$ required for charging the capacitor

$$\Delta E \ \Delta t = \frac{e^2}{2C} R_T C > h \qquad (9.20)$$

where R_T is the tunneling resistance of the potential barrier. These two tunneling conditions correspond to

$$\frac{e^2}{2C} >> k_B T \qquad (9.21a)$$

$$R_T >> \frac{h}{e^2} \qquad (9.21b)$$

where $h/e^2 = 25.813 \text{ k}\Omega$ is the quantum of resistance. When these conditions are met and the voltage across the quantum dot is scanned, the current jumps in increments every time the voltage changes by the value of Eq. (9.19), as shown by the I-versus-V (I–V) characteristic of Fig. 9.21. This is called a "Coulomb blockade" because the electrons are blocked from tunneling except at the discrete voltage change positions. The step structure observed on the I–V characteristic of Fig. 9.21 is called a "Coulomb staircase" because it involves the Coulomb charging energy $e^2/2C$ of Eq. (9.21a).

An example of single-electron tunneling is provided by a line of ligand-stabilized Au_{55} nanoparticles. These gold particles have a *structural magic number* of atoms arranged in a fcc close-packed cluster that approximates the shape of a sphere of diameter 1.4 nm, as was discussed in Section 1.2. The cluster of 55 gold atoms is encased in an insulating coating called a *ligand shell*, which is adjustable in thickness and has a typical value of 0.7 nm. Single-electron tunneling can take place between two of these Au_{55} ligand-stabilized clusters when they are in contact, with the shell acting as the barrier for the tunneling. Experiments were carried out with linear arrays of these Au_{55} clusters of the type illustrated in Fig. 9.22. An electron entering the chain at one end was found to tunnel its way through in a solitonlike manner. Estimates of the interparticle capacitance gave $C_{\text{micro}} \times 10^{-18}$F, and the estimated interparticle resistance was $R_T \times 100 \text{ M}\Omega$.[7] Section 11.4.1 discusses electron tunneling through a 500-nm-long line of gold nanoparticles connected together by conjugated organic molecules.

Another example of single-electron tunneling is the ultrasmall superconducting *Josephson junction*,[8] (to be discussed in Section 15.9.5), which consists of two superconductors separated by a thin layer of insulating material, and such an ultrasmall junction with an area $A = 0.01 \ \mu m^2$ and a thickness $d = 0.1$ nm is indeed a

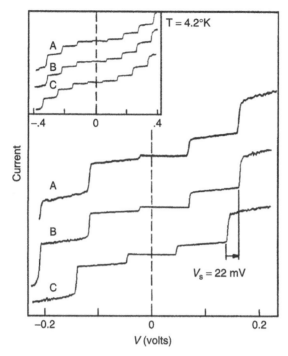

Figure 9.21. Coulomb staircase on the current I versus voltage V characteristic plot from single electron tunneling involving a 10-nm indium metal droplet. The experimental curve A was measured by a scanning tunneling microscope, and curves B and C are theoretical simulations. The peak-to-peak current is 1.8 nA. [After R. Wilkins, E. Ben-Jacob, and R. C. Jaklevic, *Phys. Rev. Lett.* **59**, 109 (1989).]

quantum dot. It has the capacitance $C = \varepsilon_0 A / d$ of about $\sim 10^{-15}$ F, and hence the change in voltage ΔV arising from the tunneling of a single electron across the insulating barrier $\Delta V = e/C = 0.16$ mV is appreciable in magnitude. This can suffice to impede the tunneling of the next electron, and the result is a Coulomb blockade.

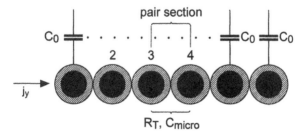

Figure 9.22. Linear array of Au_{55} ligand-stabilized nanoparticles with interparticle resistance R_T, interparticle capacitance C_{micro}, and self-capacitance C_0. The single-electron current density j_y entering from the left, which tunnels from particle to particle along the line, is indicated. (From V. Gasparian et al., in *Handbook of Nanostructured Materials and Nanotechnology*, H. S. Nalwa, ed., Academic Press, Boston, 2000, Vol. 2, Chapter 11, p. 550.)

9.6. APPLICATIONS

9.6.1. Infrared Detectors

Infrared transitions involving energy levels of quantum wells, such as the levels shown in Figs. 9.15 and 9.16, have been used for the operation of infrared photodetectors. Sketches of four types of these detectors are presented in Fig. 9.23. The conduction band is shown at or near the top of these figures, occupied and unoccupied bound-state energy levels are shown in the wells, and the infrared transitions are indicated by vertical arrows. Incoming infrared radiation raises electrons to the conduction band, and the resulting electric current flow is a measure of the incident radiation intensity. Figure 9.23a illustrates a bound-state–bound-state transition that takes place within the quantum well, and Fig. 9.23b shows a bound-state–continuum transition. In Fig. 9.23c the continuum begins at the top of the well, so the transition is referred to as a *bound-state–quasi-bound-state type*. Finally, in Fig. 9.23d the continuum band lies below the top of the well, so the transition is designated as a bound-state–miniband variety.

The responsivity of the detector is defined as the electric current [amperes, (A)] generated per watt (W) of incoming radiation. Figure 9.24 shows a plot of the dark current density (before irradiation) versus bias voltage for a GaAs/AlGaAs bound-state–continuum photodetector, and Fig. 9.25 shows the dependence of this detector's responsivity on the wavelength for normal and 45°E incidence. The responsivity reaches a peak at the wavelength $\lambda = 9.4$ μm, and Fig. 9.26 shows the dependence of this peak responsivity R_P on the bias voltage. The operating bias of 2 V was used to obtain the data of Fig. 9.25 because, as Fig. 9.26 shows, at that bias the responsivity has leveled off at a high value. This detector is sensitive for operation in the infrared wavelength range from 8.5 to 10 μm.

9.6.2. Quantum Dot Lasers

The infrared detectors described in the previous section depend on the presence of discrete energy levels in a quantum well between which transitions in the infrared

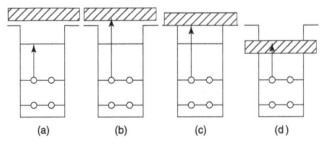

Figure 9.23. Schematic conduction band (shaded) and electron transition schemes (vertical arrows) of types: (a) bound state to bound state, (b) bound state to continuum, (c) bound state to quasibound band, and (d) bound state to miniband, for quantum well infrared photodetectors. [Adapted from S. S. Li and M. Z. Tidrow, in *Handbook of Nanostructured Materials and Nanotechnology*, H. S. Nalwa, ed., Academic Press, Boston, 2000, Vol. 4, Chapter 5, p. 563.]

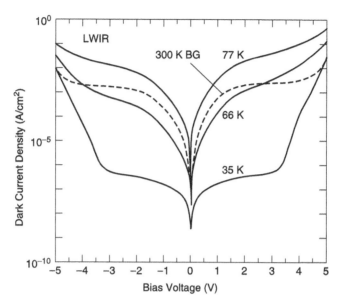

Figure 9.24. Dark current density versus bias voltage characteristics for a GaAs/AlGaAs quantum well long-wavelength infrared (LWIR) photodetector measured at three indicated temperatures. A 300-K background current plot (BG, dashed curve) is also shown. [From M. Z. Tidrow, J. C. Chiang, S. S. Li, and K. Bacher, *Appl. Phys. Lett.* **70**, 859 (1997).]

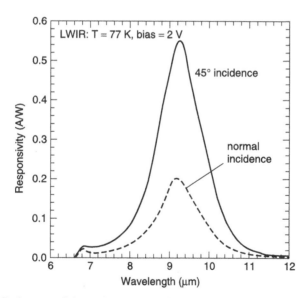

Figure 9.25. Peak responsivity versus wavelength at 77 K for a 2 V bias at normal and 45° angles of incidence. [From M. Z. Tidrow, J. C. Chiang, S. S. Li, and K. Bacher, *Appl. Phys. Lett.* **70**, 859 (1997).]

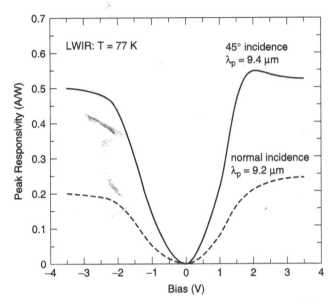

Figure 9.26. Peak responsivity versus bias voltage at 77 K at normal and 45° angles of incidence. The peak detection wavelengths λ_p are indicated. [From M. Z. Tidrow, J. C. Chiang, S. S. Li, and K. Bacher, *Appl. Phys. Lett.* **70**, 859 (1997).]

spectral region can be induced. Laser operation also requires the presence of discrete energy levels, that is, levels between which laser emission transitions can be induced. The word laser is, an acronym for *light amplification by stimulated emission of radiation*, and light emitted by a laser is both monochromatic (single-wavelength) and coherent (in-phase). Quantum well and quantum wire lasers have been constructed that make use of these laser emission transitions. These devices have conduction electrons for which the confinement and localization in discrete energy levels takes place in one or two dimensions, respectively. Hybrid-type lasers have been constructed using "dots in a well," such as InAs quantum dots placed in a strained InGaAs quantum well. Another design has employed InAs "quantum dashes," which are very short quantum wires, or from another perspective they are quantum dots elongated in one direction. The present section is devoted to a discussion of quantum dot lasers, in which the confinement is in all three spatial directions.

Conventional laser operation requires the presence of a laser medium containing active atoms with discrete energy levels between which the laser emission transitions take place. It also requires a mechanism for population inversion whereby an upper energy level acquires a population of electrons exceeding that of the lower-lying ground-state level. In a helium–neon gas laser the active atoms are Ne mixed with He, and in a Nd:YAG solid state laser the active atoms are neodymium ions substituted ($-10^{19}/cm^3$) in a yttruim aluminum garnet crystal. In the quantum dot laser to be described here, the quantum dots play role of active atoms.

Figure 9.27 provides a schematic illustration of a quantum dot laser diode grown on an N-doped GaAs substrate (not shown). The top P-metal layer has a GaAs contact layer immediately below it. Between this contact layer above and the GaAs substrate (not shown) below the diagram there are a pair of 2-μm-thick $Al_{0.85}Ga_{0.15}As$ cladding or bounding layers surrounding a 190-nm-thick waveguide consisting of $Al_{0.05}Ga_{0.95}As$. The waveguide plays the role of conducting the emitted light to the exit ports at the edges of the structure. Centered in the waveguide (dark horizontal stripe on the figure labeled QD) is a 30-nm-thick GaAs region, and centered in this region are 12 monolayers of $In_{0.5}Ga_{0.5}As$ quantum dots with a density of $1.5 \times 10^{10}/cm^2$. The inset at the bottom of the figure was drawn to represent the details of the waveguide region. The length L_C and the width W varied somewhat from sample to sample, with L_C ranging from 1 to 5 mm, and W varying between 4 and 60 μm. The facets or faces of the laser were coated with $ZnSe/MgF_2$ high-reflectivity (>95%) coatings that reflected the light back and forth inside to augment the stimulated emission. The laser light exited through the lateral edge of the laser.

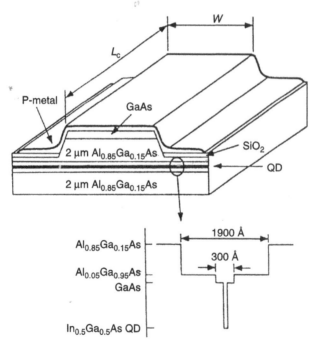

Figure 9.27. Schematic illustration of a quantum dot near-IR laser. The inset at the bottom shows details of the 190-nm-wide ($Al_{0.85}Ga_{0.15}As$-cladded) waveguide region that contains the 12 monolayers of $In_{0.5}Ga_{0.5}As$ quantum dots (indicated by QD) that do the lasing. [From G. Park, O. B. Shchekin, S. Csutak, D. L. Huffaker, and D. G. Deppe, *Appl. Phys. Lett.* **75**, 3267 (1999).]

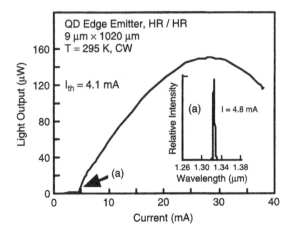

Figure 9.28. Dependence of the near-IR light output power on current for a continuous-wave, room-temperature, edge-emitting quantum dot laser of the type illustrated in Fig. 9.27. [From G. Park, O. B. Shchekin, S. Csutak, D. L. Huffaker, and D. G. Deppe, *Appl. Phys. Lett.* **75**, 3267 (1999).]

The laser output power for continuous-wave (CW) operation at room temperature is plotted in Fig. 9.28 versus the current for the laser dimensions $L_C = 1.02$ mm and $W = 9$ μm. The near-infrared output signal at the wavelength of 1.32 μm for a current setting just above the 4.1 mA threshold value, labeled point a, is shown in the inset of the figure. The threshold current density increases sharply with the temperature above 200 K, and this is illustrated in Fig. 9.29 for pulsed operation.

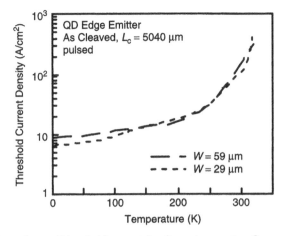

Figure 9.29. Dependence of threshold current density on temperature for a pulsed, edge-emitting quantum dot laser with the design illustrated in Fig. 9.27. [From G. Park, O. B. Shchekin, S. Csutak, D. L. Huffaker, and D. G. Deppe, *Appl. Phys. Lett.* **75**, 3267 (1999).]

PROBLEMS

9.1. For a cubic quantum dot of indium selenide, estimate the size below which there are more atoms on the surface than in the bulk.

9.2. Calculate the conduction electron energy levels of a cubic quantum dot having an edge of 10 nm assuming an infinite square well potential. What is the energy, and what is the degeneracy of the two lowest levels, including spin?

9.3. A quantum dot in the shape of a sphere with a radius of 11 nm is formed from a bulk semiconducting material doped with 10^{21} donors/cm^3. How many conduction electrons does the dot have?

9.4. If the quantum dot of Problem 9.3 were to be used as laser, what would be its wavelength?

9.5. Suggest a method for making a quantum dot laser tunable, specifically such that the wavelength can be varied. Explain your answer.

9.6. How many monomers n are in a polymer of polymethyl methacrylate $[C_5O_2H_8]_n$ with a size of 10^5 Da?

9.7. A nanoparticle in the shape of a sphere has the same surface area as another nanoparticle in the shape of a thin disk made of the same material. What is the ratio C_{sph}/C_{disk} of their capacitances?

9.8. Show that the critical surface energy is given by $\Delta G_c = (\frac{2}{3})^4 \pi r_0^3 (\alpha + \beta)$, where r_0 is the radius for which ΔG_c vanishes (i.e., $\Delta G_c = 0$). Find the ratio r_0/r_c.

9.9. To observe discrete single-election transfer to a quantum dot at temperature 0.003 K, how small does its capacitance have to be? If the quantum dot is a sphere made of InSb, how small a radius must it have?

9.10. A single electron is added to a CdTe quantum dot in the shape of a disk of radius 7 nm. What is the change in potential?

REFERENCES

1. M. Henini, Quantum dot nanostructures, *Mater. Today* 48 (June 2002).
2. L. Jacak, P. Hawrylak, and A. Wojs, *Quantum Dots*, Springer, Berlin, 1998.
3. G. Cao, *Nanostructures and Nanomaterials*, Imperial College Press, London, 2004, Chapter 3.
4. D. Heitmann and J. P. Kotthaus, The spectroscopy of quantum dot arrays, *Phys. Today* 56 (June 1993).
5. M. S. Feld and K. An, Single atom laser, *Sci. Am.* 57 (July 1998).
6. L. P. Kouwenhoven and P. L. McEuen, Single electron transport through a quantum dot, in *Nanotechnology*, G. Timp, ed., Springer Berlin, 1999, Chapter 13.
7. V. Gasparian, M. Ortuño, G. Schön, and U. Simon, in *Handbook of Nanostructured Materials and Nanotechnology*, H. S. Nalwa, ed., Academic Press, Boston, 2000, Chapter 11.
8. C. P. Poole, Jr., H. A. Farach, R. J. Creswick, and R. Prozorov, *Superconductivity*, Elsevier, Amsterdam, 2007, Chapter 15, Section VIII-G.

Carbon Nanostructures

10.1. INTRODUCTION

This chapter is concerned with various nanostructures of carbon. A separate chapter is devoted to carbon because of the important role of carbon bonding in the organic molecules of life and the unique nature of the carbon bond itself. It is the diverse nature of this bond that allows carbon to form some of the more interesting nanostructures, particularly carbon nanotubes. Possibly more than any of the other nanostructures, these carbon nanotubes have enormous applications potential, which we will discuss in this chapter.

10.2. CARBON MOLECULES

10.2.1. Nature of the Carbon Bond

In order to understand the nature of the carbon bond, it is necessary to examine the electronic structure of the carbon atom. Carbon consists of six electrons, which are distributed over the lowest energy levels of the carbon atom. The structure is designated as $(1s)^2$, $(2s)$, $(2p_y)$, $(2p_x)$, $(2p_z)$ when bonded to atoms in molecules. The lowest energy level, $1s$, with the principal quantum number $N = 1$ contains two electrons with oppositely paired electron spins. The electron charge distribution in an s state is spherically symmetric about the nucleus. These $1s$ electrons do not participate in the chemical bonding. The next four electrons are in the $N = 2$ energy state, one in a spherically symmetric s orbital, and three in p_x, p_y and p_z orbitals, which have the very directed charge distributions shown in Fig. 10.1a, oriented perpendicular to each other. This outer s orbital together with the three p orbitals form the chemical bonds of carbon with other atoms. The charge distribution associated with these orbitals mixes (or overlaps) with the charge distribution of each atom bonded to carbon. In effect, one can view the electron charge between the two atoms of a bond as the "glue" that holds the atoms together. According to this simple picture, the methane molecule (CH_4) might have the structure shown in Fig. 10.1b, where the H—C bonds are at right angles to each other. However, methane does not have this

The Physics and Chemistry of Nanosolids. By Frank J. Owens and Charles P. Poole, Jr.
Copyright © 2008 by John Wiley & Sons, Inc.

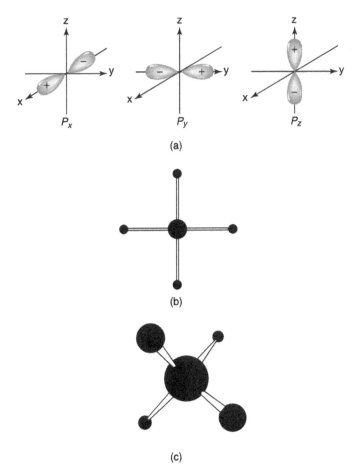

Figure 10.1. (a) Illustration of p_x, p_y, and p_z orbitals of the carbon atom; (b) structure of methane (CH_4) assuming that the valence orbitals of carbon are pure p_x, p_y, p_z orbitals; (c) actual structure of CH_4, which is explained by sp^3 hybridization.

configuration; rather, it has the tetrahedral structure shown in Fig. 10.1c, where the carbon bonds make angles of 109°28′ with each other. The explanation lies in the concept of hybridization. In the carbon atom the energy separation between the $2s$ level and the $2p$ levels is very small, and this allows an admixture of the $2s$ wavefunction with one or more of the $2p_i$ wavefunctions. The wavefunction Ψ in a valence state can be designated by the expression

$$\Psi = s + \lambda p \tag{10.1}$$

where λp indicates an admixture of p_i orbitals. With this hybridization, the directions of the p lobes and the angles between them changes. The angles will depend on the relative admixture λ of the p states with the s state. Three kinds of hybridization are

TABLE 10.1. Types of sp^n-Hybridization, Resulting Bond Angles, and Example of Molecules

Type of Hybridization	Digonal sp	Trigonal sp^2	Tetrahedral sp^3
Orbitals used for bond	s, p_x	s, p_x, p_y	s, p_x, p_y, p_z
Example	Acetylene (C_2H_2)	Ethylene (C_2H_4)	Methane (CH_4)
Value of λ	1	$2^{1/2}$	$3^{1/2}$
Bond angle (degrees)	$180°$	$120°$	$109°\,28'$

identified in Table 10.1, which shows the bond angles for the various possibilities, which are $180°$, $120°$, and $109°28'$ for the linear compound acetylene (H—C≡C—H), the planar compound benzene (C_6H_6), and tetrahedral methane (CH_4), respectively. In general, most of the bond angles for carbon in organic molecules have these values. For example, the carbon bond angle in diamond is $109°$, and in graphite and benzene it is $120°$.

Solid carbon has two main structures called *allotropic forms*, which are stable at room temperature, namely, diamond and graphite. Diamond consists of carbon atoms that are tetrahedrally bonded to each other through sp^3 hybrid bonds that form a three-dimensional network. Each carbon atom has four nearest-neighbor carbons. Graphite has a layered structure with each layer, called a *graphene sheet*, formed from hexagons of carbon atoms bound together by sp^2 hybrid bonds that form $120°$ angles with each other. Each carbon atom has three nearest-neighbor carbons in the planar layer. The hexagonal sheets are held together by weaker van der Waals forces, which were discussed in Chapter 6.

10.2.2. New Carbon Structures

Until 1964 it was generally believed that no other carbon bond angles were possible in a hydrocarbon, which are compounds containing only carbon and hydrogen atoms. In that year Phil Eaton of the University of Chicago synthesized a square carbon molecule, C_8H_8, called *cubane*, shown in Fig. 10.2a. In 1983 L. Paquette of Ohio State University synthesized a $C_{20}H_{20}$ molecule having a dodecahedron shape, shown in Fig. 10.2b, formed by joining carbon pentagons, and possessing C—C bond angles ranging from $108°$ to $110°$. The synthesis of these hydrocarbon molecules with carbon bond angle values differing from the standard hybridization values of Table 10.1, has important implications for the formation of carbon nanostructures, which would also require different bonding angles.

10.3. CARBON CLUSTERS

10.3.1. Small Carbon Clusters

Laser evaporation of a carbon substrate using the apparatus shown in Fig. 3.2 in a pulse of He gas can be used to make carbon clusters. The neutral cluster beam is

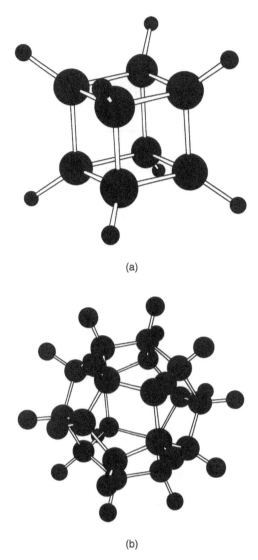

(a)

(b)

Figure 10.2. (a) Cubic structure of cubane C_8H_8; (b) carbon dodecahedron structure of $C_{20}H_{20}$.

photoionized by a UV laser and analyzed by a mass spectrometer. Figure 10.3 shows the typical mass spectral data obtained from such an experiment. For the number of atoms $N < 30$ there are clusters of every N, although some are more prominent than others. Calculations of the structure of small clusters by molecular orbital theory show that the clusters have linear or closed nonplanar monocyclic geometries, as shown in Fig. 10.4. The linear structures that have sp hybridization occur when N is odd, and the closed structure form when N is even. The open structures with 3 11 15 19 and 23 carbons have the usual bond angles and are more prominent and

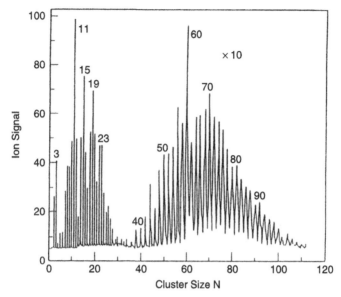

Figure 10.3. Mass spectrum of carbon clusters; the C_{60} and C_{70} fullerene peaks are evident. (With permission from S. Sugano and H. Koizuni, in *Microcluster Physics*, Springer-Verlag, Heidelberg, 1998, p. 123.)

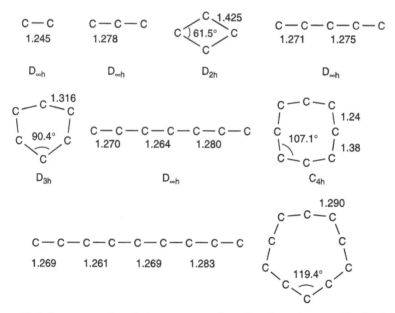

Figure 10.4. Some examples of the structures of small carbon clusters with C—C bond lengths and angles indicated. [With permission from K. Raghavacari et al., *J. Chem. Phys.* **87**, 2191 (1987).]

more stable. The closed structures have angles between the carbon bonds that differ from those predicted by the conventional hybridization concept. Notice in Fig. 10.3 that there is a very large mass peak at 60. The explanation for this peak and its structure earned a Nobel prize.

10.3.2. Buckyball

Donald Huffman of the University of Arizona and Wolfgang Kratschmer of the Max Planck Institute of Nuclear Physics in Heidelberg were interested in understanding the nature of carbon dust in outer space. They simulated the graphite dust in the laboratory and investigated light transmission through it. They created smokelike particles by striking an arc between two graphite electrodes in a helium gas environment, and then condensing the smoke on quartz glass plates. Various spectroscopic methods such as infrared (IR) and Raman spectroscopy, which can measure the vibrational frequencies of molecules, were used to investigate the condensed graphite. They did indeed obtain the spectral lines known to arise from graphite, but they also observed four additional IR absorption bands that did not originate from graphite. Although a soccer-ball-like molecule consisting of 60 carbon atoms with the chemical formula C_{60} had been envisioned by theoretical chemists for a number of years, no evidence had ever been found for its existence. Many detailed properties of the molecule had, however, been calculated by the theorists, including a prediction of what the IR absorption spectrum of the molecule would look like. To the amazement of Huffman and Kratschmer, the four bands observed in the condensed "graphite" material corresponded closely to those predicted for a C_{60} molecule. To further verify this, the scientists studied the IR absorption spectrum using carbon arcs made of the 1% abundant ^{13}C isotope, and compared it to their original spectrum, which arose from the usual ^{12}C isotope. It was well known that this change in isotope would shift the IR spectrum by the square root of the ratio of the masses, which in this case is

$$\left(\frac{13}{12}\right)^{1/2} = 1.041 \tag{10.2}$$

corresponding to a shift of 4.1%. This is exactly what was observed when the experiment was performed. The two scientists now had firm evidence for the existence of an intriguing new molecule consisting of 60 carbon atoms bonded in the shape of a sphere. Other experimental methods such as mass spectroscopy were used to verify this conclusion, and the results were published in the journal *Nature* in 1990. Professor Richard Smalley of Rice University in Houston had built the apparatus depicted in Fig. 3.2, to make small clusters of atoms using high-power pulsed lasers. In this experiment a graphite disk is heated by a high-intensity laser beam, which produces a hot vapor of carbon. A burst of helium gas then sweeps the vapor out through an opening where the beam expands. The expansion cools the atoms, which then condense into clusters. This cooled cluster beam is then narrowed by a skimmer and fed into a *mass spectrometer*, a device designed to measure the mass of molecules in the clusters. When the experiment was done using a graphite disk, the mass spectrometer yielded an unexpected result. A mass number of 720, which would consist of 60 carbon atoms, each of mass 12, was observed. Another piece of evidence for the existence a C_{60} molecule had been found! Although the data from this experiment did not give information about the structure of the carbon cluster, the scientists suggested that the molecule might be spherical, and they built a geodesiclike dome model of it.

10.3.3. The Structure of Molecular C_{60}

The C_{60} molecule has been named fullerene after the architect and inventor R. Buckminister Fuller who designed the geodesic dome that resembles the structure of C_{60}. Originally the molecule was called *buckministerfullerene*, but this name is a bit unwieldy, so it has been shortened to *fullerene*. A sketch of the molecule is shown in Fig. 10.5. It has 12 pentagonal (five-sided) and 20 hexagonal (six-sided) faces symmetrically arrayed to form a molecular ball. As can be seen from Fig. 10.5, every pentagon is surrounded by five hexagons. Each carbon atom is bonded to three other carbon atoms by sp^2 hybrid orbitals as in graphite. However, because of the curvature of the surface there is some admixture of sp^3 orbitals. The average C—C bond length of 1.44 is quite close to that of graphite. Two single bonds are located along the edge of a pentagon having a bond length of 1.46 Å. The double bonds have a length of 1.40 Å. Table 10.2 provides a summary of some of the properties of the isolated C_{60} structure. The structure has a relatively high electron affinity, and an interesting result of this is that the anion of C_{60} is slightly more stable than neutral C_{60} by about 0.74 eV as determined from molecular orbital calculations. Generally the C_{60} anion is present in any batches of C_{60} and can be observed by

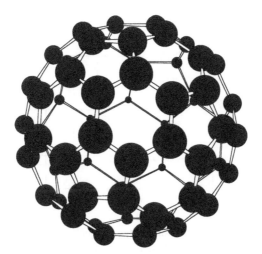

Figure 10.5. Structure of the C_{60} fullerene molecule.

TABLE 10.2. Some Properties of C_{60} Molecule

Property	Magnitude
C—C bond length in pentagon	1.46 Å
C—C bond length in hexagon	1.40 Å
Binding energy per atom	7.4 eV
Electron affinity	2.65 ± 0.05 eV
Cohesive energy per carbon	1.4 eV
Ionization potential	7.58 eV
Optical absorption edge	1.65 eV

electron paramagnetic resonance (EPR). The high electron affinity makes C_{60} a good radical scavenger, which means that it can attract and bond to free radicals. Free radicals are molecules having unpaired electrons. C_{60} is soluble in toluene. Benzyl radicals produced in solutions of toluene in which C_{60} was dissolved were shown by paramagnetic resonance and mass spectrometry to readily form such paramagnetic C_{60} adjuncts as $C_{60}(C_6H_5CH_2)_3$ and $C_{60}(C_6H_5CH_2)_5$. Figure 10.6 shows the energy levels of an isolated C_{60} molecule calculated by a relatively simple molecular orbital method called *Hückel theory*. The labels refer to the symmetry of the orbitals. The top filled level is designated h_u, and the first unfilled level is t_{1u}. Figure 10.7 shows the optical absorption spectra of C_{60} dissolved in *n*-hexane. Three sharp peaks are observed at 3.7 eV (335 nm), 4.6 eV (270 nm), and 5.8 eV (215 nm). The origin of these absorptions can be explained in terms of the energy levels shown in Fig. 10.6. According to the calculations, the likely assignments are that the 5.8 eV absorption is due to the $h_u \rightarrow g_g$ transition as shown by the vertical lines between the energy levels in Fig. 10.6. The 4.6 eV absorption is likely due to the

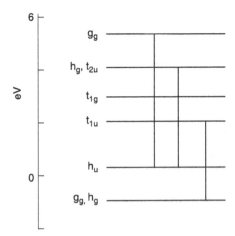

Figure 10.6. The energy levels of an isolated C_{60} molecule calculated by Hückel theory. The empty states are above the zero level and the filled states below it. [Adapted from R. C. Haddon, *Acc. Chem. Res.* **25**, 127 (1992).]

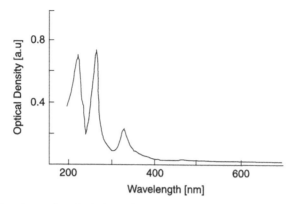

Figure 10.7. Experimental optical absorption spectrum of C_{60} in n hexane. [Adapted from S. Leach et al., *Chem. Phys.* **160**, 451 (1992).]

$g_g \rightarrow t_{1u}$ transition, and the 3.9 eV absorption is associated with the $h_u \rightarrow h_g$ excitation. Above 400 nm there are some much weaker absorptions, which have been discussed in terms of vibrational splitting of the electronic levels.

The atoms in a molecule or nanoparticle continually vibrate back and forth. Each molecule has a specific set of vibrational motions, called *normal modes of vibration*, which are determined by the symmetry of the molecule. For example, carbon dioxide (CO_2), which has the structure O=C=O, is a a linear molecule with three normal modes. One mode involves a bending motion of the molecule. Another, called the *symmetric stretch*, consists of an in-phase elongation of the two C—O bonds. The asymmetric stretch consists of out of phase stretches of the C—O bond length, where one bond length increases while the other decreases. A molecule has $3N-6$ normal modes of vibration. Thus C_{60} will have 174 normal modes of vibration. It turns out that some of these modes are *degenerate*, which means that a different set of atomic motions of the normal modes may have the same frequency. It is found that there are 46 distinct vibrational frequencies of the C_{60} molecule. The vibrational modes of an isolated C_{60} molecule can be classified according to their symmetry using the methods of group theory. It is beyond the scope of this book to go into the details of this, so we will just present the results of the analysis here (Readers interested in understanding how group theory is used to analyze molecular vibrations might consult the book by F. A. Cotton, *Chemical Applications of Group Theory*, Wiley, New York, 1990.) Consider the following formula:

$$\Gamma = 2A_g(1) + 3F_{1g}(3) + 4F_{2g}(3) + 6G_g(4) + 8H_g(5) + A_u(1) + 4F_{1u}(3)$$
$$+ 5F_{2u}(3) + 6G_u(4) + 7H_u(5) \qquad (10.3)$$

The subscripts g and u, refer to even and odd [*gerade* and *ungerade* (German), respectively] according to whether the eigenvectors describing the motion of the atoms are invariant under inversion symmetry. The capital letters are the labels of

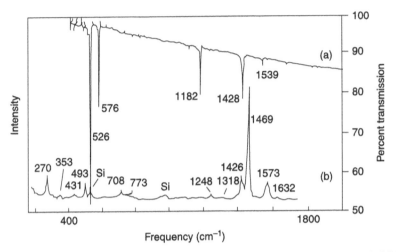

Figure 10.8. First-order infrared (a) and Raman (b) spectra of C_{60} [Adapted from A. M. Rao et al., *Science* **259**, 955 (1992).]

the irreducible representations from group theory of the icosahedral symmetry group I_h. The numbers in parentheses refer to the number of degeneracies. Figure 10.8 shows the Raman and IR spectra of C_{60}. The most intense Raman absorption at $1469 \, cm^{-1}$ is called the *pentagonal pinch mode*. The motion of the atoms in this vibration involves a tangential displacement of the carbon atoms along with a contraction of the pentagonal rings and an expansion of the hexagonal rings. The frequencies of the modes are sensitive to the bonding of other molecules to C_{60}.

10.3.4. Crystalline C_{60}

When C_{60} crystals are grown by vapor sublimination, the structure is FCC as shown in Fig. 10.9. The lattice constant is 14.17 Å. The C_{60} molecules are separated from each other by 10.02 Å. Since 26% of the volume of the unit cell is empty, many different atoms and molecules can be doped into the lattice, some of which have dramatic effects on the properties of solid C_{60}, ranging from making them superconducting to rendering them ferromagnetic. These properties are be discussed in other chapters on ferromagnetism and superconductivity. The C_{60} molecules are bonded to each other in the lattice by van der Waals forces, discussed in Section 3.4.1 The cohesive energy is 1.4 eV, greater than that of inert-gas crystals but less than that of ionic crystals. Table 10.3 lists some of the properties of solid C_{60}.

Nuclear magnetic resonance studies at room temperature indicate that the C_{60} molecules are rotating rapidly with 3 degrees of freedom. Below 261 K there is a phase transition in which the C_{60} molecules lose 2 of their 3 rotational degrees of freedom. Below 261 K the rotation is about the four diagonal axes of the cubic unit cell. The structure of the unit cell changes from a FCC structure above 261 K

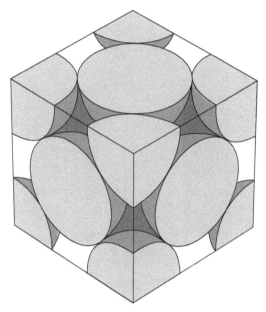

Figure 10.9. Crystal structure of face-centered cubic (FCC) phase of C_{60}.

to a simple cubic structure having a lattice constant of 14.17 Å and four C_{60} molecules per unit cell. This change in the structure of the solid results in changes of some properties of the material such as specific heat, electrical resisitivity, sound velocity, elastic properties, thermal conductivity, and lattice parameter.

Photoconductivity refers to the production of electric current by light. Usually a small voltage is applied across electrodes on the surface of the sample, and the current is measured as a function of the wavelength of the incident light. Figure 10.10 is a plot of the dependence of the magnitude of the photoinduced current on the photon energy of the light for the FCC structure of C_{60} showing the onset of significant current at 1.18 eV corresponding to the transition across the bandgap. Solid C_{60} could be called a *van der Waals semiconductor*.

TABLE 10.3. Some Properties of C_{60} Cubic Solid

Property	Magnitude
Fcc lattice constant	14.17 Å
C_{60}–C_{60} distance	10.02 Å
Density	1.72 g/cm^3
Optical absorption edge	1.7 eV
Workfunction	4.7 eV
Debye temperature	185 K
Melting temperature	1180°C
Sublimation temperature	434°C

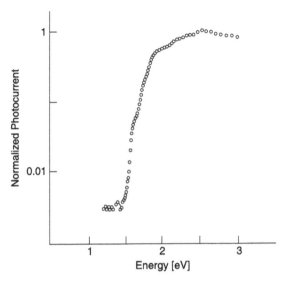

Figure 10.10. Dependence of photocurrent on photon energy of incident light in FCC phase of C_{60}. [Adapted from N. Minami and M. Sato, *Synth. Met.* **56**, 3029 (1993).]

When solid C_{60} is subjected to UV light in vacuum, it polymerizes into short polymers referred to as *oligomers*. Figure 10.11 shows the mass spectrum of a film of C_{60} that has been UV-irradiated for 12 h. The spectrum is obtained by deabsorbing the material from the film in a vacuum using a pulsed laser. The deabsorbed material

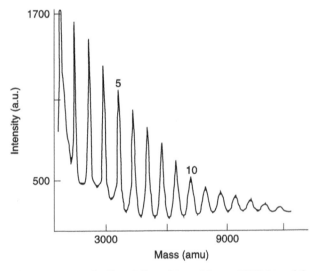

Figure 10.11. Mass spectrum of a film of C_{60} subjected first to *UV* light and then subjected to a pulse laser beam showing peaks at multiples of the mass of C_{60}. [Adapted from A. M. Rao et al., *Science* **259**, 955 (1993).]

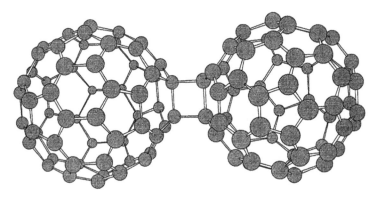

Figure 10.12. Structure of C_{60} dimer formed by photolysis of C_{60}.

is then sent to the analyzer of a mass spectrometer. The method is referred to as *laser deabsorption mass spectrometry* (LDMS). Figure 10.11 shows a series of peaks occurring at mass positions that are multiples of atomic mass 720, which is the atomic mass of C_{60}. The first largest peak is from molecular C_{60}. The second largest peak occurs at mass 1440, which corresponds to the formation of a dimer. The structure of the dimer is shown in Fig. 10.12. When the photopolymerization occurs, the Raman spectra of the pentagonal pinch mode shits from 1469 to 1458 cm^{-1} as shown in Fig. 10.13a for pristine fullerene and Fig. 10.13b for UV-irradiated fullerene. When the photopolymerized material is heated above $\sim$125°C, the intensity of the 1459-cm^{-1} mode begins to decrease and the intensity of the 1469-cm^{-1} mode begins to increase, indicating that the dimer is dissociating into two fullerene molecules. This means that the dimer phase is not stable much above 150°C. Figure 10.14 shows the temperature dependence of the two lines, which allows a determination of the kinetics of the return of the dimer phase to the monomer.

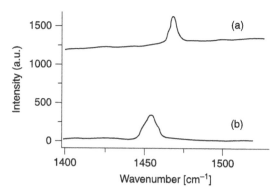

Figure 10.13. Raman spectrum of pentagonal pinch mode of C_{60} before (a) and after (b) photolysis with UV light showing downward shift of the Raman line after photolysis indicating the formation of the polymer phase.

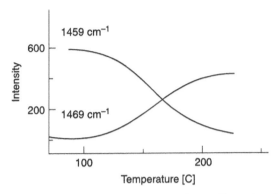

Figure 10.14. A measurement of the temperature dependence of intensity of Raman spectra of pentagonal pinch mode of C_{60} dimer at 1459 cm^{-1} showing that it decreases with increasing temperature with a concomitant growth in intensity of the 1469 cm^{-1} line of the C_{60} molecule. [Adapted from Y. Wang et al., *Chem. Phys. Lett.* **211**, 241 (1993).]

An Arrhenius plot of the ratio of the intensity of the 1459-cm^{-1} line to the intensity of the 1469-cm^{-1} line (I_{1458}/I_{4690}) yields an activation energy of 1.25 eV for the thermal dissociation process.

10.3.5. Larger and Smaller Buckyballs

Larger fullerenes such as C_{70}, C_{76}, C_{80} and C_{84} have also been synthesized. A C_{20} dodecahedral carbon molecule has also been synthesized by gas-phase dissociation of $C_{20}HBr_{13}$, and $C_{36}H_4$ has also been made by pulsed-laser ablation of graphite. A solid phase of C_{22} has been identified in which the lattice consists of C_{20} molecules bonded together by an intermediate carbon atom. One interesting aspect of the existence of these smaller fullerenes is the prediction that they could possibly be superconductors close to room temperature when appropriately doped. We will discuss this in a later chapter.

10.3.6. Buckyballs of Other Atoms

What about the possibility of buckyballs made of other materials such as silicon or nitrogen? Researchers in Japan have managed to make cage structures of silicon. However, unlike carbon atoms, pure silicon cannot form closed structures. The researchers showed that silicon can form a closed structure around a tungsten atom in the form of a hexagonal cage. Potential applications of such structures are components in quantum computers, chemical catalysts, and new superconducting materials. Nitrogen-based buckyballs have not been found, but their existence has been predicted by molecular orbital calculations. Figure 10.15 shows the predicted structure of N_{20} However, synthesis of such nitrogen buckyballs will be difficult.

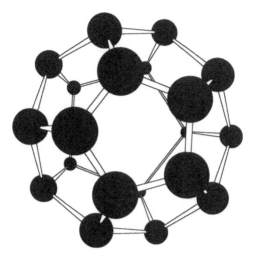

Figure 10.15. Density functional calculated structure of N_{20}. [Adapted from F. J. Owens, *J. Molec. Struct.* **623**, 197 (2003).]

10.4. CARBON NANOTUBES

Carbon nanotubes are perhaps the most interesting of all nanostructures with large application potential. One can think of a carbon nanotube as a graphene sheet rolled into a tube with atoms at the end of the sheet forming the bonds that close the tube. Figure 10.16 shows the structure of a tube formed by rolling the graphite sheet about an axis perpendicular to a C—C bond. A single-walled nanotube (SWNT) can have a diameter of 2 nm and a length of 100 μm, making it effectively a one-dimensional structure called a nanowire.

10.4.1. Fabrication

Carbon nanotubes can be synthesized by laser evaporation, carbon arc methods, and chemical vapor deposition. Figure 10.17 illustrates the apparatus for making carbon nanotubes by laser evaporation. A quartz tube containing argon gas and a graphite target are heated to 1200°C. Located in the tube, but somewhat outside the furnace, is a water-cooled trap. The graphite target contains small amounts of cobalt and nickel, which act as catalytic nucleation sites for the formation of the

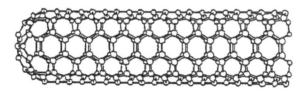

Figure 10.16. Structure of armchair form of carbon nanotubes.

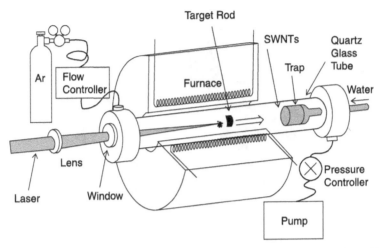

Figure 10.17. Schematic of apparatus for production of carbon nanotubes by laser ablation.

tubes. An intense pulsed-laser beam is incident on the target rod, evaporating carbon from the graphite. The argon then sweeps the carbon atoms from the high-temperature zone to the cold trap on which they condense into nanotubes. Tubes 10–20 nm wide and 100 μm long can be made by this method.

Nanotubes can also be synthesized using a carbon arc. A potential of 20–25 V is applied across carbon electrodes of 5–20 μm, diameter and separated by 1 mm at 500 torr pressure of flowing helium. Carbon atoms are ejected from the positive electrode and form nanotubes on the negative electrode. As the tubes form, the length of the positive electrode decreases, and a carbon deposit forms on the negative electrode. To produce single-walled nanotubes, a small amount of cobalt, nickel, or iron is incorporated as a catalyst in the central region of the positive electrode. Figure 10.18 sketches the apparatus of the carbon arc method. If no catalysts are used, the tubes are nested or *multiwalled nanotubes* (MWNTs), which are nanotubes within nanotubes, as illustrated in Fig. 10.19. The carbon arc method can produce SWNTs of diameters 1–5 nm with a length of 1 μm.

Chemical vapor deposition is the method with the most promise to produce large quantities of carbon nanotubes. Figure 10.20 is an illustration of the equipment needed to synthesize carbon nanotubes by chemical vapor deposition. A quartz glass tube is located in a cylindrical oven. On the bottom of the tube at the center of the oven is a wafer of silicon covered by a material such as $FeSO_4$. Carbon monoxide, hydrogen, and argon are flowed through the glass tube at very specific flow rates as the temperature of the oven is raised in a controlled manner to 1100°C. Other gases such as methane can be used as the source of carbon. The hydrogen reduces the $FeSO_4$ to iron nanoparticles. Cobalt and nickel can also be used. The incoming CO gas decomposes into atomic carbon, which assembles on the iron nanoparticles into carbon nanotubes.

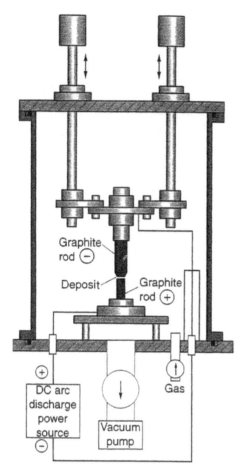

Figure 10.18. Illustration of apparatus for production of carbon nanotubes by the arc discharge method.

The nanotube growth mechanism is not totally understood. Since the metal catalyst is necessary for the growth of single-walled carbon nanotubes, the mechanism must involve the role of the Fe, Co, or Ni atoms. One proposal, referred to as the "scooter mechanism," suggests that atoms of the metal catalyst attach to the dangling bonds at the open end of the tubes, and that these atoms scoot around the rim of the tube, absorbing carbon atoms as they arrive.

Generally when nanotubes are synthesized, the result is a mix of different kinds, some metallic and some semiconducting. Methods are being developed to separate the semiconducting from the metallic nanotubes. In one approach the separation is accomplished by depositing bundles of nanotubes, some of which are metallic and some semiconducting, on a silicon wafer. Metal electrodes are then deposited over the bundle. Using the silicon wafer as an electrode, a small bias voltage is applied that prevents the semiconducting tubes from conducting, effectively making them

MULTI WALLED CARBON NANO TUBES

Figure 10.19. Illustration of a multiwalled carbon nanotube showing one tube nested inside another.

insulators. A high voltage is then applied across the metal electrodes, thereby sending a high current through the metallic tubes but not the insulating tubes. This causes the metallic tubes to vaporize, leaving behind only the semiconducting tubes.

10.4.2. Structure

Although carbon nanotubes are not actually made by rolling graphene sheets, it is possible to explain the different structures by consideration of the way graphene sheets might be rolled into tubes. A nanotube can be envisioned as being formed by rolling a graphene sheet about a vector denoted as T called the *translational vector*, as shown in Fig. 10.21a. The figure shows examples of two such vectors, T_1 and T_2, having different orientations about which the graphene sheet can be rolled to produce the armchair, the zigzag and chiral tubes. Folding about the vector denoted T_2, which is perpendicular to the edges of the hexagons, produces the armchair structure labeled (b) in the figure. Folding about vector T_1, which is parallel to the sides of the hexagons, produces the zigzag structure labeled (c) in the figure. Folding about any other direction produces chiral tubes of the type labeled

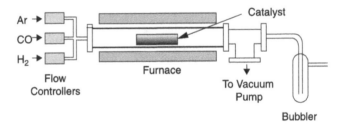

Figure 10.20. Apparatus for production of carbon nanotubes by chemical vapor deposition. (Courtesy of Z. Iqbal.)

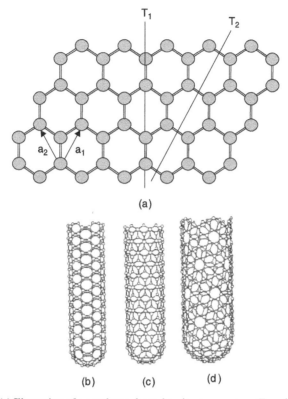

Figure 10.21. (a) Illustration of a graphene sheet showing two vectors, T_2 and T_1, about which the sheet can be folded to form the armchair tube (b) and the zigzag tube (c), respectively. The chiral tube (d) is formed by folding about any other direction.

(d) in the figure. The T_1 vector has the coordinates $(n,0)$, and the T_2 vector has the coordinates (n,n). The circumferential C_h is perpendicular to the vector T about which the sheet is folded. The vector can be written in terms of the vectors $\mathbf{a_1}$ and $\mathbf{a_2}$, which are the unit vectors of the hexagonal two-dimensional lattice of graphene and are shown in Fig. 10.21a:

$$C_h = n\mathbf{a_1} + m\mathbf{a_2} \tag{10.4}$$

The vector determines the tube diameter, which is related to the length of the vector C_h by

$$D = \frac{[3]^{1/2}a_{c-c}\{m^2 + mn + n^2\}^{1/2}}{\pi} \tag{10.5}$$

where a_{c-c} is the C—C distance, which is 1.421 Å in graphite. The chiral angle Θ is the angle between C_h and the $(n,0)$ vector T_1, shown in Fig. 10.21a, which is the direction of folding that produces the zigzag structure. Thus, when $\Theta = 0$, the structure is zigzag, and when it is $\pm 30°$, corresponding to the T_2 vector, the structure is armchair. The carbon nanotubes are metallic or semiconducting depending on the diameter and the particular T vector they are rolled about. The metallic tubes have the armchair structure shown in Fig. 10.21a and are formed from a C_h vector such that $(2n + m)/3$ is an integer. Synthesis generally results in a mixture of tubes, which are two-thirds semiconducting and one-third metallic.

10.4.3. Electronic Properties

The dispersion relationships, the dependence of the energy levels on the k vector, of carbon nanotubes can be understood in terms of the dispersion relationship of a one-dimensional sheet of graphene such as that shown in Fig. 10.21a. The tight-binding model, discussed in Section 8.3.3, can be used to obtain the dispersion relationship assuming only nearest-neighbor interactions. The unit cell of graphene, shown in Fig. 10.22, is rhombic, delimited by the lines connecting the points w,x,y,z in the figure. The fundamental lattice displacements are $\mathbf{a}_1$, which is the dashed line from a to b and has a length $1.42(3)\frac{1}{2} A = 2.46$ Å; and $\mathbf{a}_2$, the dashed line from a to e, which has the same length as $\mathbf{a}_1$. The reciprocal lattice vectors have the magnitude

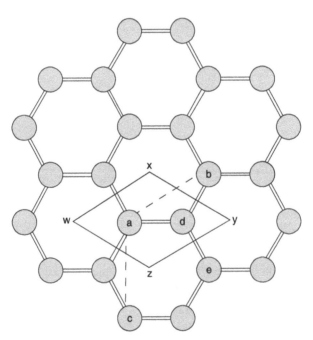

Figure 10.22. Unit cell of graphene sheet delineated by lines joining points $wxyz$. The direct hexagonal lattice vectors $w \rightarrow x$ and $w \rightarrow z$, which define the unit cell of this figure, are, respectively, in the $a \rightarrow b$ and the $a \rightarrow e$ directions. The reciprocal lattice is a triangular lattice with its reciprocal lattice vectors in the $b \rightarrow d$ and the $e \rightarrow d$ directions.

$2/a(3)^{1/2}$ and have the directions *ab* and *ae*. Thus the directions of the reciprocal lattice vectors are different from the unit cell directions in real space, which complicates the task of obtaining the dispersion relationship. The results of the tight-binding model give the dispersion relationship

$$E(k_x,k_y) = \pm t \left[1 + 4\left(\cos\left((3)^{1/2}k_x\frac{a}{2}\right)\cos\left(k_y\frac{a}{2}\right) + 4\cos^2\left(k_y\frac{a}{2}\right) \right]^{1/2} \qquad (10.6)$$

where *t* is the overlap integral. [Derivation of this relationship, which would take us a little far from the main point, is due to P. R. Wallace, *Phys. Rev.* **71**, 622 (1947).] Figure 10.23 is a plot of the dispersion relationship along K_x. Now let us look at the dispersion relationship for a sheet of graphene rolled into the armchair structure. This means that the T_2 vector in Fig. 10.21a is along X and C_h is along y. When the sheet is rolled about T the wavevector associated with C_h, the circumferential vector, can have only discrete values. Another way of looking at this is that the wavelengths of the electron moving around the circumference of the tube can only be a multiple of the circumference of the tube; otherwise they will destructively interfere with each other. The wavevector associated with the T direction, which is the x axis of the tube, remains continuous for a tube of infinite length. The periodic boundary condition for the (n,n) armchair tubes restricts the number of allowed wavevectors K_{xg} in the circumferential direction and is

$$n(3)^{1/2}k_{xq}a = 2\pi q \qquad (10.7)$$

where $q = 1\cdots 2n$. Thus the dispersion relationship for the armchair nanotubes is

$$E(k_x,k_y) = \pm t \left[1 \pm 4\left(\cos\left(\frac{q\pi}{n}\right)\cos\left(\frac{ka}{2}\right) + 4\cos^2\left(\frac{ka}{2}\right) \right]^{1/2} \qquad (10.8)$$

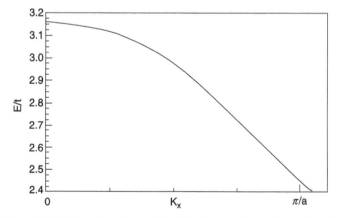

Figure 10.23. The dispersion relationship for a graphene sheet along k_x.

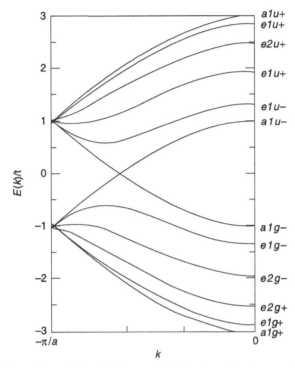

Figure 10.24. Dispersion relationships for armchair structure of carbon nanotubes. (Adapted from R. Saito, G. Dresselhaus, and M. S. Dresselhaus, *Physical Properties of Nanotubes*, Imperial College Press,1998.)

where $-\pi < Ka < \pi$ and $q = 1 \cdots 2n$. The resulting dispersion relationships for the (5,5) armchair tubes are shown in Fig. 10.24. There are six dispersion curves for the conduction and valence bands. Some of the dispersion relationships are degenerate. Those labeled a in the figure are non-degenerate; those labeled e are twofold degenerate. Thus there are 10 dispersion relationships. At the zone boundary $k = -\pi/a$ there is only one band, and it is 10-fold degenerate. This same procedure can be used to obtain the dispersion relationships for the chiral and zigzag tubes.

Figure 10.25 is a plot of the energy gap of semiconducting chiral carbon nanotubes versus the reciprocal of the diameter, showing that as the diameter of the tube increases the bandgap decreases. Scanning tunneling microscopy (STM), which is described in Section 2.3.3, has been used to investigate the electronic structure of carbon nanotubes. In this measurement the position of the STM tip is fixed above the nanotube, and the voltage V between the tip and the sample is swept while the tunneling current I is monitored. The measured conductance $G = I/V$ is a direct measure of the local electronic density of states. The density of states, which is discussed in more detail in Section 9.3.4, is a measure of how close together the energy levels are to each other. Figure 10.26 gives the STM data plotted as the differential conductance, which is $(dI/dV)/(I/V)$, versus the applied voltage

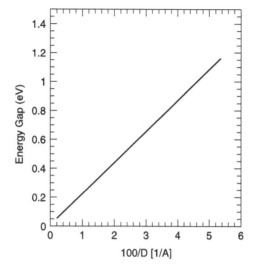

Figure 10.25. Plot of the magnitude of the energy bandgap of a semiconducting, chiral carbon nanotube versus the reciprocal of the diameter of the tube (10 Å = 1 nm). [Adapted from M. S. Dresselhaus et al., *Molec. Mater.* **4**, 27 (1994).]

between the tip and carbon nanotube for metallic (top) and semiconducting tubes. The data show clearly the energy gap in materials at voltages where very little current is observed. The voltage width of this region measures the gap, which for the semiconducting tubes is 0.7 eV.

At higher energies sharp peaks are observed in the density of states, which are referred to as *van Hove singularities*, and they are characteristic of low-dimensional conducting materials. The peaks occur at the bottom and top of a number of sub-bands. As we discussed earlier, electrons in the quantum theory can be viewed as waves. If the electron wavelength is not a multiple of the circumference of the tube, it will destructively interfere with itself, and therefore only electron wave-lengths, that are integer multiples of the circumference of the tubes are allowed. This severely limits the number of energy states available for conduction around the cylinder. The dominant remaining conduction path is along the axis of the tubes, making carbon nanotubes function as one-dimensional quantum wires. The electronic states of the tubes do not form a single wide electronic energy band, but instead split into one dimensional subbands, which are evident in the data in Fig. 10.24.

Electron transport has been measured on individual single-walled carbon nano-tubes. The measurements at a millikelvin ($T = 0.001$ K) on a single metallic nanotube lying across two metal electrodes show steplike features in the current–voltage measurements, as shown in Fig.10. 27. The steps occur at voltages that depend on the voltage applied to a third electrode that is electrostatically coupled to the nano-tube. This resembles a field effect transistor (FET) made from a carbon nanotube, which is discussed below and illustrated in Fig. 10.28. The steplike features in the

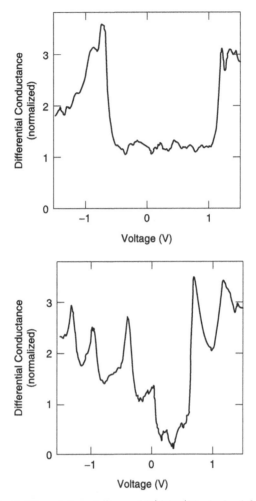

Figure 10.26. Plot of differential conductance $(dI/dV)(I/V)$ obtained from STM measurements of the tunneling current of metallic (a) and semiconducting (b) nanotubes. [With permission from C. Dekker, *Phys. Today* 22 (May 1999).]

$I-V$ curve are due to single-electron tunneling and resonant tunneling through single molecular orbitals. Single-electron tunneling occurs when the capacitance of the nanotube is so small that adding a single electron requires an electrostatic charging energy greater than the thermal energy $k_B T$. Electron transport is blocked at low voltages, which is called a *Coulomb blockade*, and this is discussed in more detail in the section on quantum dots. By gradually increasing the gate voltage, electrons can be added to the tube one by one.

In the metallic state the conductivity of nanotubes is very high. It is estimated that they can carry a billion amperes per square centimeter. Copper wire fails at one million amperes per square centimeter because resistive heating melts the wire. One reason for the high conductivity of carbon tubes is that they have very few

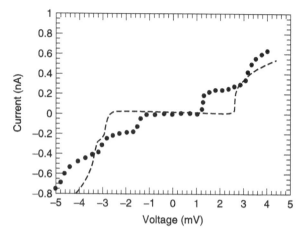

Figure 10.27. Plot of electron transport for two different gate voltages through a single metallic carbon nanotube showing steps in the *I–V* curves. [With permission from C. Dekker, *Phys. Today* 22 (May 1999).]

defects to scatter electrons, and thus a very low resistance. High currents do not heat the tubes the way they heat copper wires. Nanotubes also have a very high thermal conductivity, almost a factor of 2 more than that of diamond. This means that they are also very good heat conductors.

Because of their very high electrical conductivity, carbon nanotubes have a number of interesting application possibilities. The high electrical conductivity of carbon nanotubes means that they will be poor transmitters of electromagnetic energy. A plastic composite of carbon nanotubes could provide lightweight shielding material for electromagnetic radiation. This is a matter of much concern to the military, which is developing a highly digitized battlefield for command, control, and communication. The computers and electronic devices that are a part of this system need to be protected from weapons that emit electromagnetic pulses.

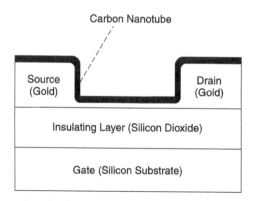

Figure 10.28. Illustration of a field effect transistor (FET) made from a carbon nanotube.

The feasibility of designing FETs, the switching components of computers, on the basis of semiconducting carbon nanotubes connecting two gold electrodes, has been demonstrated. This device is illustrated in Fig. 10.28. When a small voltage is applied to the gate, the silicon substrate, current flows through the nanotube between the source and the drain. The device is switched on when current is flowing, and off when it is not. It has been found that a small voltage applied to the gate can change the conductivity of the nanotube by a factor of $>1,000,000$, which is comparable to silicon FETs. It has been estimated that the switching time of these devices will be very fast, allowing clock speeds of a terahertz, which is 10^4 times faster than the speed of present processors. The gold sources and drains are deposited by lithographic methods, and the connecting nanotube wire is less than 1 nm in diameter. This small size should allow more switches to be packed on a chip. It should be emphasized that these devices have been built in the laboratory one at a time, and methods to produce then cheaply on large scale and connected to each other on a chip will have to be developed before they can be used in applications such as computers.

When a small electric field is applied parallel to the axis of a nanotube, electrons are emitted at a very high rate from the ends of the tube. This is called *field emission*. The effect can easily be observed by applying a small voltage between two parallel metal electrodes and spreading a composite paste of nanotubes on one electrode. A sufficient number of tubes will be perpendicular to the electrode so that electron emission can be observed. One application of this effect is the development of flat-panel displays. Television and computer monitors use a controlled electron gun to impinge electrons on the phosphors of the screen, which then emit light of the appropriate colors. Many companies are developing a flat-panel display using the electron emission of carbon nanotubes. A thin film of nanotubes is placed over control electronics with a phosphor-coated glass plate on top. Another application is to use this electron emission effect to make vacuum-tube lamps.

Magnetoresistance is a phenomenon whereby the resistance of a material is changed by the application of a dc magnetic field. Carbon nanotubes display magneto resistive effects at low temperature. Figure 10.29 shows a plot of the magnetic field dependence of the change in resistance ΔR of nanotubes at 2.3 and 0.35 K compared to their resistance R in zero magnetic field. This is a negative magnetoresistance effect because the resistance decreases with increasing dc magnetic field, so its reciprocal, the conductance $G = 1/R$, increases. This occurs because when a dc magnetic field is applied to the nanotubes, the conduction electrons acquire new energy levels associated with their spiraling motion about the field. It turns out that for nanotubes these levels, called *Landau levels*, lie very close to the topmost filled energy levels (the Fermi level). Thus there are more states available for the electrons to increase their energy, so the material is more conducting.

10.4.4. Vibrational Properties

The phonon dispersion relationships of carbon nanotubes, that is, the dependence of the vibrational frequency on the wavevector, can be obtained by folding the phonon

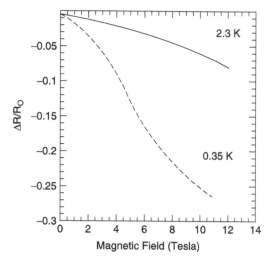

Figure 10.29. Effect of a dc magnetic field on the resistance of nanotubes at the temperatures of 0.35 and 2.3 K. (Adapted from R. Saito, G. Dresselhaus, and M. S. Dresselhaus, *Physical Properties of Nanotubes*, Imperial College Press, 1998.)

dispersion curve of a graphene sheet analogous to the way the electron energy dispersion curves were obtained in the previous section. We will not go through the details of this, but the approach is simply illustrated by Problem 10.6, which examines how the phonon dispersion relationship is modified when a linear chain of atoms is formed into a circle. Figure 10.30 shows the calculated dependence of the vibrational frequencies on the wavevector for a (5,5) armchair nanotube. Because the present synthesis methods lead to a mixture of types of tubes and tubes of different diameters, experimental measurements of phonon dispersion curves have not been made. We will therefore confine our discussion to the frequencies at $k = 0$ that can be measured by Raman and IR spectroscopy. The frequencies of the Raman and IR active vibrations depend on the kind and diameter of the nanotube. For a (10,10) armchair tube, the Raman active modes are designated by their group symmetries as

$$4A_{1g} + 4E_{1g} + 8E_{2g} \tag{10.9}$$

and the IR active modes are $A_{2u} + 7E_{1u}$. The IR active modes have not yet been observed. Figure 10.31 shows the Raman spectra of two carbon nanotube modes. Figure 10.32 illustrates some displacement vectors and the calculated vibrational frequencies of a (10,10) armchair nanotube. The low-frequency modes below 500 cm^{-1} are known as the *radial breathing modes* (RBMs), and their frequencies are sensitive to the diameter of the tube. A Raman spectrum in the radial breathing mode region is shown in Fig. 10.31b. Figure 10.33 is a plot of the A_{1g}, the strongest of the low-frequency modes, versus the radius of the nanotube. This dependence can be used to determine the radius of the nanotube. The data between 3 and 7 Å can be fit to

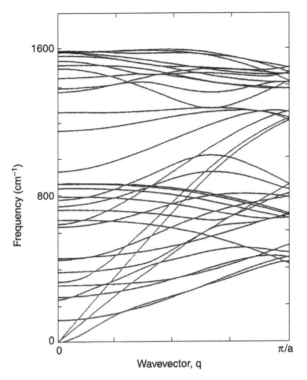

Figure 10.30. Calculated dependence of vibrational frequencies on wavevector for a (5,5) armchair carbon nanotube. (Adapted from R. Saito, G. Dresselhaus, and M. S. Dresselhaus, *Physical Properties of Nanotubes*, Imperial College Press, 1998.)

a power law of the form

$$\omega(cm^{-1}) = 165 \left[\frac{6.785}{R} \right]^{1.0017} \qquad (10.10)$$

where R is the radius in angstroms. Thus the Raman spectrum of the tubes shown in Fig. 10.31b having the A_{1g} mode at $212\,cm^{-1}$ indicates a radius of 5.3 Å. The higher-frequency modes, the E_{1g} at $1585\,cm^{-1}$, the A_{1g} at $1587\,cm^{-1}$, and the E_{2g} at 1591 cm^{-1}, are referred to as the *tangential modes*, and their frequencies are not that sensitive to the radius of the tubes, but they are sensitive to sidegroups bonded to the walls of the tubes, as we will see in the next section. The Raman spectrum of the tangential mode shown in Fig. 10.31a is strong because of the resonance Raman effect discussed earlier. The weaker lower-frequency line in the vicinity of $1320\,cm^{-1}$ is associated with defects in the carbon nanotube structure.

10.4.5. Functionalization

It has been possible to attach other atoms and molecules such as the carboxyl group, COOH, to the walls and ends of carbon nanotubes by a process called

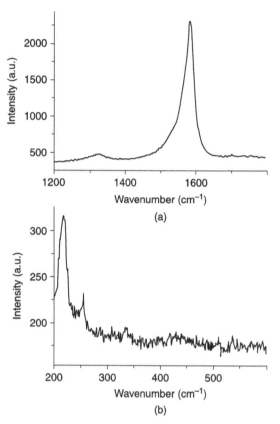

Figure 10.31. Raman spectra of carbon nanotubes showing (a) the higher-frequency tangential mode and (b) the low-frequency radial breathing mode. (F. J. Owens, unpublished.)

functionalization. A few examples of the methods used to functionalize carbon nanotubes will be discussed. This is an important area of research on carbon nanotubes because functionalized tubes are soluble or, more properly stated, form suspensions unlike nonfunctionalized tubes, in many organic solvents and even water. This allows further modification by wet chemistry techniques. Functionalized tubes may also be better in enhancing the mechanical strength of composites because of the potential of the sidegroups to bond to the matrix. Also, it has been shown that functionalized tubes in some instances tend to aggregate less than do nonfunctionalized tubes. This will be discussed in Section 10.4.5. Molecular groups can be attached to the ends of the tubes or the sidewalls. Attachment at the ends of the tubes is the least interesting because the bulk properties of the tubes do not change appreciably because of the limited number of bonding sites available. Functionalization on the sidewalls results in significant modification of properties. For example, tubes that are normally highly conducting become insulating when the sidewalls are fluorinated. The chemistry of sidewall functionalization is not easy because carbon nanotubes are not very reactive, having no available bonds for molecular groups to bind with.

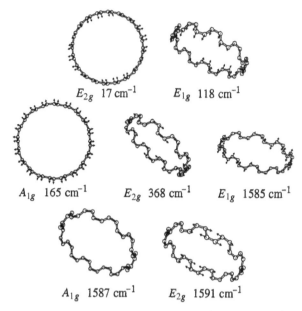

E_{2g} 17 cm^{-1} E_{1g} 118 cm^{-1}

A_{1g} 165 cm^{-1} E_{2g} 368 cm^{-1} E_{1g} 1585 cm^{-1}

A_{1g} 1587 cm^{-1} E_{2g} 1591 cm^{-1}

Figure 10.32. Illustration of normal modes of vibration of carbon nanotubes. (Adapted from R. Saito, G. Dresselhaus, and M. S. Dresselhaus, *Physical Properties of Nanotubes*, Imperial College Press, 1998.)

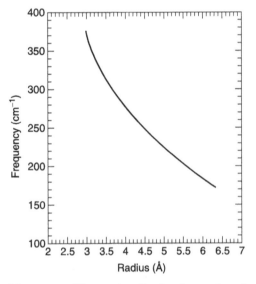

Figure 10.33. Plot of frequency of Raman A_{1g} vibrational normal mode versus radius of the nanotube (10 Å = 1 nm). (Adapted from R. Saito, G. Dresselhaus, and M. S. Dresselhaus, *Phys. Properties of Nanotubes*, Imperial College Press, 1998.)

Defects in the walls, such as a carbon vacancy, however, can play a role in functionalization by providing sites for the bonding of molecular groups. However, there are
not many of these defects on the walls of the tubes. While the synthesized tubes have
very few defects on the walls, the process of removing the Fe, Ni, and Co nanoparticles from the tubes, which involves treatment with nitric acid, can produce defects in
the sidewalls such as the rupture of C—C bonds.

A number of different methods have been used to functionalize carbon nanotubes.
Fluorination of the tubes has been accomplished by methods used to fluorinate graphite. Purified carbon nanotubes are first made into a paper. A colloidal suspension of
carbon nanotubes in a 0.2% aqueous solution of Tritonx-100, a surfactant, is vacuum-
filtered through a fine polytetrafluoroethylene (PTFE) (Teflon) filter membrane, and
then washed with methanol to remove the surfactant. The layer on the filter paper is
then peeled off, forming a paper of carbon nanotubes. The paper is then fluorinated
by vacuum baking at 1100°C for several hours to remove any residual surface con-
taminants. Fluorine gas diluted with helium is then flowed over the paper, which is
heated to 250°C. Infrared spectra of the treated paper show the C—F stretch modes
in the vicinity of $1220-1220 \, \text{cm}^{-1}$. Figure 10.34b shows the Raman spectrum of
the tangential mode in the fluorinated tubes and Fig. 10.34a, the spectra in the nonfluorinated tubes. The tangential mode in the fluorinated tubes has shifted up to
higher frequency by $21 \, \text{cm}^{-1}$. Fluorine has a high electron affinity, making it an electron acceptor, and thus pulls electrons from the tubes. Generally when carbon nanotubes are functionalized with electron acceptors, the tangential mode shifts up. Note
also that the intensity of the tangential mode in the fluorinated tubes is considerably
less than in the nonfluorinated tubes. This is the result of a reduction of resonant
enhancement because when functionalized, the tubes are less one-dimensional.
This is analogous to the reduction in intensity of the Raman active modes of polyacetylene as the chains become shorter, as discussed in Chapter 7.

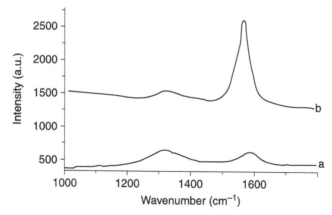

Figure 10.34. Raman spectra of the tangential mode (a) of fluorinated nanotubes and nonfluorinated (b) tubes. (F.J. Owens, unpublished.)

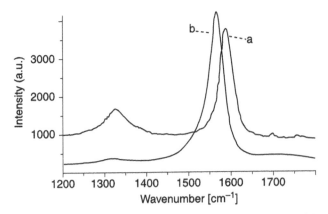

Figure 10.35. Raman spectra of carbon nanotubes (a) treated with nitric acid at 70°C while subjected to sonication and the spectra (b) of the tubes not subjected to this treatment. (F. J. Owens, unpublished.)

Nitration and addition of COOH groups has been achieved by suspending carbon nanotubes in concentrated nitric acid while applying sonication for one hour at 70°C. Figure 10.35 shows the Raman spectrum of the tangential modes of the nitric acid–treated carbon nanotubes showing an upward shift of the mode and a reduction in intensity indicative of functionalization. The IR spectra presented in Fig. 10.36 show that a high degree of COOH functionalization has occurred as evidenced by the strong COOH peaks labeled in the figure. Evidence of NO_2 vibrations is also present, indicating some nitration of the carbon nanotubes. The nitric acid–treated tubes may be soluble in water or form dispersed suspensions, providing further evidence that they are functionalized.

Electrochemical methods can also be used to functionalize carbon nanotubes. Nitration of tubes has been achieved by using carbon nanotube paper as the positive electrode in a chemical cell as illustrated in Fig. 10.37, in which the electrolyte is a 6 M aqueous solution of KNO_2 and the negative electrode is platinum or nickel. Nitration is accomplished by applying a potential of one volt across the electrodes for 3–4 h. The Raman spectra of the tangential mode of the nitrated carbon nanotubes shifted down by 3 cm^{-1} compared to the pristine tubes, and the intensity of the spectrum was reduced by a factor of 6–7 because of the increase in the dimensionality from 1 (as discussed above). Figure 10.38 shows the FTIR spectra of the SWNT paper before (a) and after treatment in the electrochemical cell having the KNO_2 solution. (b) The untreated SWNTs show no IR spectrum, whereas a number of lines are observed in the treated material. The peak at 1500 cm^{-1} can clearly be assigned to the asymmetric stretch of NO_2 providing evidence of nitration. The inset in the figure shows the results of a thermogravimetric (TGA) measurement of the nitrated carbon nanotubes as a function of increasing temperature. The TGA is a measurement of the weight loss of the sample as a function of increasing temperature. The measurement shows that there are two stages of weight loss. There is a loss starting at 150°C

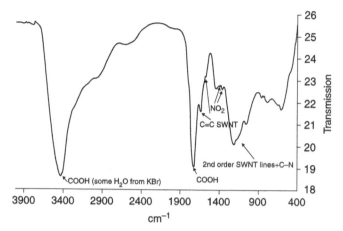

Figure 10.36. Infrared spectrum of nitric acid–treated carbon nanotubes showing evidence for large functionalization with COOH and some evidence of nitration. (F.J. Owens, unpublished.)

and a higher-temperature loss at 450°C. This can be understood by the presence of two different attachment sites for the NO_2 on the sidewall and on the ends of the tubes. Molecular orbital calculations of the bond dissociation energy for the removal of NO_2 from the tubes show that the bond dissociation energy for NO_2 at the end is substantially higher than that for removal from the sidewalls. Thus the first weight loss is likely due to removal of NO_2 from the sidewalls, and the higher-temperature reduction is due to removal of the NO_2 from the ends of the tube.

Fuel cells, which are the hope of the future for generation of electricity without emission of gases that contribute to global warming, require storage of hydrogen.

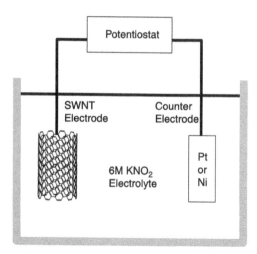

Figure 10.37. Sketch of electric chemical cell used to nitrate carbon nanotubes. [Adapted from Y. Wang et al., *Chem. Phys. Lett.* **407**, 68 (2005).]

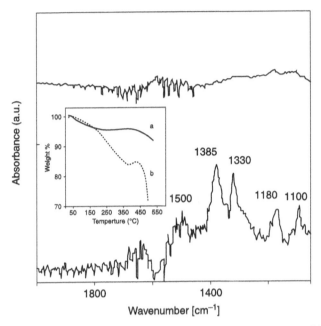

Figure 10.38. Infrared spectrum of pristine SWNT paper (a) and paper treated in an electrochemical cell in a solution of KNO$_3$ (b) The insert is a thermogravimetric measurement of weight loss in electrochemically nitrated carbon. [Adapted from Y. Wang et. al., *Chem. Phys. Lett.* **407**, 68 (2005).]

The potential of carbon nanotubes to store hydrogen is currently under investigation. Initial efforts subjected the carbon nanotubes cooled to 80 K to pressurized hydrogen gas up to 40 bars. Electrochemical methods similar to the process described above for nitrate carbon nanotubes have also been considered. The electrolyte is a 6 M aqueous solution of KOH, but the SWNT paper is now the negative electrode. Electrolysis is carried out for a few hours, generating protons that are attached to the tubes in the carbon nanopaper electrode. Raman scattering measurements of the tangential modes show a sizable drop in the intensity and a 4-cm^{-1} shift of the frequency as shown in Fig. 10.39. These results indicate that the hydrogen is bonding to the carbon nanotubes. It is unclear how the hydrogen is bonded to the tubes. Theoretical modeling has been used to gain some insight into the issue. An important question that the theoretical models would like to answer is how much hydrogen can be stored. The U.S. Department of Energy has set a goal 6.5% weight of hydrogen in the tubes to be useful as a storage system for fuel cells. While theory has not answered this question, it has provided some models for how the hydrogen bonds to the carbon nanotubes. It has been shown that the H atoms bond to the surface of the tubes in arrangements that minimize the loss of C—C π bonds. One predicted arrangement is two zigzag lines of hydrogens on the surface parallel to the axis of the tube as shown in Fig. 10.40a. Another arrangement shows the hydrogens forming an

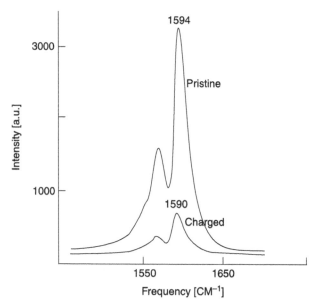

Figure 10.39. Raman spectrum of tangential mode of carbon nanotubes recorded before (pristine) and after (charged) treatment in the KOH electrochemical cell. (From Z. Iqbal, unpublished.)

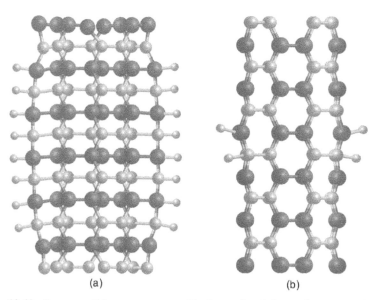

Figure 10.40. Some possible arrangements of hydrogen bonded to carbon nanotudes calculated by molecular orbital theory. [Adapted from G. F. Froudakis, *J. Phys. Condens. Matter* **14**, R453 (2002).]

armchair ring around the circumference of the tube as shown in Fig. 10.40b, which the calculation showed was the lower-energy configuration.

A field effect transistor similar to the one shown in Fig. 10.28 made of the chiral semiconducting carbon nanotubes has been demonstrated to be a sensitive detector of various gases. In one experiment, the transistor was placed in a 500-mL flask having electrical feedthrough and inlet and outlet valves to allow gases to be flowed over the carbon nanotubes of the FET, and 2–200 ppm NO_2 flowing at a rate of 700 mL/min for 10 min caused a threefold increase in the conductance of the carbon nanotubes. Figure 10.41 shows the current–voltage relationship before and after exposure to NO_2. These data were taken for a gate voltage of 4 V. The effect occurs because when NO_2 bonds to the carbon nanotube, charge is transferred from the nanotube to the NO_2, increasing the hole concentration in the carbon nanotube and enhancing the conductance.

10.4.6. Doped Carbon Nanotubes

Boron has one less electron than carbon does. If it is doped into a graphite layer, holes are introduced and the layer becomes a P-type conductor. Nitrogen, on the other hand, has one more electron than carbon does, and when it is doped into the graphite layer, it will contribute electrons, making it an N-type conductor. Thus introducing boron or nitrogen into a carbon nanotube structure should significantly alter the electrical conduction properties of the carbon nanotubes. Figure 10.42 shows the results of a calculation of the density of states in the valence and conduction band region of a

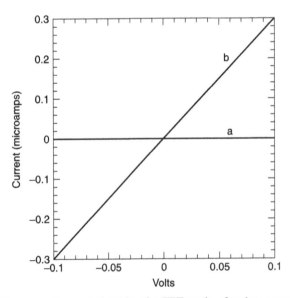

Figure 10.41. Current–voltage relationship of a FET made of carbon nanotubes before (a) and after (b) exposure to NO_2 gas.

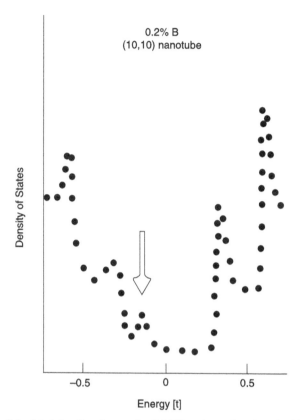

Figure 10.42. Calculated density of states for (10,10) carbon tubes doped with boron. Arrow indicates state in the bandgap introduced by doping. [Adapted from M. Terrones, *Mater. Today* (Oct. 2004).]

(10,10) armchair nanotube doped with 0.2 at % (atomic percent) of boron showing a clear peak (indicated by the arrow in the figure) just above the valence band corresponding to a hole state in the energy gap. Similarly, nitrogen doping introduces a state in the energy gap just below the conduction band.

Various methods have been used to incorporate dopants into carbon nanotubes using the methods for synthesis that were discussed above. For example, nitrogen-doped single-walled carbon nanotubes have been made by chemical vapor deposition using a system similar to that shown in Fig. 10.20. One approach deposits an ethanol solution of $FeCl_3 \cdot H_2O$ on a silicon wafer, which is heated to 60°C. These sheets are then placed at the center of the quartz tube in the oven, which is then heated to 800°C in flowing nitrogen. The oven temperature is then raised to 860°C and H_2 is introduced into the chamber in order to reduce $FeCl_3$. Then ethylenediamine is introduced into the quartz tube and decomposes, producing, among other species, nitrogen. This reaction is allowed to proceed for 10 min and then cooled down to room temperature. CN_x tubes are formed on the silicon substrates. X-Ray photoelectron spectroscopy of

Figure 10.43. High-resolution electron microscope picture of boron-doped multiwalled tubes showing bamboolike structure. [Adapted from M.Terrones, *Mater. Today* (Oct. 2004).]

the tubes confirmed the presence of nitrogen in the tubes with a nitrogen : carbon ratio of 0.02. High-resolution transmission electron microscopy (HRTEM) has been used to study the structure of nitrogen-doped multiwalled tubes made by a chemical vapor deposition (CVD) process involving the decomposition of melamine in the presence of ferrocene at 900–1050°C. Figure 10.43 is the HTREM image of the multiwalled tubes showing a bamboolike structure . The circles around the tubes are disruptions in the graphitelike structure of the tube. The multiwalled nitrogen-doped tubes made from ethylenediamine displayed enhanced electron emission, which occurred at lower fields than did undoped single walled nanotubes. Theoretical calculations have shown that CN_x multiwalled tubes have lower workfunctions than do undoped tubes. The *workfunction* is the energy needed to remove an electron from the tube. The bamboolike structure could also contribute to enhancing the emission current because it may open the graphite sheets, providing ends from which electron emission can occur.

Boron-doped tubes can be produced using CVD methods with B_2O_3 vapor introduced into the quartz tube at temperatures of 1500–1700 K. The tubes have B/C ratios of less than 0.1. The boron-doped tubes generally have the zigzag structure. The armchair structure is rarely observed. Figure 10.44 shows the Raman spectra of single-walled tubes using a 514.5-nm laser as a function of percentage of boron in the tubes. The Raman line, in the vicinity of $1334\,\mathrm{cm^{-1}}$, is not observed in high-quality undoped carbon nanotubes, but appears when the translation symmetry of the tubes is broken because of the presence of defects. It is seen that the intensity of this Raman line increases with boron concentration. The frequency of the radial breathing modes does not change much with boron doping, indicating that the radius of the tube is not affected by the concentration of boron in the tubes. However, the radial breathing modes disappear at concentrations greater than 3%. The boron-doped tubes also display better field emission properties than do the undoped tubes. Figure 10.45 is a plot of the current density of the emission versus electric field for doped and undoped multiwalled tubes showing that emission begins at lower voltages in the doped tubes.

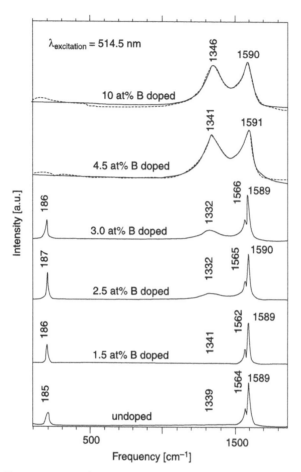

Figure 10.44. Raman spectra of boron-doped carbon nanotubes for different amounts of boron doping. [Adapted from J. Maultzsc et al., *Appl. Phys. Lett.* **81**, 2647 (2002).]

10.4.7. Mechanical Properties

Carbon nanotubes are very strong. If a weight W is attached to the end of a thin wire nailed to the roof of a room, the wire will stretch. The stress S on the wire is defined as the load, or the weight per unit cross sectional area A of the wire

$$S = \frac{W}{A} \qquad (10.11)$$

The strain e is defined as the amount of stretch ΔL of the wire per unit length L

$$e = \frac{\Delta L}{L_0} \qquad (10.12)$$

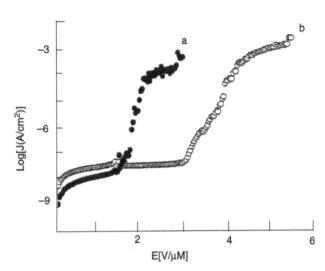

Figure 10.45. Emission current versus electric field for (a) boron-doped and (b) undoped multiwalled carbon nanotubes. [Adapted from J. C. Charlier, *Nano. Lett.* **2**, 1191 (2002).]

where L_0 is the length of the wire before the weight is attached. Hooke's law states that the increase in length of the wire is proportional to the weight at the end of the wire. More generally, we say that stress S is proportional to strain e:

$$S = Ee \qquad (10.13)$$

The proportionality constant $E = L_0 W / A \, \Delta L$ is *Young's modulus*, which is a property of a given material. It characterizes the elastic flexibility of a material. The greater the Young's modulus value, is the less flexible the material. Young's modulus of steel is about 30,000 times that of rubber. Carbon nanotubes have Young's moduli of approximately 0.64 TPa. (One terapascal (TPa) is a pressure very close to 10^7 times atmospheric pressure.) Young's modulus of steel is 0.21 TPa, which means that Young's modulus of carbon nanotubes is almost 3 times that of steel. This would imply that carbon nanotubes are very stiff and difficult to bend. However, this is not quite true because they are hollow and have very thin walls in the order of 0.34 nm.

When carbon nanotubes are bent they are very resilient. They buckle like straws but do not break, and can be straightened back without any damage. Most materials fracture on bending because of the presence of defects such as dislocations or grain boundaries. Because carbon nanotubes have so few defects in the structure of their walls, this does not occur. Another reason why they do not fracture is that as they are bent sharply, the almost hexagonal carbon rings in the walls change their structure rather than breaking. This is a unique result of the fact that the carbon–carbon bonds are sp^2 hybrids, and these sp^2 bonds can rehybridize as they are bent. The degree of change and the amount of s and p admixture both depend on the extent to which the bonds are bent.

Strength is not the same as stiffness. Young's modulus is a measure of how stiff or flexible a material is. Tensile strength is a measure of the amount of stress needed to pull a material apart. The tensile strength of carbon nanotubes is about 37 billion pascals (Pa). High-strength steel alloys break at about 2 billion Pa. Thus carbon nanotubes are about 19 times stronger than steel.

Nested nanotubes also have improved mechanical properties, but they are not as good as their single-walled counterparts. For example, multiwalled nanotubes of 200 nm diameter have a tensile strength of 0.007 TPa (i.e., 7 GPa), and a modulus of 0.6 TPa.

10.5. NANOTUBE COMPOSITES

10.5.1. Polymer–Carbon Nanotube Composites

Increasing the strengths of materials such as plastics and metals by incorporation of long carbon fibers such as polyacrylonitrile (PAN) is an established technology. The factors that determine the degree of enhancement of the strength of materials when fibers are incorporated into them are (1) the ratio of the length of the fiber to the diameter, called the *aspect ratio*; (2) the ability of the fiber to bond to the material; (3) the alignment of the fiber in the material; (4) the inherent tensile strength of the fiber itself; and (5) a good dispersion of the fibers in the material. Since carbon nanotubes have the highest tensile strength of any known fiber and the greatest length : diameter ratio, they should be the ultimate reinforcing fiber.

Thus one of the important applications of carbon nanotubes that is under investigation is their use as reinforcing fibers. Addition of carbon nanotubes to such polymers as polystyrene and poly (vinyl alcohol) has been shown to increase the yield strength, hardness, and electrical and thermal conductivity of the polymers. A polyacrylonitrile (PAN)–carbon nanotube composite has been fabricated by suspending PAN and 3–20% by weight carbon nanotubes in an organic solvent. The organic liquid is allowed to slowly evaporate while being subjected to sonication. Next the residue is dried for a few hours at 80°C to remove any remaining liquid. The residue is then pressed into a pellet. Both nonfunctionalized and functionalized carbon nanotubes have been incorporated into the polymer by this process. Raman measurements of the tangential mode of the carbon nanotube in the composite show that it has been shifted up from $1571 \, \text{cm}^{-1}$ in the pristine tubes to $1576 \, \text{cm}^{-1}$ in the composite, indicating interaction between the tube and the polymer. In composites of PAN made with fluorinated carbon nanotubes, the frequency of the tangential mode has been observed to shift down by $10 \, \text{cm}^{-1}$. Hardness is measured by placing a weighted indenter on the sample. The harder the material, the less the indenter sinks into the sample. It has been shown that hardness correlates to yield strength. Figure 10.46 is a plot of the hardness versus the percent weight of nonfluorinated and fluorinated carbon nanotubes in PAN showing that the hardness increases

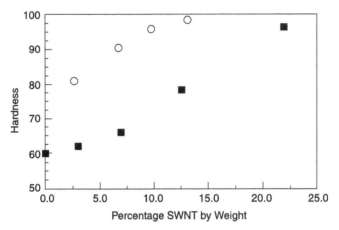

Figure 10.46. Hardness of polyacrylonitrile–carbon nanotube composites versus percent weight of fluorinated (O) and nonfluorinated carbon nanotubes (■) in composite. [F. J. Owens, *Mater. Lett.* **59**, 3720 (2005).]

with higher percentage of the nanotubes in the composite. The fluorinated tubes are more effective in increasing the hardness. The fluorinated tubes have approximately one fluorine for every two carbons, meaning that the molecular weight is 1.8 times higher than that of nonfunctionalized tubes. In a comparison of composites having the same number of tubes with nonfluorinated composites, the percentage of the fluorinated tubes was 1.8 times that of the nonfluorinated tubes. The observed increase in hardness is not optimal because the tubes have not been preferentially aligned in the polymer. Further, even though sonication is applied, the nanotubes remain bundled to some extent. This results in a reduction in the degree of enhancement of hardness because the aspect ratio of the bundles is lower than that of the isolated tubes and because there may be some slippage of tubes in the bundles under stress. The enhancement of hardness using the fluorinated tubes may be due to intertube bonding in the bundles and/or better bonding to the polymer matrix. Thus, while carbon nanotubes may show high potential to enhance the mechanical strength of polymers, optimum enhancement has not yet been achieved.

Because the nonfunctionalized tubes have high electrical and thermal conductivity, incorporating the tubes into polymers can be useful in making electrically and thermally conducting polymer materials, which can have a number of applications. Figure 10.47 shows a measurement of the current–voltage relationship of PAN with 12% by weight single-walled carbon nanotubes synthesized by the process described above. The measurement is made with four connections to a disk of the material. Two of the connections are connected to a current source, and the other two are to a voltmeter. The four-probe technique eliminates the effect of the contact resistance between the wires and the sample. The voltage–current relationship is seen as a straight line, indicating that the conductivity obeys Ohm's law. The resistivity of the sample is calculated from the data to be 5.1 Ω·cm.

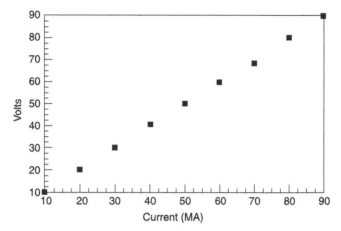

Figure 10.47. Measurement of the current–voltage relationship of carbon nanotube–polyacrylonitrile composite. [F.J. Owens, *Mater. Lett.* **59**, 3720 (2005).]

Composites made of the fluorinated polymers are not conducting because the fluorinated carbon nanotubes are not conducting. The ability to make polymer materials electrically conducting may be a near-term application of carbon nanotubes. It could be the basis of light plastic materials used to shield against electromagnetic pulses, or provide interconnects for flexible circuitry.

10.5.2. Metal–Carbon Nanotube Composites

It would be highly desirable to develop methods to incorporate single-walled or, for that matter, multiwalled carbon nanotubes into metals such as iron and aluminum and increase the yield strength of the metal without a significant increase in the weight of the metals. The applications of such stronger metals are many, ranging from stronger metals for car bodies and airplanes as well as armor on military vehicles. Progress in this area has been slow. The first successful fabrication of carbon nanotube–metal composite displaying enhanced yield strength was reported in 2006. The composites were fabricated by growing carbon nanotubes directly into low-density pellets of iron by the CVD method described earlier and shown in Fig. 10.20. The density of the iron pellets was controlled by the pressure used to press the pellets. The catalysts such as iron acetate were mixed with the iron powder used to make the pellet. The pellets were placed in the quartz tube of a CVD oven such as the one shown in Fig. 10.20. The tube was evacuated to 10^{-3} torr and heated to 500°C to decompose the iron acetate to iron oxide. Hydrogen gas was then introduced to reduce the iron oxide to nanosized iron. The temperature was then raised to 700°C and carbon monoxide was flowed through the tube for 30 min. The carbon monoxide decomposed into atomic carbon, which then formed carbon nanotubes about the iron nanoparticles in the pores of the pellet. Reference pellets without the catalysts were also subjected to the same treatment. Figure 10.48 shows the stress–strain curve of the carbon

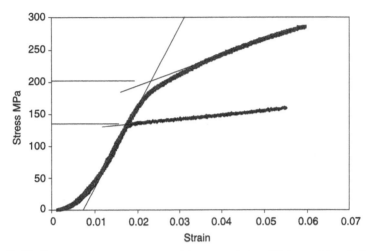

Figure 10.48. Stress–strain measurement of iron containing 2% by weight single-walled carbon nanotubes (top) curve and the reference sample (bottom) curve showing a 45% increase in yield strength for the iron containing the carbon nanotubes. [A. Goyal et al., *J. Mater. Res.* **21**, 522 (2006).]

nanotube metal composite (upper curve) and the reference pellet of iron. The yield strength is the value of stress where the stress–strain curve deviates from linearity indicated by the horizontal lines in the figure A 45% increase in yield strength compared to the reference sample is obtained in an iron pellet containing 1% by weight of single-walled carbon nanotubes. The TEM picture in Fig. 10.49 provides a possible explanation of why the strength is enhanced. It is seen that the carbon nanotubes stretch across the pore and are bonded to the walls of the pore, which contain the iron nanoparticles. It may be that the enhanced strength of the composite is due to the mechanical support provided to the pores by the bonding of the tubes to the walls of the pores.

10.6. GRAPHENE NANOSTRUCTURES

A monolayer of graphite is called *graphene*. It has been suggested that graphene sheets of nanodimensions, such as the graphene nanoribbon shown in Fig. 10.50, could have unique electronic properties comparable to those of single-walled carbon nanotubes. The graphene nanoribbon shown in Fig. 10.50 has 52 carbon atoms and is 2.3 nm long and 0.53 nm wide. The structures can be made on the (0001) face of silicon carbide by thermal deabsorption of the silicon from the surface. The surfaces are prepared by H_2 etching and heated by an electron beam in a very high vacuum to about 1000°C to remove oxide layers. The samples are then heated to 1250–1450°C for 1–20 min, producing graphene layers. The thickness of the layers is determined by the final temperature used. These nanostructures

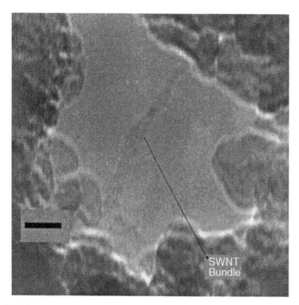

Figure 10.49. TEM image of the surface of the iron pellet showing a pore in the pellet and a carbon nanotube bundle stretching across it and bonded to the walls of the pore. The scale bar is 10 nm. [A. Goyal et al., *J. Mater. Res.* **21**, 522 (2006).]

can have very interesting electronic properties that have potential for electronic applications. Figure 10.51 is a plot of the ionization energy versus the number of carbons in the ribbon calculated using molecular orbital theory. The ionization energy drops significantly with the length of the ribbon and at the longest length calculated is less than that of single-walled carbon nanotubes. The ionization energy corresponds to the workfunction. The workfunctions of single walled carbon nanotubes are in the order of 4.8–4.6 eV, depending on the diameter of the tubes. For example, a nanoribbon having 52 carbon atoms has a calculated ionization energy of 0.67 eV, which is considerably less than that of carbon nanotubes. This suggests that the graphene ribbons could be better field-induced electron emitters than carbon nanotubes, allowing some interesting application possibilities. Figure 10.52 is a plot of the calculated bandgap

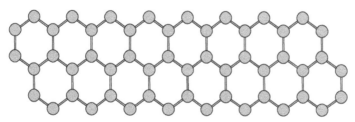

Figure 10.50. Calculated structure of a graphene nanoribbon containing 52 carbon atoms. [Adapted from F.J.Owens, *Molec. Phys.* **104**, 3107 (2006).]

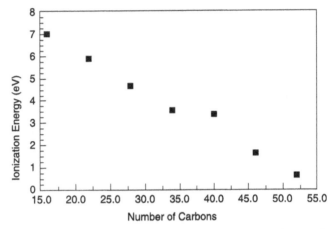

Figure 10.51. Calculated ionization energy versus number of carbons in a graphene nanoribbon. [Adapted from F. J. Owens, *Molec. Phys.* **104**, 3107 (2006).]

as a function of length of the ribbon showing that the energy gap decreases with
increasing length. The bandgap is taken as the energy separation between the
highest occupied molecular orbital (HOMO) and the lowest unoccupied molecular
orbital (LUMO). These results suggest that the length of the ribbons in the nanometer
region can be used to engineer the electronic properties of the ribbons from insulating, to semiconducting, to metallic behavior. However, isolated graphene nanorib-
bons have not yet been fabricated. As discussed above, most of the graphene
materials made to date have been on the surface of SiC, and are rarely monolayers.
Figure 10.53 shows a measurement of the derivative of the tunneling current

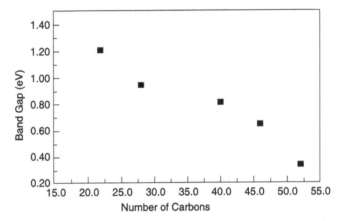

Figure 10.52. Calculated bandgap at $K = 0$ for graphene ribbon versus the number of carbons
in the ribbon. [Adapted from F. J. Owens, *Molec. Phys.* **104**, 3107 (2006).]

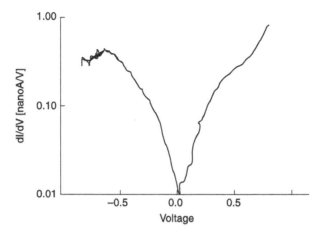

Figure 10.53. Measurement of derivative of tunneling current dI/dV using a scanning tunneling microscope on a sheet of graphene. [Adapted from C. Berger et al., *J. Phys. Chem.* **108**, 1912 (2004).]

dI/dV, obtained by a scanning tunneling microscope for a graphene layer on SiC showing a zero bandgap. The estimated size of this structure is ~ 20 nm.

If we have a rectangular slab of a conducting material whose large facet is parallel to the xy plane of a coordinate system with an electric field applied along x, the current will flow along the x direction. If a dc magnetic field is applied perpendicular to the facet along z, the conducting electrons are deflected to the bottom of the slab, which becomes negatively charged while the top is positively charged. Thus an electric field is produced by the dc magnetic field along z, perpendicular to the surface of the slab. The strength of the electric field is proportional to the strength of the dc magnetic field. This is known as the *Hall effect*. Figure 10.54 illustrates a measurement of the conductance of a graphene film consisting of three monolayers of graphene at 5 K for the dc magnetic field parallel and perpendicular to the surface of the film. The two-dimensional nature of the film is clearly shown by the fact that when the magnetic field H is parallel to the surface, (b) the magnetic field does not affect the conductance, whereas there is a significant effect (a) when the magnetic field is perpendicular to the surface.

Another approach to making graphene is by mechanically transferring thin flakes of the material from highly ordered pyrolytic graphite (HOPG) onto a smooth silicon substrate. The flakes are removed from the HOPG by scotch tape and transferred to the (111) surface of a silicon crystal. An atomic force microscope can be used to determine the number of graphene layers in the flake.

Films made in this manner were studied by Raman spectroscopy and the properties of the spectra correlated to the number of graphene layers in the film. The largest Raman line occurs in the vicinity of 1582 cm^{-1}, which is close to the frequency of the tangential mode (the G mode) in single-walled carbon nanotubes. It was found

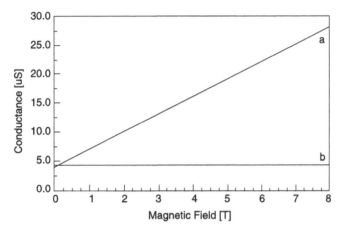

Figure 10.54. Measurement of the Hall conductance versus dc magnetic field for a triple monolayer of grapheme for the field (b) parallel and (a) perpendicular to the surface of the film. [Adapted from C. Berger et al., *J. Phys. Chem.* **108**, 1912 (2004).]

that the frequency of the G mode depended on the number of graphene layers in the material. Figure 10.55 is a plot of the frequency of the G band versus $1/N$ where N is the number of layers of graphene. This measurement can be used to estimate the number of graphene layers in a film.

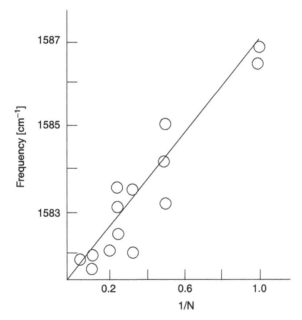

Figure 10.55. Measurement of Raman frequency of G mode versus reciprocal of number of graphene layers in the material. [Adapted from A. Gupta et al., *Nano. Lett.* **6**, 2667 (2006).]

PROBLEMS

10.1. Carbon dioxide is a linear molecule, and NO_2 is a bent by $134°$. What kind of sp^n-hybridization exists in each molecule and why?

10.2. From Table 10.1 construct a plot of $1/\lambda^2$ versus the cosine of the bond angle. From this express a general relationship between λ and the bond angle. If the bond angle of NO_2 is $134°$, what is the valence sp^n-hybrid orbital?

10.3. The cohesive energy of the face-centered cubic (FCC) structure of C_{60} is 1.6 eV. The lattice parameter is $14.17\,\text{Å}$. The nearest-neighbor C_{60} separation is $10.02\,\text{Å}$. Assuming just nearest-neighbor interactions, construct a Lennard-Jones potential for this lattice. Then use this potential to calculate the cohesive energy of the lower temperature simple cubic phase, assuming a $10.0\,\text{Å}$ separation.

10.4. Using the tight-binding model, obtain an expression for the dispersion relationship of a chain of length L with an even number of carbon atoms that forms a circle.

10.5. Consider a hypothetical two-dimensional square lattice of lithium atoms having lattice parameter a. Using the tight-binding model, calculate the dispersion relationship for E versus k. Now fold the lattice into a tube having a T vector in the $[a,0]$ direction and having a circumference of $10a$. Obtain the dispersion relationship for the tube and plot E versus k_y.

10.6. Make a drawing of the tube from Problem 10.5, then draw another tube formed by folding about the $[11]$ direction of the two-dimensional lattice

10.7. Provide a simple explanation for why the bond between an NO_2 and a carbon nanotube is stronger when the molecule is bonded to the end of the tube compared to the side of the tube. How does the bonding on the side of the tube differ from that on the end? How would a carbon vacancy on the sidewall contribute to the bonding?

10.8. It has been reported that the dispersion relationship for the band energies of graphene is linear, having the form $E(K) = \left(\frac{3}{2}\right) t\, a\, |k|$, where a is the lattice parameter and t is a constant. Obtain the group velocity and reduced mass along k_x. Discuss some of the implications of the result for the reduced mass such as the effects on the density of states at the Fermi level and on the electrical conductivity.

CHAPTER 11

Bulk Nanostructured Materials

In this chapter the properties of bulk nanostructured materials are discussed. Bulk nanostructured materials are solids having a nanosized microstructure. The basic units that make up the solids are nanoparticles. The nanoparticles can be disordered with respect to each other, where their symmetry axes are randomly oriented and their spatial positions display no symmetry. The particles can also be ordered in lattice arrays displaying symmetry. Figure 11.1a illustrates a hypothetical two-dimensional ordered lattice of Al_{12} nanoparticles, and Fig. 11.1b shows a two-dimensional bulk disordered nanostructure of these same nanoparticles.

11.1. SOLID METHODS FOR PREPARATION OF DISORDERED NANOSTRUCTURES

11.1.1. Methods of Synthesis

In this section we discuss some of the ways that disordered nanostructured solids are made. One method is referred to as *compaction and consolidation*. As an example of such a process, let us consider how nanostructured Cu–Fe alloys are made. Mixtures of iron and copper powders having the composition $Fe_{85}Cu_{15}$ are ballmilled for 15 h at room temperature. The material is then compacted using a tungsten–carbide die at a pressure of 1 GPa for 24 h. This compact is then subjected to hot compaction for 30 min at temperatures in the vicinity of 400°C and pressure ≤ 870 MPa. The final density of the compact is 99.2% of the maximum possible density. Figure 11.2 presents the distribution of grain sizes in the material showing that it consists of nanoparticles ranging in size from 20 to 70 nm with the largest number of particles having 40-nm sizes. Significant modifications of the mechanical properties of disordered bulk materials having nanosized grains is one of the most important properties of such materials. Making materials with nanosized grains has the potential to provide significant increases in yield stress, and has many useful applications such as stronger materials for automobile bodies. The reasons for the changes in mechanical properties of nanostructured materials will be discussed in the next chapter.

The Physics and Chemistry of Nanosolids. By Frank J. Owens and Charles P. Poole, Jr.
Copyright © 2008 John Wiley & Sons, Inc.

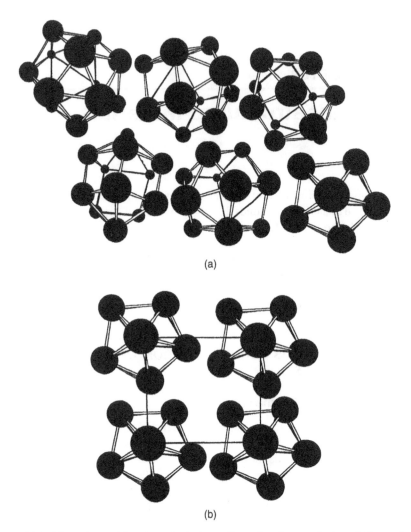

Figure 11.1. Illustrations of (b) a hypothetical two-dimensional square lattice of Al_{12} particles and (a) a two-dimensional bulk solid of Al_{12} where the nanoparticles have no ordered arrangement with respect to one another.

Nanostructured materials can be made by rapid solidification. One method, illustrated in Fig. 11.3, is called *chill block melt spinning*. Radiofrequency (RF) heating coils are used to melt a metal, which is then forced through a nozzle to form a liquid stream. This stream is continuously sprayed over the surface of a rotating metal drum under an inert-gas atmosphere. The process produces strips or ribbons ranging in thickness from 10 to 100 μm. The parameters that control the nanostructure of the material are nozzle size, nozzle–drum distance, melt ejection pressure, and speed of rotation of the metal drum. The need for lightweight, high-strength materials has lead to the development of 85–94% aluminum alloys with other metals such

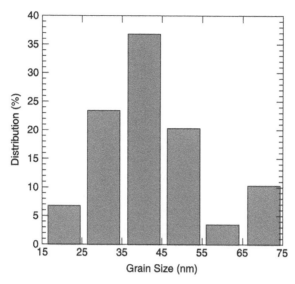

Figure 11.2. Frequency of occurrence of sizes of Fe–Cu nanoparticles made by hot compaction methods described in the text. [With permission from L. He and E. Ma, *J. Mater. Res.* **15**, 904 (2000).]

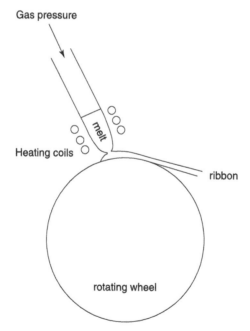

Figure 11.3. Illustration of the chill block melting apparatus for producing nanostructured materials by rapid solidification on a rotating wheel. (With permission from I. Chang, in *Handbook of Nanostructured Materials and Nanotechnology*, edited by H. S. Nalwa, ed., Academic Press, Boston, 2000, vol. 1, Chapter 11, p. 501.)

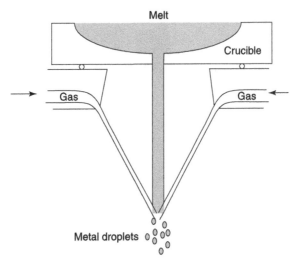

Figure 11.4. Illustration of apparatus for making droplets of metal nanoparticles by gas atomization. (With permission from I. Chang, in *Handbook of Nanostructured Materials and Nanotechnology*, H. S. Nalwa, ed., Academic Press, Boston, 2000, Vol. 1, Chapter 11, p. 501.)

as Y, Ni, and Fe made by this method. A melt spun alloy of Al–Y–Ni–Fe consisting of 10–30-nm Al particles embedded in an amorphous matrix can have a tensile strength in excess of 1.2 GPa. This high value is attributed to the presence of defect-free aluminum nanoparticles.

Another method of making nanostructured materials is *gas atomization*, in which a high-velocity inert-gas beam impacts a molten metal. The apparatus is illustrated in Fig. 11.4. A fine dispersion of metal droplets is formed when the metal is impacted by the gas, which transfers kinetic energy to the molten metal. This method can be used to produce large quantities of nanostructured powders, which are then subjected to hot consolidation to form bulk samples.

Nanostructured materials can be made by electrodeposition. For example, a sheet of nanostructured Cu can be fabricated by putting two electrodes in an electrolyte of $CuSO_4$ and applying a voltage between the two electrodes. A layer of nanostructured Cu will be deposited on the negative titanium electrode. A sheet of Cu 2 mm thick can be made by this process, having an average grain size of 27 nm, and an enhanced yield strength of 119 MPa.

11.1.2. Metal Nanocluster Composite Glasses

One of the oldest applications of nanotechnology is the colored stained-glass windows in medieval cathedrals, which are a result of nanosized metallic particles embedded in the glass. Glasses containing a low concentration of dispersed nanoclusters display a variety of unusual optical properties that have application potential. The peak wavelength of the optical absorption, which determines the color, depends on

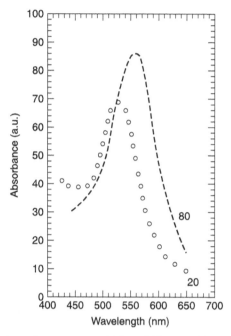

Figure 11.5. Optical absorption spectrum of 20- and 80-nm gold nanoparticles embedded in glass. [Adapted from F. Gonella et al., in *Handbook of Nanostructured Materials and Nanotechnology*, H. S. Nalwa, ed., Academic Press, Boston, 2000, Vol. 4, Chapter 2, p. 85.]

the size and the type of metal particle. Figure 11.5 shows an example of the effect of size of gold nanoparticles on the optical absorption properties of an SiO_2 glass in the visible region. The data confirm that the peak of the optical absorption shifts to shorter wavelengths when the nanoparticle size decreases from 80 to 20 nm. In Section 7.10 it was shown that in metals or semiconductors, the electron cloud oscillates with respect to the fixed positive ions of the lattice as shown in Fig. 7.24, and has a specific frequency. These plasma oscillations can be excited by an incident light beam, provided that the nanoparticle is smaller than the wavelength of the incident light. A theory developed by Mie may be used to calculate the absorption coefficient versus the wavelength of the light. The absorption coefficient α of small spherical metal particles embedded in a nonabsorbing medium is given by

$$\alpha = \frac{18\pi N_s V n_0 \varepsilon_2^3 / \lambda}{[\varepsilon_1 + 2n_0^2]^2 + \varepsilon_2^2} \tag{11.1}$$

where N_s is the number of spheres of volume V, ε_1 and ε_2 are the real and imaginary parts ($\varepsilon = \varepsilon_1 + i\varepsilon_2$) of the dielectric constant of the spheres, n_0 is the refractive index of the insulating glass, and λ is the wavelength of the incident light.

Another technologically important property of metallic glass composites is that they display nonlinear optical effects, which means that their refractive indices depend on the intensity of the incident light. The glasses have an enhanced third-order susceptibility that results in an intensity-dependent refractive index n given by

$$n = n_0 + n_2 I \qquad (11.2)$$

where I is the intensity of the light beam. Nonlinear optical effects have a potential application as optical switches, which would be a major component of photonic-based computers. When metal particles are less than 10 nm in size, confinement effects become important, and these alter the optical absorption properties. Quantum confinement is discussed in Section 9.3.6.

The earliest methods for making composite metal glasses involved mixing metal particles in molten glasses. However, it is difficult to control the properties of the glasses, such as the aggregation of the particles. More controllable processes have been developed such as ion implantation. Essentially the glasses are subjected to an ion beam consisting of atoms of the metal to be implanted, having energies ranging from 10 keV to 10 MeV. Ion exchange is also used to insert metal particles into glasses. Figure 11.6 shows an experimental setup for an ion exchange process designed to put silver particles in glasses. Monovalent surface atoms such as sodium present near the surface of all glasses are replaced by other ions such as silver. The glass substrate is placed in a molten salt bath that contains the electrodes, and a voltage is applied across the electrodes with the polarity shown in Fig. 11.6. The sodium ion diffuses in the glass toward the negative electrode, and the silver diffuses from the silver electrolyte solution into the surface of the glass.

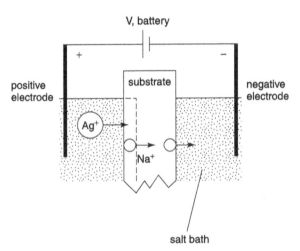

Figure 11.6. Electric-field-assisted ion exchange apparatus for doping glasses (substrate) with metals such as Ag^+ ions. [Adapted from G. De Marchi et al., *J. Non-Cryst. Solids* **196**, 79 (1996).]

11.1.3. Porous Silicon

When a wafer of silicon is subjected to electrochemical etching, the silicon wafer develops pores. Figure 11.7 is a scanning electron microscope (SEM) picture of the (100) surface of etched silicon showing pores (dark regions) of micrometer dimensions. This substance is called *porous silicon* (PoSi). By controlling the processing conditions pores of nanometer dimensions can be made. Research interest in porous silicon was intensified in 1990, when it was discovered that it was fluorescent at room temperature. *Luminescence* refers to the absorption of energy by matter and its reemission, as visible or near-visible light. If the emission occurs within 10^{-8} s of the excitation, the process is called *fluorescence*; if there is a delay in the emission, it is called *phosphorescence*. Nonporous silicon has a weak fluorescence between 0.96 and 1.20 eV in the region of the bandgap, which is 1.125 eV at 300 K. This fluorescence is due to bandgap transitions in the silicon. However, as shown in Fig. 11.8, porous silicon exhibits a strong photon-induced luminescence well above 1.4 eV at room temperature. The peak wavelength of the emission depends on the length of time that the wafer is subjected to etching, which increases the pore concentration. This observation generated much excitement because of the potential of incorporating photoactive silicon using current silicon technology, leading to new display devices or optoelectronic coupled elements. Silicon is the element most widely used to make transistors, which are the on/off switching elements in computers.

Figure 11.9 illustrates one method of etching silicon. Silicon is deposited on a metal such as aluminum, which forms the bottom of a container made of polyethylene or Teflon that will not react with the hydrogen fluoride (HF) etching solution. A voltage is applied between the platinum electrode and the Si wafer such that the Si is the positive electrode. The parameters, which influence the nature of the pores,

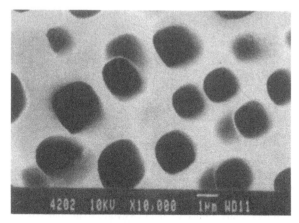

Figure 11.7. Scanning electron microscope (SEM) picture of the surface of N-doped etched silicon. The micrometer-sized pores appear as dark regions. [With permission from C. Levy-Clement et al., *J. Electrochem. Soc.* **141**, 958 (1994).]

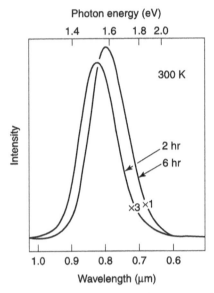

Figure 11.8. Photoluminescence spectra of porous silicon for two different etching times at room temperature. Note the change in ordinate scale for the two curves. [Adapted from L.T. Camham, *Appl. Phys. Lett.* **57**, 1046 (1990).]

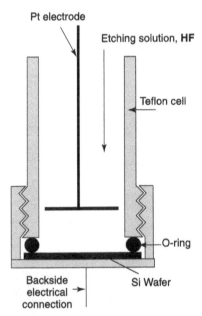

Figure 11.9. A cell for etching a silicon wafer in a hydrogen fluoride (HF) solution in order to introduce pores. (With permission from D. F. Thomas et al., in *Handbook of Nanostructured Materials and Nanotechnology*, edited by H. S. Nalwa, ed., Academic Press, Boston, 2000, Vol. 4, p. 173.)

are the concentration of HF in the electrolyte or etching solution, the magnitude of current flowing through the electrolyte, the presence of a surfactant (surface-active agent), and whether the silicon is negatively (N) or positively (P) doped.

The Si atoms of a silicon crystal have four valence electrons, and are bonded to four nearest-neighbor Si atoms. If an atom of silicon is replaced by a phosphorus atom, which has five valence electrons, four of the electrons will participate in the bonding with the four neighboring silicon atoms. This will leave an extra electron available to carry current, and thereby contribute to the conduction process. This puts an energy level in the gap just below the bottom of the conduction band. Silicon doped in this way is called an *N-type semiconductor*. If an atom of aluminum, which has three valence electrons, is doped into the silicon lattice, there is a missing electron referred to as a *hole* in one of the bonds of the neighboring silicon atoms. This hole can also carry current and contribute to increasing the conductivity. Silicon doped in this manner is called a *P-type semiconductor*. It turns out that the size of the pores produced in the silicon is determined by whether silicon is of N or P type. When P-type silicon is etched, a very fine network of pores having dimensions less than 10 nm is produced.

The effort to explain the origin of the fluorescence in porous silicon is an interesting example of how scientific understanding of phenomena develops. Because luminescence is well known to occur in molecules, it was proposed that the luminescence arose from molecules adsorbed on the surface of the pores. Such molecules as siloxane, $Si_6O_3H_6$, and an oxygen-rich siloxane, $Si_6O_{3+\delta}H_{6-\delta}$, which has a strong yellow fluorescence, were candidates. One difficulty with the molecular model is that the luminescence is still observed after the porous silicon is heated to temperatures as high as 1100°C. At these temperatures one would expect the molecules to undergo decomposition. The next model that was considered attributed the luminescence to the existence of quantum wires in the porous silicon. Quantum wires are nanostructured materials in which two dimensions are of nanometer size and the third is much larger. Carbon nanotubes, discussed earlier, are examples of quantum wires. In the case of carbon nanotubes we saw that the bandgap of the tubes increased with decreasing diameter of the tubes. In porous silicon it was observed that as the concentration of pores increased, the luminescence shifted to higher energy. The quantum wires are the solid regions between the pores, and their diameters decrease as the concentration of pores increases. If the luminescence is associated with excitation across the bandgap of a hole or electron from the valence band to the conduction band with emission occurring when the electron decays back to the valence band, then the shift of the luminescence as a function of pore concentration could be accounted for by a decrease in diameter of the quantum wires, causing an increase in the bandgap as seen in the carbon nanotubes.

However, AFM and TEM measurements showed that the solid regions between the pores in the normal direction in the solid were discontinuous and too small to be considered quantum wires. In fact, one TEM study revealed that the quantum wires were actually strings of nanosized crystallites of silicon. Figure 11.10a is a TEM picture of a 5-nm crystallite of silicon embedded in an amorphous silicon layer of porous silicon. Figure 11.10b is a greatly enlarged TEM view showing many nanosized crystallites.

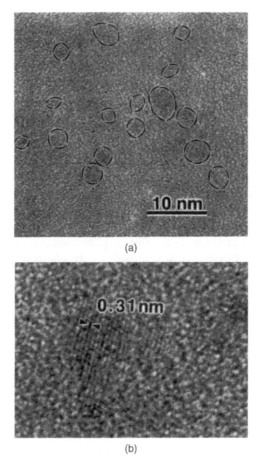

Figure 11.10. Transmission electron microscope images of porous silicon at two different magnifications showing the existence of nanosized crystallites of silicon. [Adapted from M. W. Cole et al., *Appl. Phys. Lett.* **60**, 2800 (1992).]

These results point to the possibility that quantum dots rather than wires are the source of the fluorescence. Quantum dots, which are discussed in detail in Chapter 9, are structures in which all three dimensions are nanometers. We saw in Chapter 5 that when the particle size of Si is less than 10 nm, the frequency of the optical vibrational mode of Si at 520 cm^{-1} begins to decrease, and the decrease is inversely proportional to the particle size. A measurement of the Raman spectra at different points on the surface of porous silicon using confocal micro-Raman spectroscopy, which focuses the laser beam down to a $\sim$10 μm or so cross section, and a measurement of the peak wavelength of the luminescence showed that the greater the shift of the frequency of the Raman line, the greater the shift of the peak wavelength of the luminescence. The emission shifts to higher energies as the particle size decreases because of the increase in the bandgap with decreasing particle size. Figure 11.11 is a plot of the calculated bandgap of

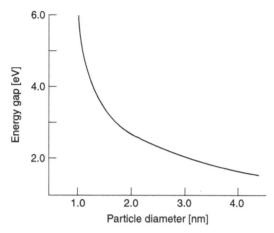

Figure 11.11. Calculation of bandgap of silicon versus particle size showing a pronounced increase below 2 nm. [Adapted from J. P. Proot et al., *Appl. Phys. Lett.* **61**, 1948 (1992).]

silicon versus particle size showing a pronounced increase below ~2 nm. These results strongly support the claim that the quantum dot of silicon is the source of the fluorescence.

11.2. NANOCOMPOSITES

In the previous section we discussed bulk materials made of grains, which all had three dimensions in the nanometer range. Here we will consider bulk nanostructured materials in which the components may have only one or two dimensions less than 100 nm, or may consist of particles in which all three dimensions are nanosized but embedded in matrices that are not nanosized, the three main types of nanocomposites are schematically illustrated in Fig. 11.12: (a) materials composed of altering layers of different materials that have nanometer thickness, (b) composites made of aligned filaments having nanometer diameters embedded in matrices that may not have nanometer dimension, and (c) composites consisting of particles where all three dimensions are of nanometer length embedded in matrices whose grain structure is not nanosized. The materials may have very different mechanical, electrical; magnetic, and optical properties compared to the same materials in the bulk. In the case of layered materials, many of the new properties are a result of the large ratio of the surface area between the layers to the volume of the layers.

11.2.1. Layered Nanocomposites

Layered nanocomposites can be synthesized by chemical vapor deposition (CVD) methods similar to the one described earlier to make carbon nanotubes. The materials that make up the layers are alternately fed into the oven, and the layers are deposited on each other systematically to obtain the desired thickness. Another process used to make nanostructured layered materials is molecular-beam epitaxy (MBE).

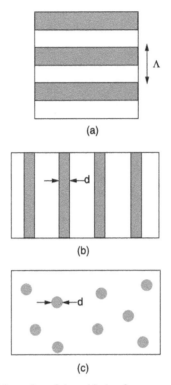

Figure 11.12. Schematics of three kinds of nanocomposite structures.

Figure 11.13 is a schematic diagram of an MBE apparatus. The system consists of a chamber, which is pumped down to a very high vacuum. The diagram shows three evaporation cells, which are the source of the vapors to be deposited. Typically the materials are heated in a boron nitride crucible at the base of the evaporation cells to produce the vapors. The vapor beams are interrupted by shutters mounted above the cells, which rotate on a shaft allowing an abrupt change of the beam material, and thus precise control of the layer thickness. The substrate on which the layers are deposited is on a molybdenum heating block and is held to the block by indium. The temperature of the substrate is precisely controlled. The electron gun shown in the figure allows an in situ monitoring of the layer-by-layer growth by probing the surface by reflected high-energy electrons. The technique is called *reflection high-energy electron diffraction* (RHEED). An electron gun produces a beam of high-energy electrons, which are incident on the substrate surface at small angles with respect to the surface, and the fluorescent screen shown in the figure detects the diffracted beams. Figure 11.14 shows a high-resolution TEM image of a layered composite of alternating layers of GaAs and GaAlAs synthesized by MBE. The period of the layers is 10 nm. Such structures are called *quantum wells*. Molecular-beam epitaxy is a relatively slow and expensive method used to synthesize nanometer-thick layered materials, but it is capable of very precise control of the thickness of

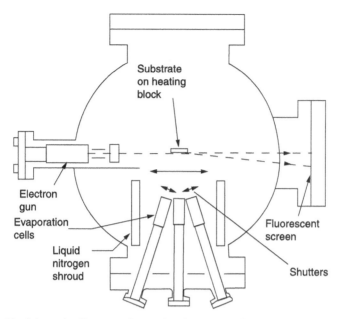

Figure 11.13. Schematic diagram of a molecular-beam epitaxy apparatus used to make nanometers thick multilayered materials. (Adapted from N. Inoue et al., *Handbook of Nanophase Materials*, A. N. Goldstein, ed., Marcel Dekker, New York, 1997, p. 83.)

the layers. Materials made of nanometer thick layers of different materials have a number of properties of technological importance. Such layered materials display enhanced mechanical strength, which will be discussed in Chapter 12, on mechanical properties. If the layers are made of magnetic materials, unusual magnetoresistive effects can be present. This will be discussed in Chapter 13, on magnetism in nanomaterials. Some of the more technologically important layered structures are made of semiconducting materials, which, because of quantum confinement effects, have very different electronic properties compared to bulk materials as explained in Chapter 9, on quantum dots. They have application as infrared detectors because the absorption of electromagnetic energy in the structures occurs in the IR and depends on the layer thickness and composition. Figure 11.15 shows the calculated dependence of the peak wavelength of the absorption on the aluminum content of a layered material made of 5 Å GaAs/25 Å $In_{.06}Ga_{.04}$/5 Å GaAs/200 Å Al_xGa_{1-x} systems. This results show how the response of the detector to different wavelengths can be altered depending on the Al composition.

11.2.2. Nanowire Composites

In Chapter 10 we discussed an example of a nanowire composite consisting of a polymer matrix in which carbon nanotubes, which are nanowires, are embedded. These nanowire polymer composites were mechanically stronger and electrically

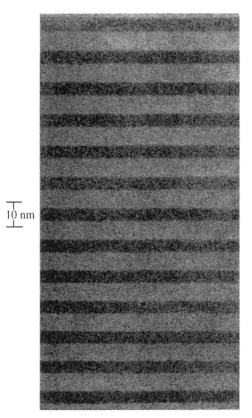

Figure 11.14. A high-resolution TEM image of a nanometer-thick multistructure of alternating layers of GaAs and GaAlAs made by molecular-beam epitaxy. [Adapted from L. L. Chang and L. Esaki, *Phys. Today* 36 (Oct. 1992).]

conducting compared to the polymers without the carbon nanotubes. Composites consisting of aligned metal nanowires have also been fabricated. A polycarbonate is first irradiated under a mask having small holes. The material is then etched, producing nanosized holes perpendicular to the surface. A thin layer of gold is deposited by sputtering on one side of the polymer. This is then used as the negative electrode in an electrochemical cell that contains a solution of, say, copper sulfate. A potential difference between the positive and negative electrodes draws the metal ions into the pores of the polymer, forming the aligned metal nanowires. The polymer can then be dissolved away, leaving a bundle of aligned metal nanowires. These aligned metal nanowires have a number of application possibilities such as electron emitters for flat-panel displays.

11.2.3. Composites of Nanoparticles

These composites generally consist of metal nanoparticles embedded in a matrix whose constituents are not of nanometer dimensions. For example, composites of

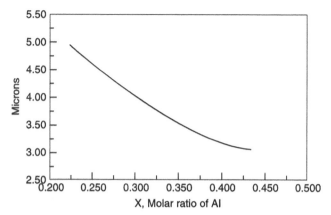

Figure 11.15. The calculated dependence of the peak wavelength of the absorption in the infrared as a function of the aluminum content of a layered structure consisting of 5 Å GaAs/25 Å $In_{.06}Ga_{.04}$/5 Å GaAs/200 Å Al_xGa_{1-x}. (Adapted from K. Choi, in *Handbook of Nanoscience and Engineering*, W. A. Goddard et al., CRC Press, 2003, pp. 9–8.)

the polymer polyisobutlyene and 40-nm copper nanoparticles have been fabricated and display increased strength. The composites are been made by dissolving the polymer in an organic solvent and suspending the metal nanoparticles in the solution. The solution is allowed to evaporate slowly while being subjected to sonication. Usually the polymer containing the nanopartiles is dried at elevated temperatures for some hours to remove any remaining solvent. Metal magnetic nanoparticle composites such as Ni/SiO_2 can be made by thin-film deposition methods where the two components are deposited on the substrate at the same time. Above ~50–60% vol%, the metal nanocomposites in the matrix may form an interconnected network, and the matrix will be electrically conducting. This volume percent is referred to as the *percolation threshold*. If the metal is magnetic, the composite will display bulk ferromagnetic behavior above the percolation threshold, and will exhibit hysteresis because of the existence of domains. Below the threshold, where the particles are separated sufficiently to be considered isolated particles, the material will display behavior characteristic of a single-domain material and have little or no hysteresis. However, we have gotten a little ahead of ourselves here and will discuss this in more detail in Chapter 13, which treats ferromagnetism.

11.3. NANOSTRUCTURED CRYSTALS

In this section we will discuss the properties of crystals made of ordered arrays of nanoparticles.

11.3.1. Natural Nanocrystals

There are some instances of what might be called "natural nanocrystals." An example is the 12-atom boron cluster, which has an icosahedral structure, shown in Fig. 11.16,

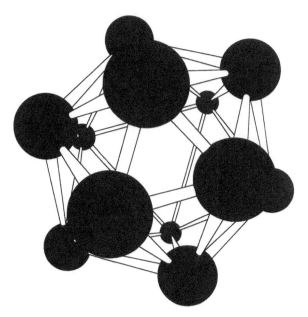

Figure 11.16. Icosahedral structure of boron nanocluster containing 12 boron atoms. This cluster is the basic unit of some boron lattices.

that exhibits 20 faces. A number of crystalline phases of solid boron contain the B_{12} cluster as a subunit. One such phase with tetragonal symmetry has 50 boron atoms in the unit cell, comprising four B_{12} icosahedra bonded to each other by an intermediary boron atom that links the clusters. Another phase consists of B_{12} icosahedral clusters arranged in a hexagonal array. Of course, there are other analogous nanocrystals such as the fullerene C_{60} compound, which forms the lattice shown in Fig. 10.9.

11.3.2. Crystals of Metal Nanoparticles

A two-phase water toluene reduction of $AuCl_4^-$ by sodium borohydride in the presence of an alkanethiol $(C_{12}H_{25}SH)$ solution produces gold nanoparticles Au_m having a surface coating of thiol, and embedded in an organic compound. The overall reaction scheme is

$$AuCl_4^-(aq) + N(C_8H_{17})_4 + (C_6H_5Me) \rightarrow N(C_8H_{17})_4 + AuCl_4^-(C_6H_5Me) \quad (11.3)$$

$$mAuCl_4^-(C_6H_5Me) + n(C_{12}H_{25}SH)(C_6H_5Me) + 3me^- \rightarrow$$
$$4mCl^-(aq) + (Au_m)(C_{12}H_{25}SH)_n(C_6H_5Me) \quad (11.4)$$

Essentially, the result of the synthesis is a chemical compound denoted as c-Au:SR, where SR is $(C_{12}H_{25}SH)_n(C_6H_5Me)$ and Me denotes the methyl radical CH_3. The

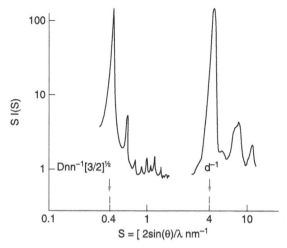

Figure 11.17. X-Ray diffraction of a lattice of aurothiol clusters showing order on two distinct length scales. The high angle pattern is from the order of the gold atoms in the clusters. The low angle pattern is from the order of the gold nanoclusters. [Adapted from R. L. Whetten et al., *Acc. Chem. Res.* **32**, 397 (1999).]

X-ray diffraction of the material shown in Fig. 11.17 indicated order on two different length scales. The scattering angle in the figure is plotted on a logarithimic scale so that both regions can be displayed together. The pattern at large angles arises from the symmetric arrangement of the gold atoms in the nanocluster. The interatomic distance is determined from the pattern to be 0.29 nm. The pattern at small angles can be accounted for by a body-centered cubic (bcc) packing structure having a repeat distance of 3 nm. In effect, a highly ordered symmetric arrangement of large gold nanoparticles had spontaneously self-assembled into a body-centered structure during chemical processing. Figure 11.18a is an illustration of the body-centered structure of the gold nanoparticles where the SR molecules are not shown. Figure 11.18b presents the structure including the SR molecules.

Superlattices of silver nanoparticles have been produced by aerosol processing. The lattices are electrically neutral, ordered arrangements of silver nanoparticles in a dense mantel of alkylthiol surfactant that is a chain molecule, $n\text{-}CH_3(CH_2)_mS$. The fabrication process involves evaporation of elemental silver above 1200°C into a flowing preheated atmosphere of high-purity helium. The flowstream is cooled over a short distance to ~400 K, resulting in condensation of the silver to nanocrystals. Growth can be abruptly terminated by expansion of the helium flow through a conical funnel accompanied by exposure to cool helium. The flowing nanocrystals are condensed into a solution of alkylthiol molecules. The material produced in this way has a superlattice structure with a FCC arrangement of silver nanoparticles with separations of <3 nm. Metal nanoparticles are of interest because of the enhancement of the optical and electrical conductance due to confinement and quantization of conduction electrons by the small volume of the nanocrystal.

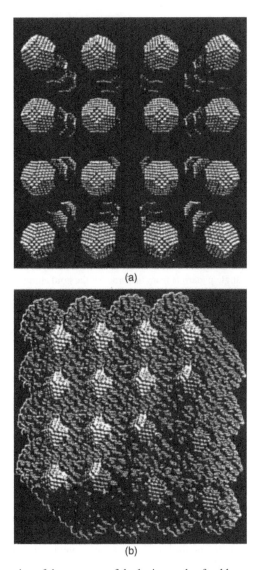

Figure 11.18. Illustration of the structure of the lattice made of gold nanoparticles without the SR molecules (a) and including the SR molecules (b). [Adapted from R. L. Whetten et al., *Acc. Chem. Res.* **32**, 397, (1999).]

When the size of the crystal approaches the order of the de Broglie wavelength of the conduction electrons, the metal clusters may exhibit novel electronic properties. They display very large optical polarizabilities, as discussed earlier, and nonlinear electrical conductance having small thermal activation energies. Coulomb blockade and Coulomb staircase current–voltage curves are observed. The Coulomb blockade

Figure 11.19. Schematic of cluster assemblies in zeolite pores. (With permission from S. G. Romanov et al., in *Handbook of Nanostructured Materials and Nanotechnology*, H. S. Nalwa, ed., Academic Press, Boston, 2000, Vol. 4, Chapter 4, p. 236.)

phenomenon was discussed in Section 9.5 Lower-dimensional lattices such as quantum dots and quantum wires have been fabricated by a subtractive approach that removes bulk fractions of the material, leaving nanosized wires and dots. These nanostructures were also discussed in Chapter 9.

11.3.3. Arrays of Nanoparticles in Zeolites

Another approach that has enabled the formation of latticelike structures of nanoparticles is to incorporate them into zeolites. Zeolites such as the cubic mineral faujasite, $(Na_2,Ca)(Al_2Si_4)O_{12} \cdot 8H_2O$, are porous materials in which the pores have a regular arrangement in space. The pores are large enough to accommodate small clusters. The clusters are stabilized in the pores by weak van der Waals interactions between the cluster and the zeolite wall. Figure 11.19 shows a schematic of a cluster assembly in a zeolite. The pores are filled by injection of the guest material in the molten state. It is possible to make lower-dimensional nanostructured solids by this approach using a zeolite material such as mordenite, which has the structure illustrated in Fig. 11.20. The mordenite has long parallel channels running through it with a diameter of 0.6 nm. Selenium can be incorporated into these channels, forming chains of single atoms. A trigonal crystal of selenium also has parallel chains, but the chains are sufficiently close together so that there is an interaction between them. In the mordenite this interaction is reduced significantly, and the electronic structure is different from that of a selenium crystal. This causes the optical absorption spectra of the selenium crystal and the selenium in mordenite to differ in the manner shown in Fig. 11.21.

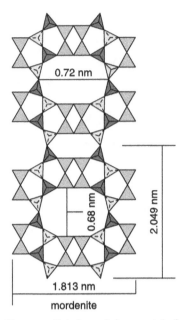

Figure 11.20. Illustration of long parallel channels in a crystal of mordenite, an orthrorhombic variety of zeolite $(Ca,Na_2,K_2)(Al_2Si_{10})O_{24} \cdot 7H_2O$. (Adapted from S. G. Romanov et al., in *Handbook of Nanostructured Materials and Nanotechnology*, H. S. Nalwa, ed., Academic Press, Boston 2000, Vol. 4, Chapter 4, p. 238.

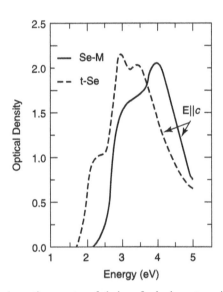

Figure 11.21. Optical absorption spectra of chains of selenium atoms in mordenite (solid line, Se–M) and in crystalline selenium (dashed line, *t*-Se) showing shift in peak of absorption and change in shape. [With permission from V. N. Bogomolov, *Solid State Commun.* **47**, 181 (1983).]

11.3.4. Nanoparticle Lattices in Colloidal Suspensions

Nanocolloidal suspensions consist of small spherical particles 10–100 nm in size suspended in a liquid. The interaction between the particles is *hard-sphere repulsion*, meaning that the center of the particles cannot get closer than the diameters of the particles. However, it is possible to increase the range of the repulsive force between the particles in order to prevent them from aggregating. This can be done by putting an electrostatic charge on the particles. Another method is to attach soluble polymer chains to the particles, in effect producing a dense brush with flexibe bristles around the particle. When the particles with these brush polymers surrounding them approach each other, the brushes compress and generate a repulsion between the particles. In both charge and polymer brush suspensions the repulsion extends over a range that can be comparable to the size of the particles. This is called *soft-sphere repulsion*. When such particles occupy over 50% of the volume of the material, the particles begin to order into lattices. The structure of the lattices is generally hexagonal close-packed, (hcp), face-centered cubic (FCC), or body-centered cubic (bcc). Figure 11.22 shows X-ray densitometry measurements on a 3 mM salt solution containing 720-nm polystyrene spheres. The dashed-line plots display the equations of state of the material, where the pressure P is normalized to the thermal energy $k_{B}T$, versus the fraction of particles in the fluid. The data show a gradual transition from a phase where the particles are disordered in the liquid to a phase where there is lattice ordering. In between there is a mixed region where there is both a fluid phase and a crystal phase. This transition is called the *Kirkwood–Alder transition*. Changing the concentration of the particles or the charge on them by removing screening ions from the fluid alters the transition. At

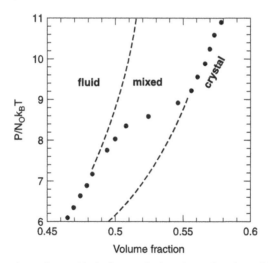

Figure 11.22. Equations of state (dashed curves) plotted as a function of fraction of 720-nm styrene spheres in a 3 mM salt solution. The constant N_{0} is Avogadro's number. [Adapted from A. P. Gast and W. B. Russel, *Phys. Today*, 24 (Dec. 1998).]

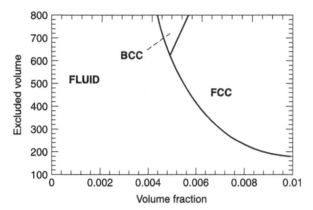

Figure 11.23. Phase diagram for soft spherical particles in suspension showing the fluid state, the body-centered cubic (BCC), and the face-centered cubic (FCC) phases. The vertical axis (ordinate) shows the volume that is inaccessible to one particle because of presence of the others. The excluded volume has been scaled to the cube of the Debye screening length, which characterizes the range of the interaction. [Adapted from A. P. Gast and W. B. Russel, *Phys. Today*, 24 (Dec. 1998).]

high concentrations or for short-range repulsions the lattice structure is FCC. Increasing the range of the repulsion, or lowering the concentration, allows the formation of the slightly less compact bcc structure. Figure 11.23 shows the phase diagram of a system of soft spherical particles. The excluded volume is the volume that is inaccessible to one particle because of the presence of the other particles. By adjusting the fraction of particles, a structural phase transition between the FCC and BCC structures can be induced. The particles can also be modified to have attractive potentials. For the case of charged particles in aqueous solutions, this can be accomplished by adding an electrolyte to the solution. When this is done, abrupt aggregation occurs.

11.3.5. Computational Prediction of Cluster Lattices

Viewing clusters as superatoms raises the intriguing possibility of designing a new class of solid materials whose constituent units are not atoms or ions, but rather clusters of atoms. Solids built from such clusters may have new and interesting properties. There have been some theoretical predictions of the properties of solids made from clusters such as $Al_{12}C$. The carbon is added to this cluster so that it has 40 electrons, which is a closed-shell configuration that makes the cluster stable. This is necessary for building solids from clusters because clusters that do not have closed shells could chemically interact with each other to form a larger cluster. Calculations of the FCC structure $Al_{12}C$ predict that it would have a very small bandgap, in the order of 0.05 eV, which means that it would be a semiconductor. The possibility of ionic solids made of KAl_{13} clusters has been considered. Since the electron affinity of Al_{13} is close to that of Cl, it may be possible for this cluster to form a structure

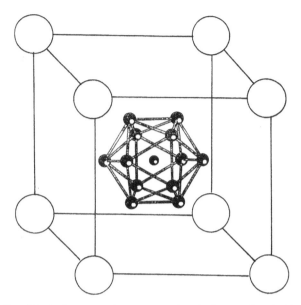

Figure 11.24. Possible body-centered structure of a lattice made of Al_{13} nanoparticles (at center) and potassium (large circles). [Adapted from S. N. Khanna and P. Jena, *Phys. Rev.* **51**, 13705 (1995).]

similar to that of KCl. Figure 11.24 shows a possible body-centered structure for this material. Its calculated cohesive energy is 5.2 eV which can be compared with the cohesive energy of KCl, which is 7.19 eV. This cluster solid is quite stable. These calculations indicate that new solids with clusters as their subunits are possible, and may have new and interesting properties; perhaps even new high-temperature superconductors will emerge. New ferromagnetic materials could result from solids made of clusters, which have a net magnetic moment.

11.4. ELECTRICAL CONDUCTION IN BULK NANOSTRUCTURED MATERIALS

In this section we discuss the electrical conduction in some bulk nanostructured materials.

11.4.1. Bulk Materials Consisting of Nanosized Grains

In order for a collection of nanoparticles to be a conductive medium, the particles must to be in electrical contact. One form of a bulk nanostructured material that is conducting consists of gold nanoparticles connected to each other by long molecules. This network is made by taking the gold particles in the form of an aerosol spray and subjecting them to a fine mist of a thiol such as dodecanethiol RSH, where R is

$C_{12}H_{25}SH$. These alkyl thiols have an endgroup —SH that can attach to a methyl —CH_3, and a methylene chain 8–12 units long that provides steric repulsion between the chains. The chainlike molecules radiate out from the particle. The encapsulated gold particles are stable in aliphatic solvents such as hexane. However, the addition of a small amount of dithiol to the solution causes the formation of a three-dimensional cluster network that precipitates out of the solution. Clusters of particles can also be deposited on flat surfaces once the colloidal solution of encapsulated nanoparticles has been formed. The in-plane electronic conduction has been measured in two-dimensional arrays of 500-nm gold nanoparticles connected or linked to each other by conjugated organic molecules. A lithographically fabricated device allowing electrical measurements of such an array is illustrated in Fig. 11.25. Figure 11.26 gives a measurement of the current versus voltage for a chain without (a) and with (b) linkage by a conjugated molecule. Figure 11.27 gives the results of a measurement of a linked cluster at a number of different temperatures. The conductance G, which is defined as the ratio of the current I to the voltage V, is the reciprocal of the resistance: $R = V/I = 1/G$. The data in Fig. 11.26 show that linking the gold nanoparticles substantially increases the conductance. The temperature dependence of the low-voltage conductance is given by

$$G = G_0 \exp \frac{-E}{k_B T} \qquad (11.5)$$

where E is the activation energy. The conduction process for this system can be modeled by a hexagonal array of single-crystal gold clusters linked by resistors, which are the connecting molecules, as illustrated in Fig. 11.28. The mechanism of conduction is electron tunneling from one metal cluster to the next. The tunneling process is a quantum-mechanical phenomenon where an electron can pass through an energy barrier higher than the kinetic energy of the electron. Thus, if a sandwich is constructed consisting of two similar metals separated by a thin insulating material,

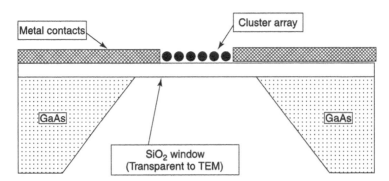

Figure 11.25. Cross-sectional view of a lithographically fabricated device to measure the electrical conductivity in a two-dimensional array of gold nanoparticles linked by molecules. (With permission from R. P. Andreas et al., in *Handbook of Nanostructured Materials and Nanotechnology*, H. S. Nalwa, ed., Academic Press, Boston, 2000, Vol. 3, Chapter 4, p. 217.)

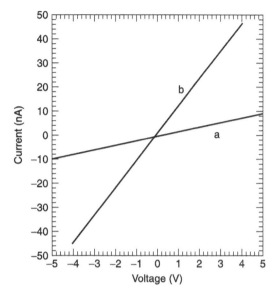

Figure 11.26. Room-temperature current–voltage relationship for a two-dimensional cluster array: (a) without linkage and (b) with the particles linked by a $(CN)_2C_{18}H_{12}$ molecule. [Adapted from D. James et al., *Superlat. Microsruct.* **18**, 275 (1995).]

as shown in Fig. 11.29a, an electron, under certain conditions, can pass from one metal to the an next. For the electron to tunnel from one side of the junction to the other, there must be available unoccupied electronic states on the other side. For two identical metals at $T = 0$ K, the Fermi energies will be at the same level,

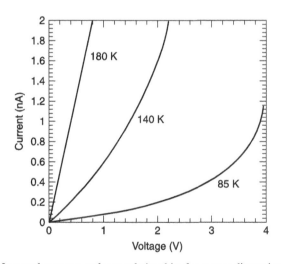

Figure 11.27. Measured current–voltage relationship for a two-dimensional linked cluster array at temperatures 85, 140, and 180 K. [Adapted from D. James et al., *Superlat. Microsruct.* **18**, 275 (1995).]

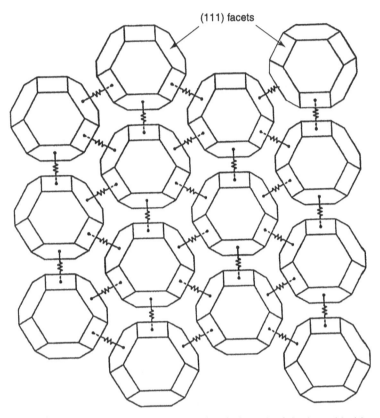

Figure 11.28. Sketch of a model to explain the electrical conductivity in an ideal hexagonal array of single-crystal gold clusters with uniform intercluster resistive linkage provided by resistors connecting the molecules. (With permission from R. P. Andreas et al., in *Handbook of Nanostructured Materials and Nanotechnology,* H. S. Nalwa, ed., Academic Press, Boston, 2000, Vol. 3, Chapter 4, p. 221.)

and there will be no states available, as shown in Fig. 11.29b, and tunneling cannot occur. The application of a voltage across the junction increases the electronic energy of one metal with respect to the other by shifting one Fermi level relative to the other. The number of electrons that can then move across the junction from left to right (Fig. 11.29c) in an energy interval dE is proportional to the number of occupied states on the left and the number of unoccupied states on the right as follows

$$N_1(E - eV)f(E - eV)[N_2(E)(1 - f(E)]$$ (11.6)

where N_1 is the density of states in metal 1, N_2 is the density of states in metal 2, and $f(E)$ is the Fermi–Dirac distribution of states over energy. The net flow of current I

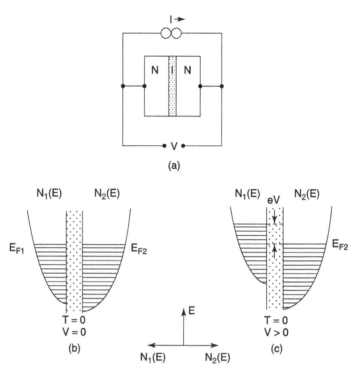

Figure 11.29. (a) Metal–insulator–metal junction; (b) density of states of occupied levels and Fermi level before a voltage is applied to the junction; (c) density of states and Fermi level after application of a voltage. Diagrams (b) and (c) plot the energy vertically and the density of states horizontally, as indicated at the bottom center of the figure. Levels above the Fermi level that are not occupied by electrons are not shown.

across the junction is the difference between the currents flowing to the right and the left as follows

$$I = K \int N_1(E - eV)N_2(E)[f(E - eV) - f(E)]dE \qquad (11.7)$$

where K is the matrix element, which gives the probability of tunneling through the barrier. The current across the junction will depend linearly on the voltage . If the density of states is assumed constant over an energy range eV, then, for small V and low T, we obtain

$$I = KN_1(E_f)N_2(E_f)eV \qquad (11.8)$$

which can be rewritten in the form

$$I = G_{nn}V \qquad (11.9)$$

where

$$G_{nn} = KN_1(E_f)N_2(E_f)e \qquad (11.10)$$

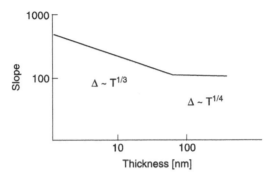

Figure 11.30. Plot of log of slope of dependence of conductivity on temperature versus log of thickness of an amorphous germanium film. [Adapted from A. Arbouet et al., *Phys. Rev. Lett.* **90**, 177401 (2003).]

and G_{nn} is identified as the conductance. The junction, in effect, behaves in an ohmic manner, that is, with the current proportional to the voltage.

11.4.2. Nanometer-Thick Amorphous Films

The electrical conductivity depends on the thickness of a film in the nanometer regime. For bulk three-dimensional amorphous materials the conduction mechanism involves charge hopping and depends on the temperature as $T^{1/4}$. An amorphous material is one in which the atoms do not have long-range crystalline order. In a two-dimensional amorphous material it has been shown that the conductivity depends on the temperature, as $T^{1/3}$. Figure 11.30 is a plot of the log of the slope of the conductivity dependence on $T^{1/4}$ and $T^{1/3}$ versus the log of the film thickness of amorphous films of germanium. The plot shows the crossover from two-dimensional to three-dimensional behavior occurring at 50 nm. Notice that only below 50 nm does the slope depend on the film thickness.

11.5. OTHER PROPERTIES

While the emphasis of the previous discussion has been on the effect of nanosized microstructure on electrical properties and optical properties, many other properties of bulk nanostructured materials are also affected. For example, the magnetic behavior of a bulk ferromagnetic material made of nanosized grains is quite different from the same material made with conventional grain sizes. Because of its technological importance relating to the possibility of enhancing magnetic information storage capability, this will be discussed in more detail in Chapter 13. Mechanical properties are also affected, which will be discussed in the next chapter.

In Chapter 3 it was seen that the inherent reactivity of nanoparticles depends on the number of atoms in the cluster. Such behavior might also be expected to be manifested in bulk materials made of nanostructured grains, providing a possible way to protect against corrosion and the detrimental effects of oxidation, such as the formation of the black silver oxide coating on silver. Indeed, there have been some advances in this area. The nanostructured alloy $Fe_{73}B_{13}Si_9$ has been found to have enhanced resistance to oxidation at temperatures between 200°C and 400°C. The material consists of a mixture of 30-nm particles of Fe(Si) and Fe_2B. The enhanced resistance is attributed to the large number of interface boundaries, and the fact that atom diffusion occurs more rapidly in nanostructured materials at high temperatures. In this material the Si atoms in the FeSi phase segregate to interface boundaries, where they can then diffuse to the surface of the sample. At the surface the Si interacts with the oxygen in the air to form a protective layer of SiO_2, which hinders further oxidation.

The melting temperature of nanostructured materials is also affected by the grain size. More recently it has been shown that the melting temperature of indium containing 4-nm nanoparticles can be lowered by 110 K. Superconducting properties can also be affected, and this will be discussed in Chapter 15.

In Chapter 3 we saw that the optical absorption properties of nanoparticles arising from transitions to excited states depend on their size and structure . In principle it should therefore be possible to engineer the optical properties of bulk nanostructured materials. A high-strength transparent metal would have many application possibilities.

PROBLEMS

11.1. Design an optical switch based on the dependence of the index of refraction on the intensity of the light. (*Hint*: the angle of reflection from a surface depends on the index of refraction.)

11.2. The table below relates the peak wavelength of the optical absorption of gold nanoparticles to the radius r of the particle. Plot the wavelength versus the square of the radius of the particle, and write an equation $\lambda = f(r)$ that approximately fits the points on the graph.

Wavelength, nm	Radius, nm
520	20
550	80
600	120
625	140

If you wish to make a stained-glass window having two panels, one red and one green, using gold nanparticles, what size particles would you put in each panel?

11.3. One crystalline form of boron consists of a hexagonal lattice of boron icosahedra (illustrated in Fig. 11.16). Given the nearest-neighbor separation of the icosahedra of 4.908 A and the lattice energy of 0.19 eV, construct a potential that would account for the cohesive energy of the lattice.

11.4. The structures of the lattice of $(Al_{13})^-$ nanoparticles and potassium ions K^+ are shown in Fig. 11.24. Construct a potential that would allow calculation of the cohesive energy by assuming that the negatively charged Al_{13} group is spherical. Given a center to-center distance between K and Al_{13} of 4.87 A and assuming that Al_{13} has the same electron affinity as Cl, calculate the cohesive energy per ion pair. Compare your result with the cohesive energy of KCl.

CHAPTER 12

Mechanical Properties of Nanostructured Materials

This chapter is concerned with how nanostructuring affects the mechanical properties of materials. We start with a brief overview of the mechanical properties of macro-materials, especially metals, and the reasons why they mechanically fail. Then we examine the effect on mechanical properties of reducing the grain size of bulk consolidated materials to nanometer dimensions as well as nanodimension layered materials. Finally the mechanical and dynamical properties of some devices of nanodimensions are discussed. Some of the challenges that need to be overcome in developing application of nanosized devices such as gas detectors will be considered.

12.1. STRESS–STRAIN BEHAVIOR OF MATERIALS

In Section 10.4.8 stress S was defined as the weight per unit area applied to the material

$$S = \frac{W}{A} \qquad (12.1)$$

The strain e is the change in length ΔL divided by the unloaded length L_0. The change in length could be a decrease produced by a compressive stress or an elongation as in the case of a wire having a weight on the end. For elastic deformation the strain is proportional to the stress [Eq. (10.13)]

$$S = Ee \qquad (12.2)$$

where E is the modulus of elasticity or Young's modulus. This is basically Hooke's law. In the elastic region when the stress is released, the material returns to its original length. However, it does not return instantaneously; rather, it takes some time to return, albeit small, to its original length. This time dependence is referred to as

The Physics and Chemistry of Nanosolids. By Frank J. Owens and Charles P. Poole, Jr.
Copyright © 2008 John Wiley & Sons, Inc.

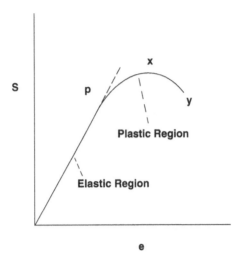

Figure 12.1. Typical stress–strain curve for a metal showing where the elastic–plastic transition occurs (*P*), which measures the yield strength, and also showing where the tensile strength (*X*) and fracture point (*Y*) are located.

anelasticity. When a compressive load is applied perpendicular to the surface of the material, there will also be strains parallel to this surface. Eventually a stress is achieved where elastic deformation is no longer possible. For most metals this generally takes place at a value of $\Delta L/L_0$ of 0.005. When this occurs, stress is no longer proportional to strain and the material does not obey Hooke's law. In this region of stress, referred to as the *plastic region*, the stress–strain plot is no longer linear. Figure 12.1 presents a typical stress–strain curve for a metal showing the elastic–plastic transition at point *P*. The yield strength of the material is the point at which the stress–strain plot deviates from linearity. After the elastic–plastic transition the stress needed to continue deformation in metals increases to a maximum, point *X* in Fig. 12.1, and then decreases to the fracture point *Y*. The tensile stress is the stress at the maximum of the stress–strain curve. This is the maximum stress that can be sustained by the material. If this stress is applied and held for some time the material will break; such a rupture is called *fracture*. Table 12.1 summarizes some of the mechanical properties of various metals. Another important property

TABLE 12.1. Mechanical Properties of Some Common Metals

Metal	Yield Strength, MPa	Tensile Strength, MPa
Al	35	90
Cu	69	200
Fe	130	262
Ni	138	480
Ti	450	520

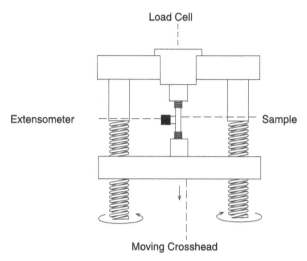

Figure 12.2. Illustration of apparatus to measure a tensile stress–strain curve. The sample is elongated by moving the crosshead. The load cell and extensometer measure the applied load and the amount of elongation.

Figure 12.3. A device used to measure the hardness of a sample.

is *ductility*, which is a measure of the amount of deformation a material can undergo before fracture. Figure 12.2 is an illustration of the apparatus used to measure the stress–strain curve of a material in the elongation mode. In the compression mode typically a cylinder of the material is compressed in a controlled fashion with a specific time dependence. The reduction of the length of the cylinder is measured versus the load. Hardness is another property used to characterize the strength of a material. It is a measure of the material's resistance to indentation. The depth or size of the indentation is measured. The deeper the indentation, the lower the hardness. Hardness measurements are relative, and a number of arbitrary scales exist to specify hardness such as the Knop, Brinell, and Mohs scales. Both tensile strength and hardness are a measure of a metal's resistance to plastic deformation. Generally there is a correlation between hardness and tensile strength. Figure 12.3 shows a device used to measure hardness. It consists of a fixed weight on the top of a penetrator that presses down on the material to be measured. A meter reads the depth of penetration.

12.2. FAILURE MECHANISMS OF CONVENTIONAL GRAIN-SIZED MATERIALS

In order to understand how nanosized grains affect the bulk structure of materials, it is necessary to discuss how conventional grain-sized materials fail mechanically. A brittle material fractures before it undergoes an irreversible elongation. Fracture occurs because of the existence of cracks in the material. Figure 12.4 shows an example of a crack in a two-dimensional lattice. A crack is essentially a region of a material where there is no bonding by adjacent atoms of the lattice. If such a material is subjected to tension, the crack interrupts the flow of stress. The stress accumulates at the bond at the end of the crack, making the stress at that bond very high, perhaps exceeding the bond strength. This results in a breaking of the bond at the end of the crack, and a resulting lengthening of the crack. Then the stress builds up on the next bond at the bottom of the crack, and it also breaks. This process of crack propagation continues until eventually the material separates

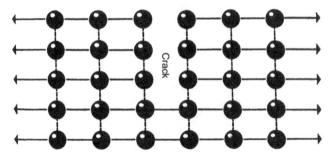

Figure 12.4. A crack in a two-dimensional rectangular lattice.

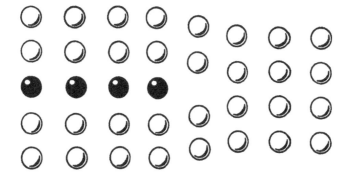

Figure 12.5. An edge dislocation in a two-dimensional rectangular lattice.

at the crack. A crack provides a mechanism whereby a weak external force can break stronger bonds one by one. This explains why the stresses that induce fracture are actually weaker than the bonds that hold the atoms of the metal together. Another kind of mechanical failure discussed above is the elastic–plastic transition, where the stress–strain curve deviates from linearity, as seen in Fig. 12.1. In this region the material irreversibly elongates before fracture. When the stress is removed after the elastic–plastic transition, the material does not return to its original length. The transition to ductility is a result of another kind of defect in the lattice called a *dislocation*. Figure 12.5 illustrates an edge dislocation in a two-dimensional lattice. There are also other kinds of dislocations such as a screw dislocation. Dislocations are essentially regions where lattice deviations from a regular structure extend over a large number of lattice spacings. Unlike cracks, the atoms in the region of the dislocation are bonded to each other, but the bonds are weaker than in the normal regions. In the ductile region one part of the lattice is able to slide across an adjacent part of the lattice. This occurs between sections of the lattice located at dislocations where the bonds between the atoms along the dislocation are weaker. One method of increasing the stress at which the brittle–ductile transition occurs is to impede the movement of the dislocations by introducing tiny particles of another material into the lattice. This process is used to harden steel, where particles of iron carbide are precipitated into the steel. The iron carbide particles block the movement of the dislocations.

12.3. MECHANICAL PROPERTIES OF CONSOLIDATED NANOGRAINED MATERIALS

The intrinsic elastic modulus of a consolidated nanostructured material is essentially the same as that of the bulk material having micrometer-sized grains until the grain size becomes very small, less than 5 nm. The larger the value of Young's modulus, the less elastic the material. Figure 12.6 is a plot of the ratio of Young's modulus E in nanograined iron to its value in conventional-grain-sized iron E_0 as a function of grain size. We see from the figure that below ∼20 nm Young's modulus begins to decrease from its value in conventional-grain-sized materials.

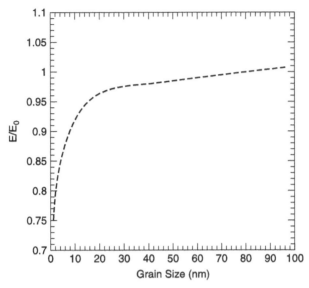

Figure 12.6. Plot of the ratio of Young's modulus E in nanograin iron to its value E_0 in conventional granular iron as a function of grain size.

The yield strength σ_y of a conventional-grain-sized material is related to the grain size by the Hall–Petch equation

$$\sigma_y = \sigma_0 + Kd^{-1/2} \tag{12.3}$$

where σ_0 is the frictional stress opposing dislocation movement, K is a constant, and d is the grain size in micrometers. Hardness can also be described by a similar equation. Figure 12.7 plots the measured yield strength of Fe–Co alloys as a function of $d^{-1/2}$, showing the linear behavior predicted by Eq. (12.3). Assuming that the equation is valid for nanosized grains, a bulk material having a 50 nm grain size would have a yield stress of 4.14 GPa. The reason for the increase in yield strength with smaller grain size is that materials having smaller grains have more grain boundaries that block dislocation movement. However, deviations from the Hall–Petch behavior have been observed for materials made of particles less than 20 nm in size. The deviations involve no dependence on particle size. It is believed that conventional-dislocation-based deformation is not possible in bulk nanostructured materials with sizes less than 30 nm because mobile dislocations are unlikely to occur. Examination of small-grained bulk nanomaterials by transmission electron microscopy (TEM) during deformation does not show any evidence for mobile dislocations.

Most bulk nanostructured materials are quite brittle and display reduced ductility under tension, typically having elongations of a few percent for grain sizes less than 30 nm. For example, conventional coarse-grained annealed polycrystalline copper is very ductile, having elongations of ≤60%. Measurements in samples with grain sizes

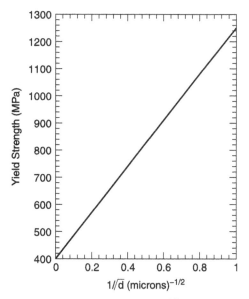

Figure 12.7. Yield stress of Fe–Co alloys versus $1/d^{1/2}$, where d is the size of the grain. [Adapted from Chang-He Shang et al., *J. Mater. Res.* **15**, 835 (2000).]

less than 30 nm yield elongations no more than 5%. Most of these measurements have been performed on consolidated particulate samples, which have large residual stress, and flaws due to imperfect particle bonding, which restricts dislocation movement. However, nanostructured copper prepared by electrodeposition displays almost no residual stress and has elongations up to 30% as shown in Fig. 12.8. These results emphasize the importance of the choice of processing procedures, and the effect of flaws and microstructure on measured mechanical properties. In general, the results of ductility measurements on nanostructured bulk materials are mixed because of sensitivity to flaws and porosity, both of which depend on the processing methods. The porosity or density of a particulate compact can strongly influence the mechanical properties. It is important to note that the Hall–Petch relationship between particle size and yield strength applies to a compacted material of different grain sizes but the same porosity. It has been shown that Young's modulus of aluminum oxide composite can change by 200% in going from a volume fraction of porosity of 0.4–0.1. Figure 12.9 is a plot of Young's modulus versus the volume fraction of pores in a compact of aluminum oxide. It has been found empirically that the modulus E depends on the volume fraction of pores in the material by the equation

$$E = E_0(1 - 1.9P + 0.9P^2) \tag{12.4}$$

where E_0 is the modulus for the materials with no porosity. The line through the data in Fig. 12.9 is a fit to Eq. (12.4).

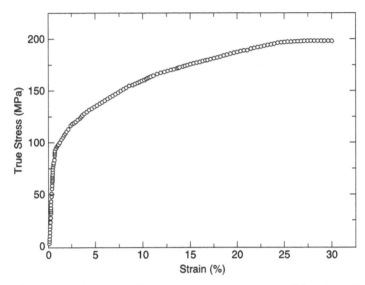

Figure 12.8. Stress–strain curve of nanostructured copper prepared by electrodeposition. [With permission from L. Lu et al., *J. Mater. Res.* **15**, 270 (2000).]

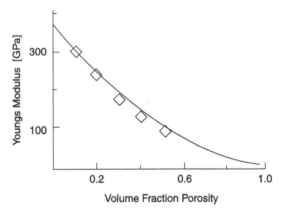

Figure 12.9. Plot of Young's modulus of elasticity versus the volume fraction of a compact of aluminum oxide that is porous. The points are experimental, and the line is a fit to Eq. (12.4) of the text. [Adapted from R. L. Cobel et al. *J. Am. Ceram. Soc.* **39**, 381 (1956).]

12.4. NANOSTRUCTURED MULTILAYERS

Another kind of bulk nanostructure consists of periodic layers of nanometer thickness of different materials such as alternating layers of TiN and NbN. These layered materials can be fabricated by various vapor-phase methods such as sputter deposition and chemical vapor phase deposition (CVD) as discussed in Sections 10.4.1 and 11.2.1. Electrochemical and molecular-beam epitaxy (MBE) can also be used to fabricate them. The materials have very large interface area densities. This means that the density of atoms on the planar boundary between two layers is very high. For example, 1 cm^2 of a 1-μm-thick multilayer film having layers of 2 nm thickness each has an interface area of 1000 cm^2. Since the material has a density of $\sim$6.5 g/cm^3, the interface area is 154 m^2/g, which is comparable to that of typical heterogeneous catalysts. The interfacial regions have a strong influence on the properties of these materials.

Figure 12.10 shows a plot of the hardness of a TiN/NbN nanomultilayered structure as a function of the bilayer period (or thickness) of the layers, showing that as the layers get thinner in the nanometer range, there is a significant enhancement of the hardness until about 30 nm, where it appears to level off and become constant. The reason for this is that the thinner the layers, the more interfaces are present to block dislocation movement, assuming that the total thickness of the laminate is not changed. It has been found that a mismatch of the crystal structures between the layers actually enhances the hardness. The compounds TiN and NbN both have the same rock salt or NaCl structure with the respective lattice constants 0.4235 and 0.5151 nm, so the mismatch between them is relatively large, as is the hardness. Interestingly, multilayers in which the alternating layers have different crystal structures were found to be even harder. In this case dislocations move less easily between the layers, and essentially became confined in the layers, resulting in an increased hardness.

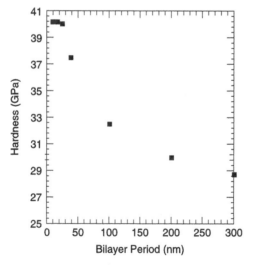

Figure 12.10. Plot of the hardness of TiN/NbN multilayer materials as a function of the thickness of the layers. [Adapted from B. M. Clemens, MRS *Bull.* 20 (Feb. 1999).]

12.5. MECHANICAL AND DYNAMICAL PROPERTIES OF NANOSIZED DEVICES

12.5.1. General Considerations

In order to develop machines with nanosized components, it is necessary to understand the mechanical properties of nanosized objects and the factors that influence mechanical behavior. Machines having micrometer-sized components have been developed, and because it is a much more mature technology than nanosized systems, it is useful to examine what has been learned about how micrometer sizing affects mechanical properties. The extensive fabrication infrastructure developed for the manufacture of silicon integrated circuits has made possible the development of machines and devices having components of micrometer dimensions. Lithographic techniques, combined with metal deposition processes, are used to make micrometer devices. Microelectromechanical systems (MEMS) involve a mechanical response to an applied electrical signal, or an electrical response resulting from a mechanical deformation. The most common component in such systems are beams clamped at both ends or at one end such as the cantilever, which supports the tip on the atomic force. microscope (AFM).

The major advantages of MEMS devices are miniaturization, multiplicity, and the ability to directly integrate the devices into microelectronics. *Multiplicity* refers to the large number of devices and designs that can be rapidly manufactured, lowering the price per unit item. For example, miniaturization has enabled the development of micrometer-sized accelerometers for activating airbags in automobiles. Previously an electromechanical device, the size of a soda can, weighing several pounds and costing about $15, triggered airbags. The currently used accelerometers based on MEMS devices are the size of a dime, and cost only a few dollars. The size of MEMS devices, which is comparable to electronic chips, allows their integration directly on the chip. Below we will present a few examples of MEMS devices and describe how they work. But before we do this, let us examine what has been learned about the difference between the mechanical behavior of machines in the macro- and micro-scale worlds.

In the microscale world the ratio of the surface area to the volume of a component greatly exceeds that in conventional-sized devices. This makes friction more important than inertia. In the macroscale world a pool ball continues to roll after being struck because friction between the ball and the table is less important than the inertia of its forward motion. In the microscale regime the surface area : volume ratio is so large that surface effects are very important. In the microscale world, mechanical behavior can be altered by a thin coating of a material on the surface of a component. We will describe MEMS sensors that take advantage of this property. Another characteristic of the microscale world is that molecular attractions between micrometer-sized objects can exceed mechanical restoring forces. Thus the elements of a microscale device, such as an array of cantilevers, micrometer-sized boards fixed at one end, could become stuck together when deflected. To prevent this, the elements of micromachines may have to be coated with special nonstick coatings. In the case of large motors and machines, electromagnetic forces are utilized, and electrostatic forces have little impact. In contrast to this, electromagnetic forces become too small when the elements of the motors have micrometer dimensions, while electrostatic forces become large. Electrostatic actuation is often used in micromachines, which means that the elements are charged, and the repulsive electrostatic force between the elements causes them to move. Many of these differences between micromachines and macromachines may become more pronounced in the nanorange. Figure 12.11 illustrates the principle behind a MEMS accelerometer used to activate airbags in automobiles. Figure 12.11a shows the device, which consists of a horizontal bar of silicon a few micrometers in length attached to two vertical hollow bars, having flexible inner surfaces. The automobile is shown moving from left to right in the figure. When the car suddenly comes to a halt because of impact, the horizontal bar is accelerated to the right in the figure, which causes a change in the separation between the plates of the capacitor, as shown in Fig. 12.11b. This changes the value of the electrical capacitance of the capacitor, which in turn electronically triggers a pulse of current through a heating coil embedded in sodium azide (NaN$_3$). The instantaneous heating causes a rapid decomposition of the azide material, thereby producing nitrogen gas (N$_2$) through the reaction $2NaN_3 \rightarrow 2Na + 3N_2$, which inflates the airbag.

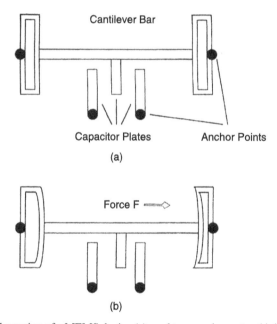

Figure 12.11. Illustration of a MEMS device (a) used to sense impact and initiate expansion of airbags. The automobile is moving from left to right. On impact (b) the horizontal cantilever bar is accelerated to the right and changes the separation of the capacitor plates, thereby triggering a pulse of electric current that activates the bag expansion mechanism. (Adapted from M. Gross, *Travels to the Nanoworld*, Plenum Press, New York, 1999, p. 169.)

Coated cantilever beams are the basis of a number of sensing devices employing MEMS. A cantilever is a small supported beam. The simplest of such devices consist of arrays of singly supported polysilicon cantilevers having various length : width ratios in the micrometer range. The beams can be made to vibrate by electrical or thermal stimuli. Optical reflection techniques are used to measure the vibrational frequency. As we will see, this frequency is very sensitive to the length of the beam. Thermal sensors have been developed using these supported micrometer-sized cantilevers by depositing on the beams a layer of a material, which has a different coefficient of thermal expansion than the polysilicon cantilever itself. When the beam is heated, it bends because of the different coefficients of expansion of the coating and the silicon, and the resonant frequency of the beam changes. The sensitivity of the device is in the microdegree range, and it can be used as an infrared (IR) sensor. A similar design can be used to make a sensitive detector of dc magnetic fields. In this case the beam is coated with a material, which displays *magnetorestrictive effects*, meaning that the material changes its dimensions when a dc magnetic field is applied. This causes the beam to bend and change its resonant frequency. These devices can detect magnetic fields as small as 10^{-5} G (10^{-9} T). In principle similar devices could be made with nanosized components and in many instances may even be more sensitive.

12.5.2. Nanopendulum

Let us examine how nanosizing effects the vibrational frequency of a pendulum. A pendulum consists of a mass M on the end of a light nonextendable string of length L. If it is pulled to one side, it swings back and forth. Figure 12.12 illustrates how the gravitational force Mg affects the pendulum. The restoring force that brings the pendulum back to the vertical position is the tangential component of Mg, which is directed toward the earth and is given by

$$F = Mg \sin \theta \qquad (12.5)$$

where g is the acceleration due to gravity. For small θ, the sin of θ (in radians) is approximately equal to θ and the displacement along the arc is $L\theta$. Thus

$$F = -Mg\theta = \frac{Mgx}{L} = kx \qquad (12.6)$$

where $k = Mg/L$. The frequency f of a harmonic oscillator is,

$$f = \frac{(k/M)^{1/2}}{2\pi} = \frac{(g/L)^{1/2}}{2\pi} \qquad (12.7)$$

The frequency of the pendulum is independent of the mass M on the end. Now let us examine how the frequency is affected when the length of the pendulum L is

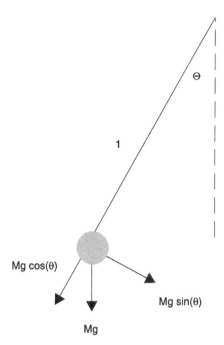

Figure 12.12. Illustration of a simple pendulum showing the forces acting on it.

reduced to nanometers. Figure 12.13 is a plot of the frequency as a function of the length of the pendulum in the nanometer range. The plot shows that below $\sim$40 nm there is a substantial increase in frequency. This increase is an important characteristic of components of nanosized machines. When the frequency increases

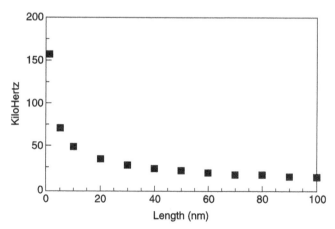

Figure 12.13. Plot of the oscillation frequency of a simple pendulum versus the length of the pendulum in nanometers.

as we saw in Section 7.8, the amplitude of vibration decreases, and as we will see, this introduces some difficulties in the ability to detect these high frequencies.

12.5.3. Vibrations of a Nanometer String

The theory of vibrating strings is one of the oldest branches of applied mathematics, dating back to the Greek mathematician Pythagoras. The reason, of course, is that many musical instruments produced sound by vibrations of strings of different lengths and tensions. Here we are interested in examining how the vibrations of the string behave when the string approaches nanometer length.

The equation of motion of a string of length L clamped at both ends is

$$\frac{d^2y}{dx^2} = \frac{[\mu/T]d^2Y}{dt^2} \tag{12.8}$$

where μ is the linear density of the string and T is the tension in the string. The string vibration is confined to the xy plane, where x is parallel to the string and $[T/\mu]^{1/2}$ is given in units of velocity, v. Assuming a solution of the form

$$v(x,t) = f(x)\cos(\omega t) \tag{12.9}$$

and substituting into Eq. (12.8), we get

$$\frac{d^2f}{dx^2} = -\frac{\omega^2 f}{v}. \tag{12.10}$$

Since at $x = 0$ where the string is clamped, there will be no displacement along y, so an acceptable solution is

$$f(x) = A\sin\frac{\omega x}{v}. \tag{12.11}$$

At $x = L$ the displacement is also zero, which can be satisified by making

$$\frac{\omega L}{v} = N\pi \tag{12.12}$$

Thus we have

$$\omega = \frac{N\pi}{L}\left(\frac{T}{\mu}\right)^{1/2} \tag{12.13}$$

Thus a vibrating string of length L tied down at both ends will have a number of vibrational frequencies. The frequencies are such that the wavelength of the vibration is an integral multiple of the length of the string. The $N = 1$ mode has the largest amplitude and in the case of a string instrument determines the frequency of the sound heard. This frequency is referred to as the *fundamental frequency*. The higher N vibrations, referred to as *overtones*, have smaller amplitudes of vibration decreasing with increasing N. In music they are, however, important in contributing to the quality of the sound. Equation (12.13) shows us that the frequency of vibration of the string increases as the length of the string decreases. For a string of 40 nm having a tension of 70 N and a linear density of 0.0038 g/cm^2, the vibrational frequency of the fundamental mode will 3.4×10^9 cycles per second, far beyond our ability to hear it.

12.5.4. The Nanospring

If a particle vibrates about an equilibrium position under the influence of a force that is proportional to the distance from the equilibrium position, the particle will have simple harmonic motion and the equation of motion will be

$$\frac{Md^2x}{dt^2} + Kx = 0 \qquad (12.14)$$

Assuming a time-dependent solution of the form $x = x_0 \exp[-i\omega t]$ and substituting into Eq. (12.14) yields an expression for the vibrational frequency

$$f = \frac{1}{2\pi} \left(\frac{K}{M}\right)^{1/2} \qquad (12.15)$$

Consider a spring with mass M on the end which is a rectangular block of length L and cross-sectional area A having sides of length a. The block has a density ρ. The frequency can be rewritten

$$f = \frac{1}{2\pi} \left(\frac{K}{\rho La^2}\right)^{1/2} \qquad (12.16)$$

Ignoring the length and mass of the spring, it is seen that if we reduce the length of the mass M, the frequency will increase as $1/L^{1/2}$. However, if both a and L are reduced, the frequency will scale as $1/a(L)^{1/2}$. This illustrates another important point, namely, that the increase in frequency also depends on the number of dimensions of the vibrating body that are reduced to nanometer size.

The rectangular block oscillating at the end of the spring will encounter resistance from the fluid through which it is moving, perhaps air. This resistive force is given by

$$F \sim \frac{A\eta \, dx}{dt} \qquad (12.17)$$

where η is the viscosity of the fluid, A is the cross sectional area of the block and dx/dt is the velocity. This damping force will cause an attenuation of the amplitude of vibration. The equation of motion of the damped spring is

$$\frac{M\,d^2x}{dt^2} + \frac{b\,dx}{dt} + kx = 0 \tag{12.18}$$

The solution $X(t)$ to this equation for a small damping factor b is

$$X(t) = A\exp\left(\frac{-bt}{2M}\right)\cos\left(\omega t + \delta\right) \tag{12.19}$$

with the frequency ω given by

$$\omega = \left[\left(\frac{k}{M}\right) - \left(\frac{b}{2M}\right)^2\right]^{-1/2} \tag{12.20}$$

Equation (12.19) describes a system oscillating at a fixed frequency ω with an amplitude exponentially decreasing in time. The displacement as a function of time is plotted in Fig. 12.14a.

If one applies an external oscillating force $F_0\cos(\omega' t)$ to a damped harmonic oscillator, a very large increase in amplitude occurs when the frequency of the applied force ω' equals the natural resonant frequency ω_0 of the oscillator. This is called *resonance*. The increase of the amplitude depends on the magnitude of the damping term b in Eq. (12.18), which is the cause of dissipation. Figure 12.14b shows how the magnitude of the damping factor affects the amplitude at resonance for a vibrating mass on a spring. Notice that the smaller the damping factor, the narrower the resonance peak, and the greater the increase in amplitude. The quality factor Q for the resonance given by the expression $Q = \omega_0/\Delta\omega$, where $\Delta\omega$ is the width of the resonance at half-height, and ω_0 the resonant frequency. The quality factor is the energy stored divided by the energy dissipated per cycle, so the inverse of the quality factor $1/Q$ is a measure of the dissipation of energy. Nanosized oscillators can have very high Q values and dissipate little energy as they oscillate. Such devices will be very sensitive to external damping, which is essential to developing sensing devices. High-Q devices also have low thermomechanical noise, which means significantly less random mechanical fluctuations.

12.5.5. The Clamped Beam

While springs and pendulums serve to illustrate how dynamical behavior is affected when an oscillating body has nanometer dimensions, the most common element in a

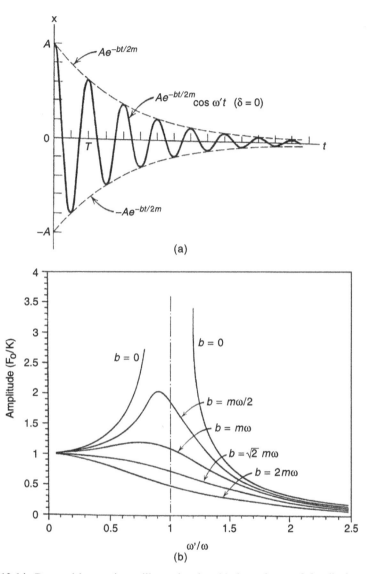

Figure 12.14. Damped harmonic oscillator showing (a) dependence of the displacement of a free-running (no driving force) oscillator on the time and (b) dependence of the amplitude of a driven oscillator versus the ratio of the driving frequency ω' to the undamped natural frequency ω_0 for several damping constants b, where F_0 is the amplitude of the driving force.

nanomachine will likely be a beam clamped at one or both ends. Figure 12.15 is an electron microscopic image picture of an array of beams having micrometer lengths, which is the basis of a number of applications such as thermal and magnetic sensors discussed above. These beams are clamped at one end. The equation of motion of the

Figure 12.15. A MEMS cantilever array.

clamped beam is*

$$\frac{\rho A d^2 U}{dt^2} + \frac{EI d^4 U}{dx^4} = q(x) \tag{12.21}$$

where U is the displacement, ρ the density, A the cross-sectional area, E Young's modulus, I the moment of inertia, and $q(x)$ a distributed applied load. The natural frequency of the first mode of vibration is given by

$$f = \frac{1.758}{\pi L^2} \left(\frac{EI}{\rho A}\right)^{1/2} \tag{12.22}$$

where L is the length of the beam. The natural frequency of a clamped beam scales as the square of the inverse of the length of the beam, which is a much stronger dependence on length than the pendulum or spring discussed in the previous sections. Figure 12.16 is a plot of the measured frequency dependence of a micrometer-sized beam versus $1/L^2$ showing that the scaling is as predicted by the theory. Notice that in the micrometer range the frequencies are in the hundreds of kilohertz ($>10^5$ cycles/s).

Now, a beam having a length of 10 nm and thickness of 1 nm will have a resonant frequency 10^5 times greater, of the order of 20–30 GHz (2–3×10^{10} cycles/s). As the frequency increases the amplitude of vibration decreases, and in this frequency range, beam displacements can range from a picometer (10^{-12} m) to a femtometer (10^{-15} m).

These high frequencies and small displacements are very difficult, if not impossible, to detect. Optical reflection methods such as those used in the micrometer range on the cantilever tips of scanning tunneling microscopes (STMs) are not applicable because of the diffraction limit. This occurs when the size of the object from

---

[1]It would take us too far afield to go through the derivation of this equation. The derivation can be found in any basic textbook on solid mechanics.

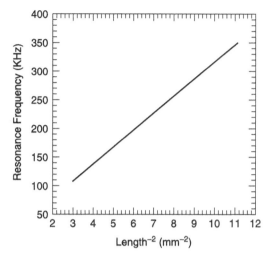

Figure 12.16. Plot of the resonance frequency of a MEMS cantilever versus the square of the reciprocal of the length of the beam. [Adapted from R. C. Benson, *Johns Hopkins Tech. Digest* **16**, 311 (1955).]

which light is reflected becomes smaller than the wavelength of the light. Transducers are generally used in MEMS devices to detect motion. The MEMS accelerometer shown in Fig. 12.11 is an example of the detection of motion using a transducer. In the accelerometer mechanical motion is detected by a change in capacitance, which can be measured by an electric circuit. It is not clear whether such a transducer sensor can be built that can detect displacements as small as 10^{-15}–10^{-12} meters, and do so at frequencies up to 30 GHz. These issues present significant obstacles to the development of NEMS devices.

12.5.6. The Challenges and Possibilities of Nanomechanical Sensors

Because of the potential for increased sensitivity, sensors of nanometer dimensions are of much interest. For example, a gas detector can be designed around a cantilever analogous to the micrometer-sized one shown in Fig. 12.15. The adsorbed gas can be sensed by the change in resonant frequency of the cantilever. The cantilever is a beam clamped at one end, and the resonant frequency of the beam is given by

$$f_0 = \frac{a}{6.154L^2}\left[\frac{Ea^2L}{M}\right]^{1/2} \tag{12.23}$$

where a is the thickness of the beam, L the length, M the mass, and E Young's modulus of the material. So a 10-nm-long silicon beam with 2 nm thickness will weigh 93.16×10^{-27} kg and have a resonant frequency of 75.5 GHz. If we want to make a humidity sensor out of this beam, we might coat the beam with a water-absorbing material. If one water molecule were to absorb on the cantilever, the resonant frequency of the beam would change to 69.1 GHz. So, in principle, such a device is capable of detecting one water molecule, which makes it a very sensitive detector. However, things are not so simple. The amplitude of vibration of the beam can be calculated from Eq. (7.30) and is 4.44 Å, a very small value. The common method of measuring the *deflection* of micrometer beams is by reflection of laser light as in the AFM system in the atomic force microscope (AFM). The deflection is detected by interference of the incident light and the reflected light. However, the wavelength of the laser light must be smaller than the amplitude of deflection of the beam for this to work. The light of a helium−neon laser has a wavelength of 6320 Å, which is much greater than the vibration amplitude of the nanobeam and therefore will not be able to sense the motion. So a challenge is to figure out a way to measure such small amplitude vibrations. Another issue is that the probing light used to measure the position of the beam may actually disturb the beam. In other words, the Heisenberg uncertainty principle may limit the ability to detect the motion and therefore limit the resolution. Finally, there is the matter of thermal fluctuations. We saw in Chapter 3 that SEM observations of 2–20-nm gold nanoparticles over time showed fluctuations in geometry of the particles. Thermal fluctuations have also been observed on the smooth crystalline surfaces of nanosized objects. The surface fluctuations consist of ridges and depressions that come and go with time. These effects have been observed by scanning tunneling microscopy (STM). Figure 12.17 shows the presence of these ridges and depressions imaged by STM at one instant of time on a 11×11-μm silicon surface. Because the atoms in step formation are bound to fewer atoms than they would be on a normal surface, atoms in step formation are more susceptible to thermal fluctuations. These distortions come and go in intervals of a second or so. Such effects will influence the performance of a nanosized beam such as a gas detector. Therefore much more work needs to be done before nanosized sensors are developed.

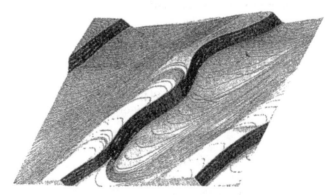

Figure 12.17. STM image of the step and depression formations on the surface of an 11×11-μm piece of silicon at a given instant of time. The surface depressions and formations change with time. [Adapted from Z. Toroczkai and E. D. Williams, *Phys. Today* 25 (Dec. 1999).]

12.5.7. Methods of Fabrication of Nanosized Devices

Nanomechanical machines and devices are in the early stages of development, and many are still in conceptual stages. Numerous ideas have been proposed and computer simulations of possibilities have been created. It turns out that nature is far ahead of humans in its ability to produce nanosized machines. Nanomotors exist in biological systems such as the flagellar motor of bacteria. Flagellae are long thin bladelike structures that extend from the bacterium. The motion of these flagellae propel the bacterium through water. These whiplike structures are made to move by a biological nanomotor consisting of a highly structured conglomerate of protein molecules anchored in the membrane of the bacterium. The motor has a shaft and a structure about the shaft resembling an armature. However, the motor is not driven by electromagnetic forces, but rather by the breakdown of adenosine triphosphate (ATP) energy-rich molecules, which cause a change in the shape of the molecules. Applying the energy gained from ATP to a molecular ratchet enables the protein shaft to rotate. Perhaps the study of biological nanomachines will provide insights that will enable us to improve the design of mechanical nanomachines.

Optical lithography is an important manufacturing tool in the semiconductor industry. However, to fabricate semiconductor devices less than 100 nm, ultraviolet light of short wavelengths (193 nm) is required, but this will not work because the materials are not transparent at these wavelengths. Electron-beam and X-ray lithography, discussed in previous chapters, can be used to make nanostructures, but these processes are not amenable to the high rate of production that is necessary for large-scale manufacturing. Electron-beam lithography uses a finely focused beam of electrons, which is scanned in a specific pattern over the surface of a material. It can produce a patterned structure on a surface having a 10 nm resolution. Because it requires the beam to hit the surface point by point in a serial manner, it cannot produce structures at sufficiently high rates to be used in assembly-line manufacturing processes. X-Ray lithography can produce patterns on surfaces having 20 nm resolution, but its mask technology and exposure systems are complex and expensive for practical applications.

A technique called *nanoimprint lithography* has now been developed that may provide a low-cost, high-production-rate manufacturing technology. Nanoimprint lithography patterns a resist by physically deforming the resist shape with a mold having a nanostructure pattern on it, rather than by modifying the resist surface by radiation, as in conventional lithography. A *resist* is a coating material that is sufficiently soft that it can be imprinted by a harder material. A schematic of the process is illustrated in Fig. 12.18. A mold having a nanoscale structured pattern on it is pressed into a thin resist coating on a substrate (Fig. 12.18a), creating a contrast pattern in the resist. After the mold is lifted off (Fig. 12.18b), an etching process is used to remove the remaining resist material in the compressed regions (Fig. 12.18c). The resist is a thermoplastic polymer, which is a material that softens on heating. It is heated during the molding process to soften the polymer relative to the mold. The polymer is generally heated above its glass transition temperature, thereby allowing it to flow and conform to the mold pattern. The

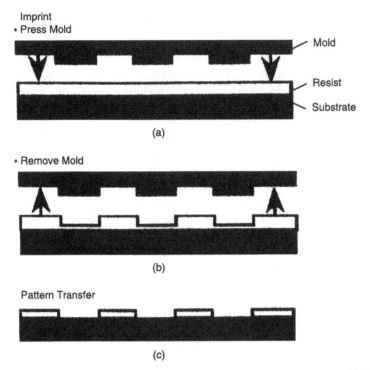

Figure 12.18. Schematics of steps in a nanoimprint lithography process: (a) a mold of a hard material made by electron-beam lithography is pressed into a resist, a softer material, to make an imprint; (b) the mold is then lifted off; (c) the remaining soft material is removed by etching from the bottom of the groves. [With permission from S. Y. Chou et al., *Science* **272**, 85 (1996).]

mold can be a metal, insulator, or semiconductor fabricated by conventional lithographic methods. Nanoimprint lithography can produce patterns on a surface having 10 nm resolution at low cost and high rates because it does not require the use of a sophisticated radiation beam generating patterns for the production of each structure.

The scanning tunneling microscope (STM), described in detail in Section 2.3.3, uses a narrow tip to scan across the surface of the material about a nanometer above it. When a voltage is applied to the tip, electrons tunnel from the surface of the material and a current can be detected. If the tip is kept at a constant distance above the surface, then the current will vary as the tip scans the surface. The amount of detected current depends on the electron density at the surface of the material, and this will be higher were the atoms are located. Thus mapping the current by scanning the tip over the surface produces an image of the atomic or molecular structure of the surface. An alternate mode of operation of the STM is to keep the current constant, and monitor the deflection of the cantilever on which the tip is held. In this mode the recorded cantilever deflections provide a map of the atomic structure of the surface.

The STM has been used to build nanosized structures atom by atom on the surface of materials. An adsorbed atom is held on the surface by chemical bonds with the atoms of the surface. When such an atom is imaged in an STM, the tip has a trajectory of the type shown in Fig. 12.19a. The separation between the tip and the adsorbed atom is such that any forces between them are small compared to the forces binding the atom to the surface, and the adsorbed atom will not be disturbed by the passage of the tip over it. If the tip is moved closer to the adsorbed atom (Fig. 12.19b) such that the interaction of the tip and the atom is greater than that between the atom and the surface, then the atom can be dragged along by the tip. At any point in the scan the atom can be reattached to the surface by increasing the separation between the tip and the surface. In this way adsorbed atoms can be rearranged on the surfaces of materials, and structures can be built on the surfaces atom by atom. The surface of the material has to be cooled to liquid helium temperatures in order to reduce thermal vibrations, which may cause the atoms to diffuse thermally, thereby disturbing the arrangement of atoms being assembled. Thermal diffusion is a problem because this method of construction can be carried out only on materials in which the lateral or in-plane interaction of the adsorbed atom and

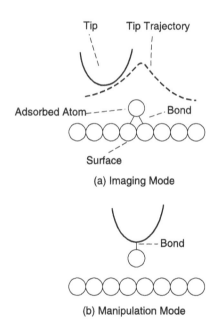

Figure 12.19. Illustration of a trajectory of a scanning tunneling microscope (STM) tip over an atom adsorbed on the surface of a material (a) when the tip is not in contact with the surface but is close enough to obtain an image and (b) when the tip is sufficiently close to the surface such that the adsorbed atom bonds to the tip. (Adapted from D. Eigler, in *Nanotechnology*, G. Timp, ed., AIP Press, New York, 1998, p. 425.)

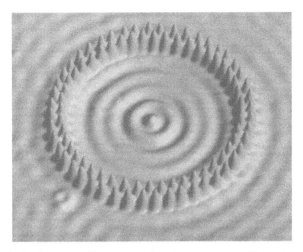

Figure 12.20. A circular array of iron atoms placed on a copper surface using an STM tip, as described in the text, forming a so-called quantum corral. The ripples inside the corral represent the surface distribution of electron density that arises from the quantum-mechanical energy levels of a circular two-dimensional potential well. [With permission from D. Eigler in Nanotechnology, G. Timp, ed., Springer Verlag Heidelberg, 1998, p. 443.]

the atoms of the surface is not too large. The manipulation also has to be done performed under ultrahigh vacuum in order to keep the surface of the material clean.

Figure 12.20 shows a circular array of iron atoms on a copper surface, called a *quantum corral*, assembled by STM manipulation. The wavelike structure inside the corral is the surface electron density distribution inside the well corresponding to three quantum states of this two-dimensional circular potential well, in effect providing a visual affirmation of the electron density predicted by quantum theory. This image is taken using an STM with the tip at such a separation that it does not move any of the atoms. The adsorbed atoms in this structure are not bonded to each other. The atoms will have to be assembled in three-dimensional arrays and be bonded to each other to use this technique to build nanostructures. Because the construction three-dimensional structures has not yet been achieved, the slowness of the technique together with the need for liquid helium cooling and high vacuum all indicate that STM manipulation is a long way from becoming a practical fabrication technique for large-scale production of nanostructures. It is important, however, in that it demonstrates that building nanostructures atom by atom is feasible, and it can be used to build interesting structures such as the quantum corral in order to study their physics.

PROBLEMS

12.1. Explain why a nanostructured multilayer is stronger if there is a mismatch between the lattice parameters of the adjacent layers.

12.2. It is not clear that the Hall–Petch equation (12.3) applies to all materials with nanosized grains. Let us assume that it does for a compact of brass alloy of Cu and Zn where the parameters of the equation are $\sigma_0 = 25$ Mpa and $K = 18.75$ Mpa$/D^{1/2}$, where D is in micrometers. Calculate the yield strength for grain sizes of 10 and 50 nm. What are some technological implications of this result?

12.3. Consider a cubic object made of silicon spheres, each having a radius of 5 nm, and all in direct contact with each other either via a simple cubic structure or via a FCC structure. Calculate the volume fraction of each structure that is empty. Which structure will have the higher yield strength? Why?

12.4. One way to make a bulk material consisting of grains is to compress the grains at high pressure and temperature. It has been found that compacts that have two different grain sizes such as 100 and 1000 nm are mechanically stronger. The word *bimodal* is used to characterize such materials. Explain why a bimodal bulk material might be stronger.

12.5. According to Stokes' law, the force needed to move a sphere of radius R through a fluid having a viscosity η is $F = 6\pi\,\eta R\,dx/dt$. Consider a spherical ball made of silicon at the end of a spring. The spring has a force constant of 200 N/m. The ball is immersed in water, which has a viscosity of 1.8×10^{-3} Pa·s. Calculate the frequency of vibration as a function of the radius of the ball from 10 to 100 nm in 10-nm intervals. Make a plot of the frequency versus the radius of the ball.

12.6. An approximate expression for the vibrational frequency of a beam clamped at both ends is $f = [E/\rho]^{1/2}[a/L^2]$, where a is the thickness of the beam, L is the length, ρ is the density, and E is Young's modulus. For the case of a silicon beam 5 nm thick, calculate the frequency versus the length of the beam from 10 to 100 nm in 10-nm intervals. The density of silicon is 2.329 gm/cm^3, and E is 130 GPa.

12.7. Consider a clamped beam of silicon with a very thin layer of material that would absorb an explosive molecule such as TNT. The beam has dimensions $5 \times 1 \times 1$ nm. Calculate the change in frequency given that one TNT molecule, $[C_7H_6(NO_2)_3]$, is absorbed on the beam. What are some applications of this result?

12.8. Given a silicon beam 10 nm long with a square cross section of 1×1 nm, calculate the number of atoms in the beam. What percentage of the atoms are on the large surface of the beam? How will the number of atoms on the surface influence the mechanical properties?

12.9. Let us say that we have a nanosized pendulum with a 1×1-nm cube on the end. A 25.0-mW beam of light from a helium–neon laser is focused to a spot of 1 μm^3 and is incident on a side of the cube perpendicular to the length of the pendulum. The pendulum is in an evacuated glass container. Will the pendulum be deflected by the incident light? If so, calculate the initial velocity of the cube. Neglect the weight of the string holding the ball. What are the implications of this result for measuring the vibrational frequencies of nanosized cantilevers with laser light?

CHAPTER 13

Magnetism in Nanostructures

13.1. BASICS OF FERROMAGNETISM

In this chapter we discuss the effect of nanostructuring on the properties of ferromagnets, and how the size of nanograins that make up bulk magnet materials affect ferromagnetic properties. We will examine how control of nanoparticle size can be used to design magnetic materials for various applications. In order to understand the role of nanostructuring in ferromagnetism, we will present a brief overview of the properties of ferromagnets. In Section 3.2.7 we saw that certain atoms whose energy levels are not totally filled have a net magnetic moment, and in effect behave like small bar magnets. The value of the magnetic moment of a body is a measure of the strength of the magnetism that is present. Atoms in the various transition series of the periodic table have unfilled inner energy levels in which the spins of the electrons are unpaired, giving the atom a net magnetic moment. The iron atom has 26 electrons circulating about the nucleus, 18 of which are in filled energy levels that constitute the argon atom inner core of the electron configuration. The d level of the $N = 3$ orbit contains only six of the possible 10 electrons that would fill it, so it is incomplete to the extent of four electrons. This incompletely filled electron d shell causes the iron atom to have a strong magnetic moment.

When crystals such as bulk iron are formed from atoms having a net magnetic moment, a number of different situations can occur relating to how the magnetic moments of the individual atoms are aligned with respect to each other. Figure 13.1 illustrates some of the possible arrangements that can occur in two dimensions. The point of the arrow is the north pole of the tiny bar magnet associated with the atom. If the magnetic moments are randomly arranged with respect to each other, as shown in Fig. 13.1a, then the crystal has a zero net magnetic moment, and this is referred to as the *paramagnetic state*. The application of a dc magnetic field aligns some of the moments, giving the crystal a small net moment. In a ferromagnetic crystal these moments all point in the same direction, as shown in Fig. 13.1b, even when no dc magnetic field is applied, so the whole crystal has a magnetic moment and behaves like a bar magnet producing a magnetic field outside of it. If a crystal is composed of two types of atoms, each having a magnetic moment of a different

The Physics and Chemistry of Nanosolids. By Frank J. Owens and Charles P. Poole, Jr.
Copyright © 2008 John Wiley & Sons, Inc.

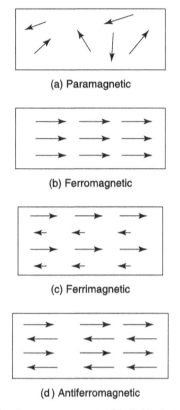

(a) Paramagnetic

(b) Ferromagnetic

(c) Ferrimagnetic

(d) Antiferromagnetic

Figure 13.1. Illustration of various arrangements of individual atomic magnetic moments that constitute paramagnetic, ferromagnetic, ferrimagnetic, and antiferromagnetic materials.

strength (indicated in Fig. 13.1c by the length of the arrow), the arrangement shown in Fig. 13.1c can occur, and it is called *ferrimagnetic*. Such a crystal will also have a net magnetic moment, and behave like a bar magnet. In an *antiferromagnet* the moments of identical atoms are arranged in an antiparallel scheme, that is, opposite to each other, as shown in Fig. 13.1d, and hence the material has no net magnetic moment. In the present chapter we are concerned mainly with ferromagnetic and to a lesser extent antiferromagnetic ordering.

Now let us consider the question of why the individual atomic magnets align in some materials and not others. When a dc magnetic field is applied to a bar magnet, the magnetic moment tends to align with the direction of the applied field. In a crystal each atom having a magnetic moment has a magnetic field about it. If the magnetic moment is large enough, the resulting large dc magnetic field can force a nearest neighbor to align in the same direction provided the interaction energy is higher than the thermal vibrational energy $k_B T$ of the atoms in the lattice. The interaction between atomic magnetic moments is of two types: the so-called exchange interaction and the dipolar interaction. The exchange interaction is a purely quantum-mechanical effect, and is generally the stronger of the two interactions.

In the case of a small particle such as an electron that has a magnetic moment $g\mu_B$, the application of a dc magnetic field forces its spin vector to align such that it can have only two projections in the direction of the dc magnetic field, which are $\pm\frac{1}{2}$ $g\mu_B$, where μ_B is the unit magnetic moment called the *Bohr magneton*, and $g = 2.0023$ is the dimensionless gyromagnetic ratio of a free electron. The wavefunction representing the state $+\frac{1}{2}g\mu_B$ is designated α, and for $-\frac{1}{2}g\mu_B$ it is β. The numbers $\pm\frac{1}{2}$ are called the *spin quantum numbers* m_s. For a two-electron system it is not possible to specify which electron is in which state. The Pauli exclusion principle does not allow two electrons in the same energy level to have the same spin quantum numbers m_s. Quantum mechanics deals with this situation by requiring that the wavefunction of the electrons be antisymmetric, that is, change sign if the two electrons are interchanged. The form of the wavefunction that meets this condition is $(\frac{1}{2})^{-1/2}[\Psi_A(1)\Psi_B(2) - \Psi_A(2)\Psi_B(1)]$. The electrostatic energy for this case is given by the expression

$$E = \int \left[\frac{\frac{1}{2}e^2}{r_{12}}\right][\Psi_A(1)\Psi_B(2) - \Psi_A(2)\Psi_B(1)]^2 dV_1 \, dV_2 \qquad (13.1)$$

which involves carrying out an integration over the volume. Expanding the square of the wavefunctions gives two terms;

$$E = \int \left[\frac{e^2}{r_{12}}\right][\Psi_A(1)\Psi_B(2)]^2 dV_1 \, dV_2$$
$$- \int \left[\frac{e^2}{r_{12}}\right]\Psi_A(1)\Psi_B(1)\Psi_A(2)\Psi_B(2) dV_1 \, dV_2. \qquad (13.2)$$

The first term is the normal Coulomb interaction between the two charged particles. The second term, called the *exchange interaction*, represents the difference in the Coulomb energy between two electrons with spins that are parallel and antiparallel. It can be shown that under certain assumptions the exchange interaction between the spins S_1 and S_2 can be written in a much simpler form as $J \, S_1 \cdot S_2$, where J is called the *exchange integral*, or *exchange interaction constant*. This is the form used in the Heisenberg model of magnetism. For a ferromagnet J is negative, and for an antiferromagnet it is positive. The exchange interaction, because it involves the overlap of orbitals, is primarily a nearest-neighbor interaction, and it is generally the dominant interaction. The other interaction, which can occur in a lattice of magnetic ions, called the *dipole–dipole interaction*, has the form

$$\frac{(\mu_1 \cdot \mu_2)}{r^3} - 3(\mu_1 \cdot r)\frac{(\mu_2 \cdot r)}{r^5} \qquad (13.3)$$

where **r** is a vector along the line separating the two magnetic moments μ_1 and μ_2 and r is the magnitude of this distance.

The magnetization M of a bulk sample is defined as the total magnetic moment per unit volume. It is the vector sum of all the magnetic moments of the magnetic atoms in the bulk sample divided by the volume of the sample. It increases strongly at the Curie temperature T_c, the temperature at which the sample becomes ferromagnetic, and the magnetization continues to increase as the temperature is lowered further below T_c. It has been found empirically that below the Curie temperature the magnetization depends on the temperature as (Bloch equation)

$$M(T) = M(0)(1 - cT^{3/2}) \tag{13.4}$$

where $M(0)$ is the magnetization at zero degrees Kelvin and c is a constant. The susceptibility χ of a sample is defined as the ratio of the magnetization at a given temperature to the applied field H, that is, $\chi = M/H$.

Generally for a bulk ferromagnetic material below the Curie temperature, the magnetic moment M is less than the moment the material would have if every individual atomic moment were aligned in the same direction. The reason for this is because of the existence of domains. *Domains* are regions in which all the atomic moments point in the same direction so that within each domain the magnetization is saturated, attaining its maximum possible value. However, the magnetization vectors of different domains in the sample are not all parallel to each other. Thus the total sample has a magnetization less than the value for the complete alignment of all moments. Some examples of domain configurations are illustrated in Fig. 13.2a. They exist because the magnetic energy of the sample is lowered by the formation of domains.

Applying a dc magnetic field can increase the magnetic moment of a sample in one of two ways: (1) the first process occurs in weak applied fields when the volume of the domains that are oriented along the field direction increases; (2) the second process dominates in stronger applied fields that force the domains to rotate toward the direction of the field. Both of these processes are illustrated in Fig. 13.2b. Figure 13.3 shows a schematic plot of the magnetization curve of a ferromagnetic material. It is a plot of the total magnetization of the sample M versus the applied dc field strength H. In the mks (millimeter–kilogram–second) system the units of both H and M are amperes per meter; in the cgs (centimeter–gram–second) system the units of M are emu/g (electromagnetic units per gram), and the units of H are oersteds (Oe). Initially as H increases, M increases until a saturation point M_s is reached. When H is decreased from the saturation point, M does not decrease to the same value it had earlier when the field was increasing; rather, it is higher on the curve of decreasing field. This is called *hysteresis*. It occurs because the domains that were aligned with the increasing field do not return to their original orientations when the field is lowered. When the applied field H is returned to zero, the magnet still has a magnetization, referred to as the *remnant*

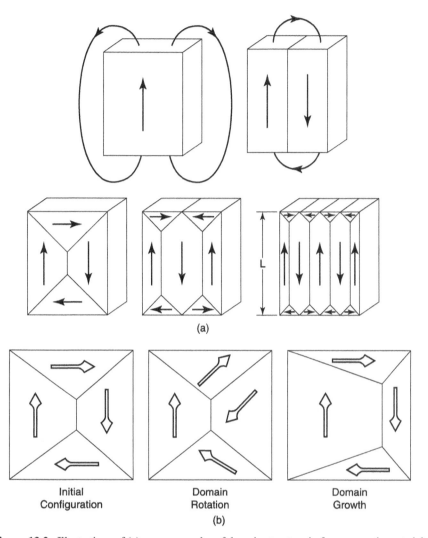

Figure 13.2. Illustrations of (a) some examples of domain structure in ferromagnetic materials and (b) ways in which the domains can change by the mechanisms of rotation and growth when a dc magnetic is applied.

magnetization M_r. In order to remove the remnant magnetization, a field H_c has to be applied in the opposite direction to the initial applied field, as shown in Fig. 13.3. This field, called the *coercive field*, causes the domains to rotate back to their original positions. The properties of the magnetization curve of a ferromagnetic have a strong bearing on the use of magnet materials, and there is much ongoing research to design permanent magnets having different kinds of magnetization curves.

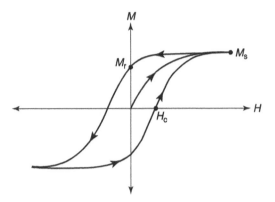

Figure 13.3. Plot of the magnetization M versus an applied magnetic field H for a hard ferromagnetic material, showing the hysteresis loop with the coercive field H_c, the remnant magnetization M_r, and the saturation magnetization M_s, indicated.

13.2. BEHAVIOR OF POWDERS OF FERROMAGNETIC NANOPARTICLES

13.2.1. Properties of a Single Ferromagnetic Nanoparticle

The existence of domains in magnetic materials as described above leads to a hysteresis of the magnetization when it is measured as a function of increasing and decreasing applied dc magnetic field. When the size of an individual magnetic particle is reduced below some critical length, it is no longer energetically favorable to form domains, and the particles exist as a single domain. For example, Fe_2O_3 nanoparticles become single-domain below $\sim 100\,nm$ and $CoFe_2O_4$ magnetic nanoparticles are single-domain below 50 nm. The existence of single-domain nanoparticles results in the phenomenon of *superparamagnetism*, where the magnetization as a function of increasing and decreasing magnetic field displays no hysteresis. In a single-domain ferromagnetic nanoparticle the magnetic moment of the individual atoms are aligned in the same direction, giving the particle a total magnetic moment, which is the sum of the individual moments of the constituent atoms. When a dc magnetic field is applied to a collection or powder of monodomain ferromagnetic nanoparticles, the total magnetic moment of each particle aligns with the dc magnetic field. As we will discuss below, this alignment can be eliminated at some higher temperature where the thermal energy k_BT of the particles exceeds the energy of interaction $\boldsymbol{\mu}\cdot\mathbf{H}$ between the magnetic moment of the particles and the applied dc magnetic field.

For a monodomain nanoparticle, the total magnetic moment of the particle will be directly proportional to the number of magnetic atoms N, in the particle, that is, $N\mu_i$, where μ_i is the magnetic moment of the individual atoms. For the case of an iron nanoparticle, the increase in total magnetic moment as a function of particle diameter can be obtained using the relationship between the number of atoms and the diameter of the particle given in Table 1.1 (of Chapter 1). Figure 13.4 is a plot of the relative

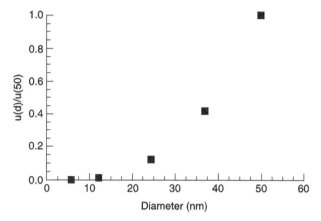

Figure 13.4. Relative change of the total magnetic moment μ of an iron nanoparticle versus the diameter d of the particle.

change of the total magnetic moment of an iron nanoparticle as a function of particle diameter in the monodomain region. However, when the particles become multi-domain, the increase will begin to level off. The plot in Fig. 13.4 does not account for the reduction of the net moment of the particle because of disordering of the atomic moments in the surface layers of the particles. This effect will be discussed in a later section. As discussed above, the interaction between the magnetic atoms of the particles that produces the net alignment of the atomic moments is the exchange interaction $JS_1 \cdot S_2$, where the exchange integral J involves an overlap of the outer wavefunctions of nearest-neighbor atoms. The strength of this interaction will determine the transition temperature. A stronger interaction means that more thermal energy will be needed to cause the moments to become disordered and the ferromagnetism to disappear, resulting in a higher Curie temperature. Most ferromagnetic materials are metals. We saw earlier that the lattice parameters of metals do not change much until the particle size is quite small, at which point it decreases slightly, meaning that there should be a slight increase in the Curie temperature. However, if the magnetic metal nanoparticles have an oxide layer on them as most do, the lattice parameters of the metal may change with particle size because of interfacial stress due to the mismatch of the lattice parameter of metal and the metal oxide. For example, the lattice parameter of nickel is 0.3524 nm, and that of nickel oxide is 0.4203 nm. It has been shown by high-resolution TEM (HRTEM) that nickel nanoparticles have a 2–3-nm layer of nickel oxide on their surfaces. The mismatch of the lattice parameters produces an interfacial stress that causes the lattice parameters of the nickel to expand at small particle sizes. Figure 13.5 is a plot of the measured lattice parameter of nickel versus the reciprocal of particle size. The effect occurs only at very small particle sizes, less than $\sim$40 nm. This means that there will be a reduction in the value of the exchange integral and therefore the magnetization with reduced particle size in a given dc magnetic field.

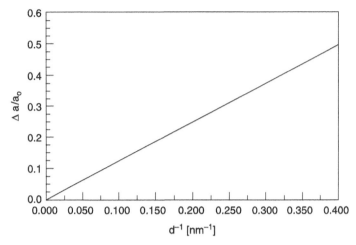

Figure 13.5. Plot of increase in lattice parameter a of nickel nanoparticles versus inverse of particle size due to the interfacial stress caused by the mismatch of the lattice parameters of the Ni lattice and the NiO surface layer. [Adapted from B. Rellinghaus et al., *Eur. Phys. J.* **16**, 249 (2001).]

13.2.2. Dynamics of Individual Magnetic Nanoparticles

The study of magnetic materials, particularly of films made of nanomagnets, sometimes called *mesoscopic magnetism*, is driven by the desire to increase storage space on magnetic storage devices such as hard drives in computers. The basic information storage mechanism involves alignment of the magnetization in one direction of a very small region on the magnetic tape called a *bit*. To achieve a storage of 10 gigabits (10^{10} bits) per square inch, a single bit would be approximately 1 μm long and 70nm wide. The film thickness could be about 30nm. Existing magnetic storage devices such as hard drives are based on tiny crystals of cobalt chromium alloys. One difficulty that arises when bits become less than 10nm in size is that the magnetization vector can be flipped by random thermal vibrations, in effect erasing the memory. One solution to this is to use nanosized grains, which have higher saturation magnetizations, and hence stronger interactions between the grains. The interaction between the grains is the dipolar interaction given by Eq. (13.3), where μ becomes the total magnetic moment of the particle. This interaction is relatively short-ranged and will occur only when the nanoparticles are densely packed.

As discussed above, when nanoparticles reach a certain small diameter, the magnetic vectors of the individual atoms of the particle become aligned in the ordered pattern of a single domain in the presence of a dc magnetic field, eliminating the complication of domain walls and regions having magnetization in different directions. The Stoner–Wohlfarth (SW) model has been used to account for the dynamical behavior of small nanosized elongated magnetic grains. Elongated grains are generally of the type used in magnetic storage devices. The SW model postulates that in the absence of a dc magnetic field ellipsoidal magnetic particles can have only two stable orientations for their magnetization, either up or down with respect to the long axis of the magnetic particles, as illustrated in Fig. 13.6. The energy versus orientation of the vectors is a symmetric double-well potential with a barrier between the two orientations. The particle may flip its orientation by thermal activation, due to an Arrhenius process where the probability P for reorientation is given by

$$P \sim \exp\frac{-E}{k_B T} \tag{13.5}$$

where E is the height of the energy barrier between the two orientations. The particle can also flip its orientation by a much lower probability process called *quantum-mechanical tunneling*. This can occur when the thermal energy $k_B T$ of the particle is much less than the barrier height. This process is a purely quantum-mechanical effect resulting from the fact that solution to the wave equation for this system predicts a small probability for the up state of the magnetization to change to the down state. If a magnetic field is applied the shape of the potential changes, as shown by the dashed line in Fig. 13.6, and one minimum becomes unstable at the coercive field.

The SW model provides a simple explanation for many of the magnetic properties of small magnetic particles, such as the shape of the hysteresis loop. However, the model has some limitations. It overestimates the strength of the coercive field because it allows only one path for reorientation. The model assumes that the magnetic energy of a particle is a function of the collective orientation of the spins of the magnetic atoms in the particle, and the effect of the applied dc magnetic field. This implies that the magnetic energy of the particle depends on its volume.

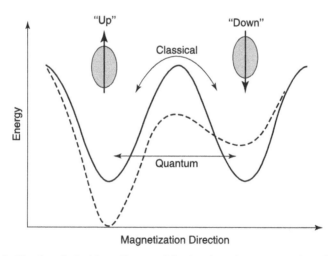

Figure 13.6. Sketch of double-well potential showing the energy plotted versus the orientation of the magnetization for up and down orientations of magnetic nanoparticles in the absence (—) and presence (- - - -) of an applied magnetic field. [With permission from D. D. Awschalom and D. P. DiVincenzo, *Phys. Today* 44 (April 1995).]

However, when particles are in the order of 6 nm in size, most of their atoms are on the surface, which means that they can have magnetic properties very different from those of larger grain particles. It has been shown that treating the surfaces of nanoparticles of α-Fe that are 600 nm long and 100 nm wide with various chemicals can produce variations in the coercive field by as much as 50%, underlining the importance of the surface of nanomagnetic particles in determining the magnetic properties of the grain. Thus the dynamical behavior of very small magnetic particles is somewhat more complicated than predicted by the SW model, and remains a subject of continuing research.

13.2.3. Measurements of Superparamagnetism and the Blocking Temperature

Consider a powder of ferromagnetic nanoparticles cooled to a very low temperature, say, 4.2 K, in zero magnetic field. The direction of the total magnetic moment of each particle is frozen in a random arrangement and the collection of particles will have a very low or zero net magnetization. If a dc magnetic field is applied and the susceptibility is measured as a function of increasing temperature, some fraction of the particles will begin to align with the field, and the susceptibility will increase with increasing temperature. If the particles are cooled to this low temperature in a dc magnetic field, the direction of maximum magnetization of the particles (often called the "easy direction") will align with the applied dc field. When the temperature is raised, a temperature is reached called the *blocking temperature* T_B, where the thermal energy $k_B T$ of the particles is greater than the energy of interaction $\mathbf{\mu} \cdot \mathbf{H}$ of the moments of the particle $\mathbf{\mu}$ with the dc magnetic field $\mathbf{H}$. At this temperature the direction of the magnetic moments of the particles begins to fluctuate about the direction of the dc magnetic field and the moments become disordered. The average time τ between reorientations is given by

$$\tau = \tau_0 \exp \frac{\mu H}{k_B T}. \tag{13.6}$$

Thus the blocking temperature T_B is a temperature below which the magnetic moments of the particles are locked in the direction of the magnetic field and above which they are randomly oriented. The blocking temperature can be measured by comparing the magnetization versus increasing temperature with zero-field-cooled (ZFC) and field-cooled (FC) particles. Figure 13.7 is a plot of the susceptibility of 8.5-nm particles of $CoFe_2O_4$ as a function of increasing temperature in field-cooled and zero-field-cooled material in different applied magnetic fields. The temperature where the FC and ZFC susceptibilities diverge is the blocking temperature T_B. This temperature corresponds to the peak in magnetization versus increasing temperature data for ZFC material. Figure 13.8 is a plot of the magnetization versus increasing temperature for ZFC $CoFe_2O_3$ nanoparticles of different sizes measured in a 100-G applied field. Figure 13.9 is a plot of the blocking temperature versus

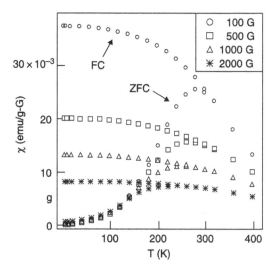

Figure 13.7. A measurement of the susceptibility of 8.5-nm particles of $CoFe_2O_4$ in field-cooled (FC) and zero-field-cooled (ZFC) material as a function of increasing temperature. [Adapted from C. R. Vestal and Z. J. Zhang *Int. J. Nanotechnol.* **1**, 240 (2004).]

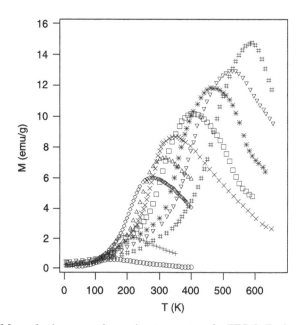

Figure 13.8. Magnetization versus increasing temperature for ZFC $CoFe_2O_4$ nanoparticles of different sizes measured in a 100-G field. The particle sizes range from 5 nm for the bottom curve to 30 nm for the top curve. [Adapted from C. R. Vestal and Z. J. Zhang, *Int. J. Nanotechnol.* **1**, 240 (2004).]

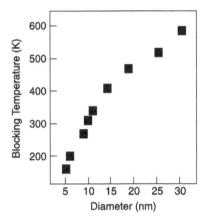

Figure 13.9. Blocking temperature versus particle size for $CoFe_2O_4$ nanoparticles obtained from the data in Fig.13.8. [Adapted from C. R. Vestal and Z. J. Zhang, *Int. J. Nanotechnol.* **1**, 240 (2004).]

particle size for the $CoFe_2O_3$ nanoparticles. The blocking temperature increases with particle size because the total magnetic moment of the particles increases with size. Figure 13.10 shows how the saturation magnetization is affected as the particle size increases. The decrease in magnetization with reduced particle size is a result of the decrease in the number of magnetic atoms in the particle with reduced size and the increased number of ferromagnetic atoms on the surface of the particles. The magnetic atoms on the surface of a ferromagnetic nanoparticle have weaker exchange interactions because they have fewer nearest neighbors. Further, the atoms on the surface are less tightly bound than the atoms in the interior of the particle, and hence can more easily fluctuate about their equilibrium positions. As a result of this, the magnetic

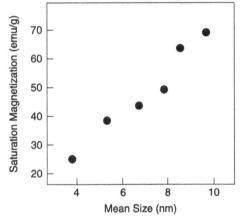

Figure 13.10. Saturation magnetization versus particle size for $CoFe_2O_4$ nanoparticles. [Adapted from C. R. Vestal and Z. J. Zhang, *Int. J. Nanotechnol.* **1**, 240 (2004).]

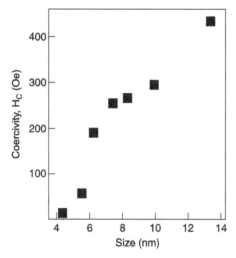

Figure 13.11. Coercivity versus temperature in the irreversible temperature region for nano-particles of $MnFe_2O_4$. [Adapted from C. R. Vestal and Z. J. Zhang, *Int. J. Nanotechnol.* 1, 240 (2004).]

moment of the surface atoms may become randomly oriented, causing a decrease in the total magnetic moment of the nanoparticle. The coercivity in the irreversible temperature range also increases with particle size as shown in Fig. 13.11.

13.2.4. Nanopore Containment of Magnetic Particles

Nanoferromagnets have also been studied in the empty spaces of porous substances. In fact, there are actually naturally occurring materials having molecular cavities filled with nanosized magnetic particles. Ferritin is a biological molecule, 25% iron by weight, which consists of a symmetric protein shell in the shape of a hollow sphere having an inner diameter of 7.5 nm and an outer diameter of 12.5 nm. The molecule plays the role in biological systems as a means of storing Fe^{3+} for an organism. In the human body, 25% of the iron is in ferritin and 70% is in hemoglobin. The ferritin cavity is normally filled with a crystal of iron oxide $5Fe_2O_3 \cdot 9H_2O$. The iron oxide can be synthesized into the cavity from solution, with the number of iron atoms per protein controlled from a few to a few thousand per protein molecule. The magnetic properties of the molecule depend on the number and kind of particles in the cavity, and the system can be engineered to be ferromagnetic or antiferromagnetic. The particles in the pores also display blocking temperature behavior. Figure 13.12 shows that the blocking temperature decreases with a decrease in the number of iron atoms in the cavity. Ferritin also displays magnetic quantum tunneling at very low temperatures. In zero magnetic field and the very low temperature of 0.2 K, the magnetization tunnels coherently back and forth between two minima. This effect appears as a resonance line in frequency-dependent magnetic susceptibility data. Figure 13.13 shows the results of a measurement of the resonant frequency

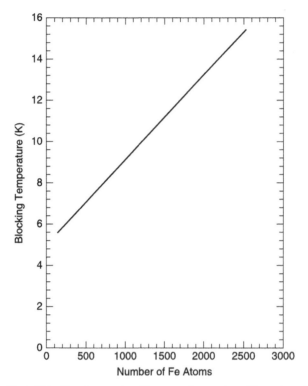

Figure 13.12. Plot of blocking temperature T_B versus the number of iron atoms in the cavity of ferritin. [Adapted from D. D. Awschalom, *Phys. Today*, p. 44 (April 1995).]

of this susceptibility versus the number of iron atoms per molecule. We see that the frequency decreases from 3×10^8 Hz for 800 atoms to 10^6 Hz for 4600 atoms. The resonance disappears when a dc magnetic field is applied, and the symmetry of the double-walled potential is broken.

13.3. FERROFLUIDS

Ferrofluids, also called *magnetofluids*, are colloids consisting typically of 10-nm magnetic particles coated with a surfactant to prevent aggregation, and suspended in a liquid such as transformer oil or kerosene. The nanoparticles are single-domain magnets, and in zero magnetic field, at any instant of time, the magnetization vector of each particle is randomly oriented so that the liquid has a zero net magnetization. When a dc magnetic field is applied the magnetizations of the individual nanoparticles, all align with the direction of the field, and the fluid acquires a net magnetization. Typically ferrofluids employ nanoparticles of magnetite (Fe_3O_4). Figure 13.14 shows the magnetization curve for a ferrofluid made of 6-nm Fe_3O_4 particles exhibiting almost immeasurable hysteresis. Ferrofluids are soft magnetic

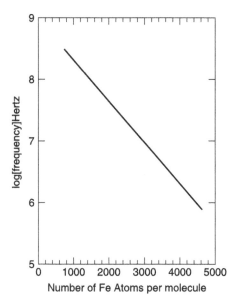

Figure 13.13. Resonance frequency of ferritin as a function of the number of atoms in the cavity of the molecule. [Adapted from D. D. Awschalom, and D. P. DiVincenzo *Phys. Today*, p. 44 (April 1995).]

materials that are superparamagnetic. Interestingly, suspensions of magnetic particles in fluids have been used since the 1940s in magnetic clutches, but the particles were larger, having micrometer dimensions. Application of a dc magnetic field to this fluid causes the fluid to congeal into a solid mass, and in the magnetic state the material is

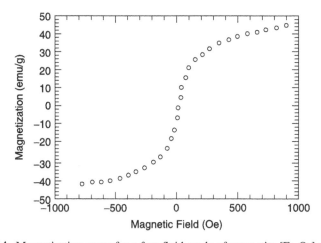

Figure 13.14. Magnetization curve for a ferrofluid made of magnetite [Fe_3O_4] nanoparticles showing the soft (nonhysteretic) magnetic behavior. (1 Oe = 10^{-4}T). [Adapted from D. K. Kim, *J. Magn. Magn. Mater.* **225**, 30 (2001).]

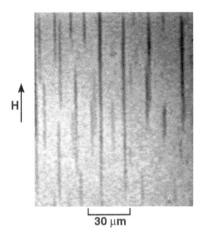

H

30 μm

Figure 13.15. Photograph taken through an optical microscope of chains of magnetic nanoparticles formed in a film of a ferrofluid when the dc magnetic field is parallel to the plane of the film. [With permission from H. E. Hornig et al., *J. Phys. Chem. Solids* **62**, 1749 (2001).]

not a liquid. A prerequisite for the existence of a ferrofluid is that the magnetic particles have nanometer sizes. Such fluids will not congeal into solids when a dc magnetic field is applied.

Ferrofluids have a number of interesting properties, such as magnetic-field-dependent anisotropic optical properties. Analogous properties are observed in liquid crystals, which consist of long molecules having large electric dipole moments, which can be oriented by the application of an electric field in the fluid phase. Electric-field-modulated bifringence or double refraction of liquid crystals is widely used in optical devices, such as liquid crystal displays in digital watches, and screens of portable computers. This suggests a potential application of ferrofluids employing magnetic-field-induced bifringence. To observe the behavior, the ferrofluid is sealed in a glass cell having a thickness of several micrometers. When a dc magnetic field is applied parallel to the surface and the film is examined by an optical microscope, it is found that some of the magnetic particles in the fluid agglomerate to form needle-like chains parallel to the direction of the magnetic field. Figure 13.15 shows an image of the chains taken through an optical microscope. As the magnetic field increases, more particles join the chains, and the chains become thicker and longer. The separation between the chains also decreases. Figures 13.16a and 13.16b show plots of the chain separation and the chain width as a function of the dc magnetic field. When the field is applied perpendicular to the face of the film, the ends of the chains arrange themselves in the pattern shown in Fig. 13.17, which is also a photograph taken through an optical microscope. Initially at low fields the ends of the chains are randomly distributed in the plane of the fluid. As the field increases, a critical field is reached where the chain ends become ordered in a two-dimensional hexagonal array as shown in Fig. 13.17. This behavior is

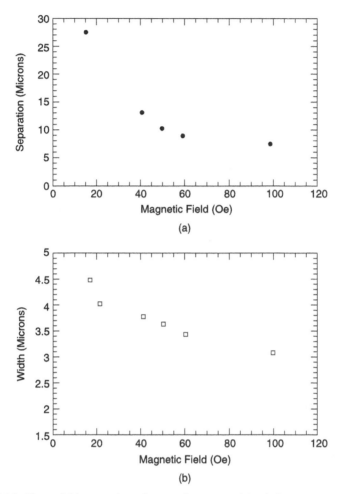

Figure 13.16. Plots of (a) separation of magnetic nanoparticle chains versus strength of a magnetic field applied parallel to the surface of the film and (b) thickness of the chains as a function of dc magnetic field strength (1 Oe = 10^{-4} T). [Adapted from H. E. Hornig et al., *J. Phys. Chem. Solids* **62**, 1749 (2001).]

analogous to the formation of the vortex lattice in type II superconductors, which will be discussed in Section 15.4.

The formation of the chains in the ferrofluid film when a dc magnetic field is applied makes the fluid optically anisotropic. Light, or more generally electromagnetic waves, have oscillatory magnetic and electric fields perpendicular to the direction of propagation of the beam. Light is linearly polarized when these vibrations are confined to one plane perpendicular to the direction of propagation, rather than having random transverse directions. When linearly polarized light is incident on a magnetofluid film to which a dc magnetic field is applied, the light emerging from

Figure 13.17. Optical microscopic image of the ends of chains of magnetic nanoparticles in a ferrofluid film when the dc magnetic field is perpendicular to the surface of the film. The field strength is high enough to form the hexagonal lattice configuration. [With permission from H. E. Hornig et al., *J. Phys. Chem. Solids* **62**, 1749 (2001).]

the other side of the film is elliptically polarized. Elliptically polarized light occurs when the *E* and *H* vibrations around the direction of propagation are confined to two mutually perpendicular planes, and the vibrations in each plane are out of phase. This is called the Cotton–Mouton effect. An experimental arrangement to investigate this effect is shown in Fig. 13.18. A He–Ne laser beam linearly polarized by a polarizer is incident on the magnetofluid film. A dc magnetic field is applied parallel to the plane of the film. To examine the polarization of the light emerging from

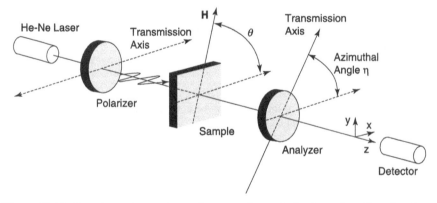

Figure 13.18. Experimental arrangement for measuring optical polarization effects in a ferrofluid film that has a dc magnetic field *H* applied parallel to its surface. [With permission from H. E. Hornig et al., *J. Phys. Chem. Solids* **62**, 1749 (2001).]

the film, another polarizer called an *analyzer* is placed between the film and the light detector, which is a photomultiplier. The intensity of the transmitted light is measured as a function of the orientation of the polarizing axis of the analyzer given by the angle η in the figure. Figure 13.19 shows that the transmitted light intensity depends strongly on the angle η. These effects could be the basis of optical switches, where the intensity of transmitted light is switched on and off using a dc magnetic field, or the orientation of a polarizer.

Ferrofluids can also form magnetic field tunable diffraction gratings. Diffraction is the result of interference of two or more light waves of the same wavelength traveling paths of slightly different lengths before arriving at a detector such as a photographic film. When the path length differs by half a wavelength, the waves destructively interfere, resulting in a dark band on the film. When the path lengths differ by a wavelength, then the waves constructively interfere, producing a bright band on the film. A diffraction grating consists of small slits separated by distances of the order of the wavelength of the incident light. We saw above that when a dc magnetic field of sufficient strength is applied perpendicular to a magnetofluid film, an equilibrium two-dimensional hexagonal lattice is formed with columns of nanoparticles occupying the lattice sites. This structure can act as a two-dimensional optical diffraction grating that diffracts incoming visible light. Figure 13.20 shows the chromatic (colored) rings of light and darkness resulting from the diffraction and interference when a focused parallel beam of white light is passed through a magnetic fluid

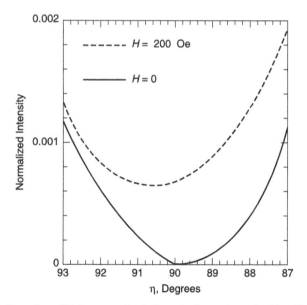

Figure 13.19. Intensity of light transmitted through the analyzer in Fig. 13.18 versus the azimuthal angle η of the light beam in zero field, and in a 200-Oe (0.02-T) applied magnetic field *H*. [Adapted from H. E. Hornig et al., *J. Phys. Chem. Solids* **62**, 1749 (2001).]

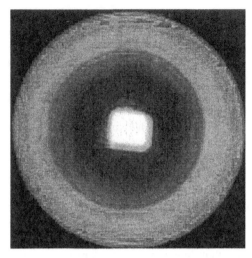

Figure 13.20. Chromatic rings resulting from the diffraction and interference of a beam of white light incident on a ferrofluid film in a perpendicular dc applied magnetic field (the original figure was in color). [With permission from H. E. Hornig et al., *J. Phys. Chem. Solids* **62**, 1749 (2001).]

film that has a magnetic field applied perpendicular to it. The diffraction pattern is determined by the equation

$$d \sin\Theta = n\lambda \tag{13.7}$$

where d is the distance between the chains of nanoparticles, Θ is the angle between the outgoing light and the direction normal to the film, n is an integer, and λ is the wavelength of the light. We saw earlier that the distance between the chains d depends on the strength of the applied dc magnetic field. In effect, we have a tunable diffraction grating, which can be adjusted to a specific wavelength by changing the strength of the dc magnetic field.

Ferrofluids have a number of present commercial uses. They are employed as contaminant exclusion seals on hard drives of personal computers, and vacuum seals for high-speed high-vacuum motorized spindles. In this latter application the ferrofluid is used to seal the gap between the rotating shaft and the pole piece support structure, as illustrated in Fig. 13.21. The seal consists of a few drops of ferrofluid in the gap between the shaft and a cylindrical permanent magnet that forms a collar around it. The fluid forms an impermeable O-ring around the shaft, while allowing rotation of the shaft without significant friction. Seals of this kind have been utilized in a variety of applications. Ferrofluids are used in audiospeakers in the voice gap of the driver to dampen moving masses. Nature even employs ferrofluids. For example, it is believed that ferrofluids play a role in the directional sensing of trout fishes. Cells near the nose of the trout are thought to contain a ferrofluid of magnetite nanoparticles. When the trout changes its orientation with respect to the earth's

Magnetically
Permeable Pole Piece

Ferrofluid

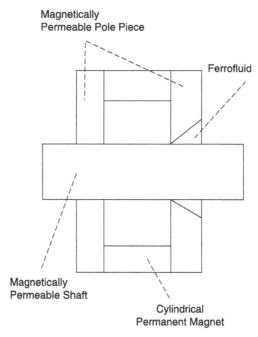

Magnetically
Permeable Shaft

Cylindrical
Permanent Magnet

Figure 13.21. Illustration of the use of a ferrofluid as a vacuum seal on a rotating magnetically permeable shaft mounted on the pole pieces of a permanent magnet.

magnetic field, the direction of magnetization in the ferrofluid of the cells changes, and this change is processed by the brain to give the fish orientational information.

13.4. BULK NANOSTRUCTURED MAGNETIC MATERIALS

In this section we discuss solid magnetic materials whose grain size is in nanometers.

13.4.1. Effect of Nanosized Grain Structure on Magnetic Properties

The diverse applications of magnets require the magnetization curve to have different properties. Magnets used in transformers and rotating electrical machinery are subjected to rapidly alternating ac magnetic fields, so they repeat their magnetization curves many times per second, causing a loss of efficiency and a rise in the temperature of the magnet. The rise in temperature is due to frictional heating from domains as they continuously vary their orientations. The amount of loss during each cycle, meaning the amount of heat energy generated during each cycle around a hysteresis loop, is proportional to the area enclosed by the loop. In these applications small or zero coercive fields are required to minimize the enclosed area. Such magnets are called *soft magnetic materials*. On the other hand, in the case of permanent magnets used as a part of high-field systems, large coercive fields are required, and the widest possible hysteresis loop is desirable. Such magnets are called *hard magnets*. High-saturation magnetizations are also needed in permanent magnets.

Nanostructuring of bulk magnetic materials can be used to design the magnetization curve. Amorphous alloy ribbons having the composition $Fe_{73.5}Cu_1Nb_3Si_{13.5}B_9$ prepared by a roller method, and subjected to annealing at 673–923 K for one hour in inert-gas atmospheres, were composed of 10-nm iron grains in solid solutions. Such alloys had a saturation magnetization M_s of 1.24 T, a remnant magnetization M_r of 0.67 T, and a very small coercive field H_C of 0.53 A/m. Nanoamorphous alloy powders of $Fe_{69}Ni_9CO_2$ having grain sizes of 10–15 nm prepared by decomposition of solutions of $Fe(CO)_5$, $Ni(CO)_4$, and $Co(NO)(CoO)_3$ in the hydrocarbon solvent decalin ($C_{10}H_{18}$) under an inert-gas atmosphere showed almost no hysteresis in the magnetization curve. Figure 13.22 presents the magnetization curve for this material. As discussed earlier, a magnetic material with grain-sized magnetic moments, which has no hysteresis at any temperature, is said to be superparamagnetic.

The strongest known permanent magnets are made of neodymium, iron, and boron. They can have remnant magnetizations as high as 1.3 T and coercive fields

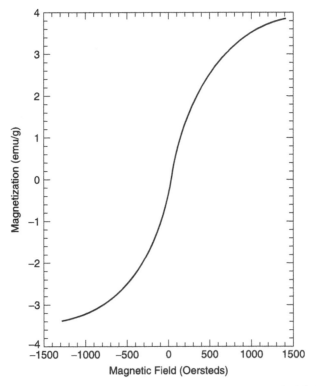

Figure 13.22. Reversible magnetization curve for nanosized powders of a Ni–Fe–Co alloy that exhibits no hysteresis (1 Oe = 10^{-4} T). [Adapted from K. Shafi et al., *J. Mater. Res.* **15**, 332 (2000).]

as high as 1.2–T. The effect of the size of the nanoparticle grain structure on $Nd_2Fe_{14}B$ has been investigated. The results, shown in Figs 13.23 and 13.24, indicate that in this material the coercive field decreases significantly below $\sim$40 nm and the remnant magnetization increases. Another approach to improving the magnetization curves of this material has been to make nanoscale compositions of hard $Nd_2Fe_{14}B$ and the soft α phase of iron. Measurements of the effect of the presence of the soft iron phase mixed in the hard material confirm that the remnant field can be increased by this approach. This is believed to be due to the exchange coupling between the hard and soft nanoparticles, which forces the magnetization vector of the soft phase to be rotated in the direction of magnetization of the hard phase.

The size of magnetic nanoparticles has also been shown to influence the value M_S at which the magnetization saturates. Figure 13.25 shows the effect of particle size on the saturation magnetization of zinc ferrite, illustrating how the magnetization increases significantly below a grain size of 20 nm. Thus, decreasing the particle size of a granular magnetic material can considerably improve the quality of magnets fabricated from it. It should be noted that this increase in magnetization with reduced grain size of the bulk is opposite to what is observed for individual nanoparticles. Perhaps this is a result of the dense packing of the grains in bulk

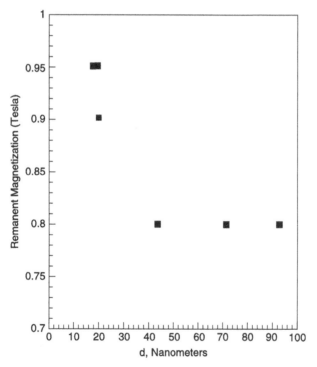

Figure 13.23. Dependence of the remnant magnetization M_r on the particle size d of the grains that form the structure of a Nd–B–Fe permanent magnet. [Adapted from A. Manaf et al., *J. Magn. Magn. Mater.* **101**, 360 (1991).]

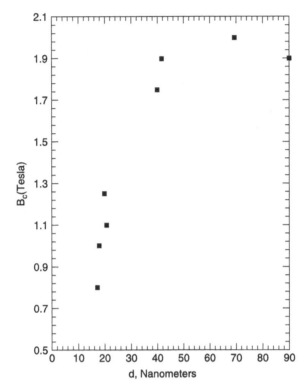

Figure 13.24. Dependence of the coercive field B_C (i.e., H_C) on granular particle size d of a Nd–B–Fe permanent magnet. [Adapted from A. Manaf et al., *J. Magn. Magn. Mater.* **101**, 360 (1991).]

materials, which does not allow the moments of the surface atoms to become disordered because of interparticle interactions.

13.4.2. Magnetoresistive Materials

Magnetoresistance is a phenomenon where the application of a dc magnetic field changes the resistance of a material. The phenomenon has been known for many years in ordinary metals, and is due to the conduction electrons being forced to move in helical trajectories about an applied magnetic field. The effect becomes evident only when the magnetic field is strong enough to curve the electron within a length equal to its mean free path. The *mean free path* is the average distance an electron travels in a metal when an electric field is applied before it undergoes a collision with atoms, defects, or impurity atoms. The resistance of a material is the result of the scattering of electrons out of the direction of current flow by these collisions. In this case the resistance increased and is thus a positive magnetoresistance effect. The magnetoresistance effect occurs in metals only at very high magnetic fields and low

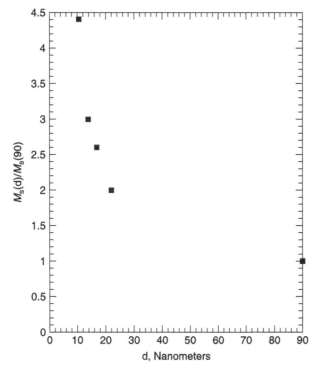

Figure 13.25. Dependence of the saturation magnetization M_s of zinc ferrite on the granular particle size d normalized to the value $M_s(90)$ for a 90-nm grain. [Adapted from C. N. Chinnasamy, *J. Phys. Condens. Matter* **12**, 7795 (2000).]

temperatures. For example, in pure copper at 4 K a field of 10 T produces a factor of 10 change in the resistance.

Because of the large magnetic field and low-temperature requirements, magneto-resistance in metals has few potential application possibilities. In the 1960s a negative magnetoresistance was observed when dc magnetic fields as low as 100 G were applied to iron whiskers, which are thin threads of iron. The effect was explained by the absence of domain structure in the whiskers. Domain boundaries called *Bloch walls* are impediments to the flow of current. A monodomain material does not have these boundaries, and the resistance is lower. In 1988 giant magnetoresistance (GMR) was observed in materials synthetically fabricated by depositing on a substrate alternate layers of nanometer-thick ferromagnetic materials and nonferromagnetic metals. A schematic of the layered structure and the alternating orientation of the magnetization in the ferromagnetic layer is shown in Fig. 13.26a. The effect was first observed in films made of alternating layers of iron and chromium, but since then other layered materials composed of alternating layers of cobalt and copper have been made that display much higher magnetoresistive effects. Figure 13.27 shows the effect of a dc magnetic field on the resistance of the

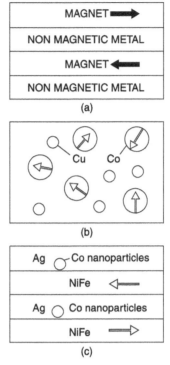

Figure 13.26. Three arrangements for producing colossal magnetoresistance: (a) layers of nonmagnetic material alternating with oppositely magnetized (arrows) ferromagnetic layers; (b) randomly oriented ferromagnetic cobalt nanoparticles (large circles) in a nonmagnetic copper matrix (small circles); (c) hybrid system consisting of cobalt nanoparticles in a silver (Ag) matrix sandwiched between nickel–iron (NiFe) magnetic layers, with alternating magnetizations indicated by arrows.

iron–chromium multilayered system. The magnitude of the change in the resistance depends on the thickness of the iron layer, as shown in Fig. 13.28, and it reaches a maximum at a thickness of 7 nm. The effect occurs because of the dependence of electron scattering on the orientation of the electron spin with respect to the direction of magnetization. Electrons whose spins are not aligned along the direction of the magnetization M are scattered more strongly than are those with their spins aligned along M. The application of a dc magnetic field parallel to the layers forces the magnetization of all the magnetic layers to be in the same direction. This causes the magnetizations pointing opposite to the direction of the applied magnetic field to become flipped. The conduction electrons with spins aligned opposite to the magnetization are more strongly scattered at the metal–ferromagnet interface, and those aligned along the field direction are less strongly scattered. Because the two spin channels are in parallel, the lower-resistance channel determines the resistance of the material.

The magnetoresistance effect in these layered materials is a sensitive detector of dc magnetic fields, and is the basis for the development of a new more sensitive reading

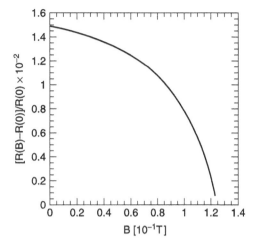

Figure 13.27. Dependence of the electrical resistance $R(B)$, relative to its value $R(0)$ in zero field, of a layered iron–chromium system on a magnetic field B applied parallel to the surface of the layers. [Adapted from R. E. Camley, *J. Phys. Condens. Matter* **5**, 3727 (1993).]

head for magnetic disks. Prior to this, magnetic storage devices have used induction coils to both induce an alignment of the magnetization in a small region of the tape (write mode) and sense the alignment of a recorded area (read mode). The magnetoresistive reading head is considerably more sensitive than the inductive coil method.

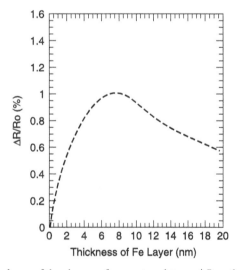

Figure 13.28. Dependence of the change of magnetoresistance ΔR on the thickness of the iron magnetic layer in a constant dc magnetic field for the Fe–Cr layered system. [Adapted from R. E. Camley, *J. Phys. Condens. Matter* **5**, 3727 (1993).]

Materials made of single-domain ferromagnetic nanoparticles with randomly oriented magnetizations embedded in nonmagnetic matrices also display giant magnetoresistance. Figure 13.26b shows a schematic of this system. The magnetoresistance in these materials, unlike that in the layered materials, is isotropic. The application of the dc magnetic field rotates the magnetization vector of the ferromagnetic nanoparticles parallel to the direction of the field, which reduces the resistance. The magnitude of the effect of the applied magnetic field on the resistance decreases with the strength of the magnetic field and as the size of the magnetic nanoparticles decreases. Figure 13.29 shows a representative measurement at $100\,K$ of a film consisting of Co nanoparticles in a copper matrix. Hybrid systems consisting of nanoparticles in metal matrices sandwiched between metal magnetic layers, as illustrated schematically in Fig. 13.26c, have also been developed and exhibit similar magnetoresistance properties.

Subsequently materials have been discovered having much larger magnetoresistive effects than the layered materials, and this phenomenon is called *colossal magnetoresistance* (CMR). These materials also have a number of application possibilities, such as in magnetic recording heads, or as sensing elements in magnetometers. The perovskitelike material $LaMnO_3$ has manganese in the Mn^{3+} valence state. If the La^{3+} is partially replaced with ions having a valence of $+2$, such as Ca, Ba, Sr, Pb, or Cd, some Mn^{3+} ions transform to Mn^{4+} to preserve the electrical neutrality. The result is a mixed-valence system of Mn^{3+}/Mn^{4+}, with the presence of many mobile charge carriers. This mixed-valence system has been shown to exhibit very large magnetoresistive effects. The unit cell of the crystal is sketched in Fig. 13.30. The particular system $La_{0.67}Ca_{0.33}MnO_x$ displays more than a 1000fold change in resistance with the application of a 6-T dc magnetic field. Figure 13.31

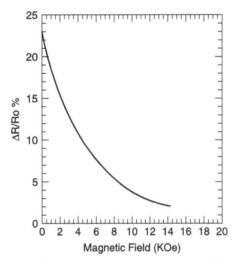

Figure 13.29. Dependence of the change of magnetoresistance ΔR versus the applied magnetic field for a thin film of Co nanoparticles in a copper matrix ($1\,k$ Oe $= 0.1\,T$). [Adapted from A. E. Berkowitz, *Phys. Rev. Lett.* **68**, 3745 (1992).]

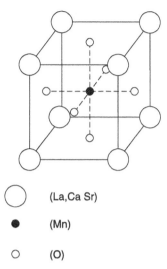

Figure 13.30. Crystal structure of $LaMnO_3$, which displays colossal magneto resistance when the La site is doped with Ca or Sr. (from F. J. Owens and C. P. Poole, Jr., *Electromagnetic Absorption in Copper Oxide Superconductors*, Academic/Plenum Publishers, 1999, p. 89.)

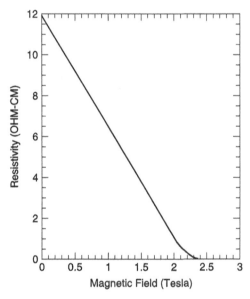

Figure 13.31. Dependence of the resistivity (normalized magnetoresistance) of La–Ca–Mn–O on an applied magnetic field in the neighborhood of the Curie temperature at 250 K. (from F. J. Owens and C. P. Poole, Jr., *Electromagnetic Absorption in Copper Oxide Superconductors*, Kluwer Academic/Plenum Publishers, 1999, p. 90.)

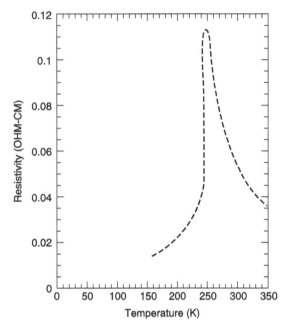

Figure 13.32. Temperature dependence of resistivity (normalized resistance) in sintered samples of La–Ca–Mn–O in zero dc magnetic field. (from F. J. Owens and C. P. Poole, Jr., *Electromagnetic Absorption in Copper Oxide Superconductors*, Kluwer Acdemic/ Plenum Publishers, 1999a, p. 89.)

shows how the normalized resistance, called the *resistivity* (normalized magnetoresistance), of a thin film of the material exhibits a pronounced decrease with increasing values of the dc magnetic field. The temperature dependence of the resistivity also displays the unusual behavior shown in Fig. 13.32 as the temperature is lowered through the Curie point. When the grain sizes of these materials are nanometers, there are dramatic changes in the magnetoresistance. Figure 13.33 is a plot of the magnetoresistance of $La_{0.9}Te_{0.1}MnO_3$ at 100 K in a 0.5-T dc magnetic field as a function of the grain size. The magnetoresistance MR is defined as

$$MR(\%) = \frac{\Delta\rho}{\rho(0)} = \frac{[\rho(0) - \rho(H)]}{\rho(0)} \times 100\% \qquad (13.8)$$

where $\rho(0)$ is the resisitivity when no magnetic field is applied and $\rho(H)$ is the resisitivity when a field H is applied. Figure 13.33 shows that when the grain size of the CMR is below 50 nm, there is a very pronounced increase in the magnetoresisitance. Similar results have been obtained in other CMR materials such as $La_{2-2x}Ca_{1+2x}Mn_2O_7$. It was also observed that the resistivity in zero field, $\rho(0)$ increased with reduced grain size. The Curie temperature and the magnetization at a given temperature are also observed to increase as the grain size decreases.

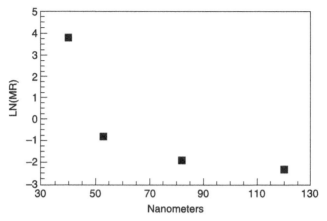

Figure 13.33. Plot of the logarithm of the magnetoresistance of $La_{0.9}Te_{0.1}MnO_3$ at 100 K in a 0.5-T dc magnetic field as a function of particle size.

A possible explanation for this is given in Fig. 13.34, which is a plot of the volume of the unit cell measured by X-ray diffraction versus the grain size. This indicates that the Mn–Mn separation decreases and therefore the interatomic exchange integral J increases resulting in an increase in the magnetization and the Curie temperature. In Section 9.5 we discussed the conductivity of a linear array of gold nanoparticles and showed that tunneling between the particles was involved in the mechanism of conductivity. The large increase in magnetoresistance below 50 nm may be due to the formation of nanosized magnetic tunnel junctions formed by the contact of the grains of the CMR material. Magnetic tunnel junctions consist of two ferromagnetic metal electrodes separated by a small insulating barrier as shown in Figure 13.35. The application of a small voltage between the electrodes causes electrons to tunnel from the negative electrode to the positive electrode. When small magnetic fields are

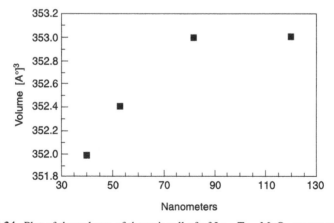

Figure 13.34. Plot of the volume of the unit cell of of $La_{0.9}Te_{0.1}MnO_3$ versus particle size.

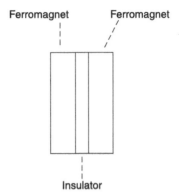

Figure 13.35. Illustration of a magnetic tunnel junction consisting of two ferromagnetic materials separated by a thin insulating barrier layer.

applied parallel to the electrodes, the tunneling conductance in the junction depends on the relative orientation of the magnetization in the two electrodes. If the magnetization vectors are parallel, the conductance is maximum or the resistance is minimum. When the magnetization vectors are antiparallel, the resistance is maximum. The *tunneling magnetoresistance ratio* (TMR) is defined as

$$TMR = \frac{R_{ap} - R_p}{R_p} \tag{13.9}$$

where R_{ap} is the junction resistance when the magnetizations are antiparallel and R_p is the resistance when they are parallel. Tunnel junctions have been fabricated from CMR materials where both electrodes are $La_{0.66}Sr_{0.3}MnO_3$. With a small voltage of 1.0 mV across the electrodes, a TMR ratio of 1800% was observed at 4.2 K. It was also observed that as the magnetoresistance effect increased, the size of the junction decreased. The qualitative similarity of the magnetoresistive behavior in the fabricated junctions and in CMR materials consisting of nanosized grains suggests that the effects are due to the existence of tunnel junctions formed by contact of the grains. The mechanism of the magnetoresistance effect in magnetic tunnel junctions will be discussed in Section 14.2, on spintronics.

13.4.3. Carbon Nanostructured Ferromagnets

As discussed in Section 10.4, the presence of iron or cobalt particles is necessary for the nucleation and growth of carbon nanotubes fabricated by pyrolysis. In the preparation of aligned carbon nanotubes by pyrolyzing iron(II) phthalacyanide (FePc), it has been demonstrated that nanotube growth involves two iron nanoparticles. A small iron particle serves as a nucleus for the tube to grow on, and at the other end of the tube larger iron particles enhance the growth. Aligned nanotubes are prepared on quartz glass by pyrolysis of FePc in an argon and hydrogen atmosphere.

Figure 13.36. Scanning electron microscope image of iron particles (light spots) on the tips of aligned carbon nanotubes. [With permission from Z. Zhang et al., *J. Magn. Magn. Mater.* **231**, L9 (2001).]

Figure 13.36 shows a scanning electron microscope (SEM) image of iron particles on the tips of aligned nanotubes, and Fig. 13.37 shows the magnetization curve at 5 and 320 K for the magnetic field parallel to the tubes showing that the hystersis is greater at 5 K. Figures 13.38 and 13.39 show plots of the temperature dependence of the coercive field H_C and ratio of remnant magnetization M_r to saturation magnetization M_S. We see that the coercive field increases by a factor of >3 when the temperature is reduced from room temperature (300 K) to liquid helium temperature (4 K). These iron particles at the tips of aligned nanotubes could be the basis of high-density

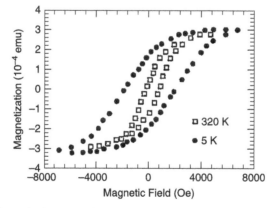

Figure 13.37. Magnetization curve hystersis loops for iron particles on the tips of aligned nanotubes at the temperatures of 5 and 320 K, for a magnetic field H applied parallel to the tubes (1 Oe $= 10^{-4}$ T). [Adapted from Z. Zhang et al., *J. Magn. Magn. Mater.* **231**, L9 (2001).]

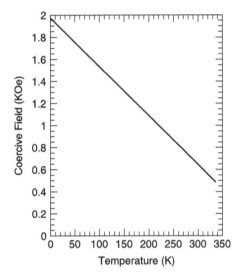

Figure 13.38. Plot of coercive field H_c versus temperature T for iron particles on the tips of aligned nanotubes ($1\,\text{k Oe} = 0.1\,\text{T}$). [Adapted from Z. Zhang et al., *J. Magn. Magn. Mater.* **231**, L9 (2001).]

magnetic storage devices. The walls of the tubes can provide nonmagnetic separation between the iron nanoparticles, ensuring that the interaction between neighboring nanoparticles is not too strong. If the interaction is too strong, the fields required to flip the orientation would be too large.

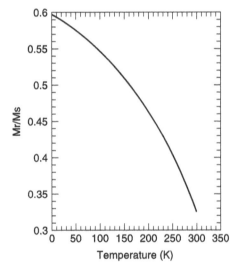

Figure 13.39. Ratio of remnant magnetization M_r to saturation magnetization M_s versus temperature T for iron nanoparticles on the tips of aligned carbon nanotubes. [Adapted from Z. Zhang et al., *J. Magn. Magn. Mater.* **231**, L9 (2001).]

Efforts to synthesize a nonpolymeric organic ferromagnet have received increased interest because of the possibility that such a magnet can be engineered by chemical modification of the molecules of the system. The C_{60} molecule discussed in Chapter 10 has a very high electron affinity, which means that it strongly attracts electrons. On the other hand, the molecule $C_2N_2(CH_3)_8$ or tetrakisdimethylaminoethylene (TDAE) is a strong electron donor, which means that it readily gives its electron to another molecule. When C_{60} and TDAE are dissolved in a benzene–toluene solvent, a precipitate of a new material forms that consists of the TDAE molecule complexed with C_{60}, and this 1–1 complex has a monoclinic crystal structure. A measured large increase in the susceptibility at 16 K indicated that this material had become ferromagnetic, and up until very recently (as of 2007) this was the highest Curie temperature for an organic ferromagnet.

In Section 10.3.4 we saw that when C_{60} is irradiated with UV light in a vacuum, it forms short polymers of nanometer length. The mass spectra data presented in Fig. 10.11 show that the dimer of C_{60} was the most likely to be produced, but that chains of 20 C_{60} molecules are also formed that are approximately 32 nm long. In a rather surprising result it was found that when the photolysis is done in the presence of oxygen or air, a ferromagnetic phase is formed. Figure 13.40a shows a measurement of the magnetization versus dc magnetic field using a SQUID (superconducting quantum interference device) magnetometer at a number of temperatures in photolyzed C_{60} showing clear evidence of the existence of ferromagnetism at 300 K Figure 13.40b is a plot of the temperature dependence of the magnetization in a 300-G magnetic field. As discussed in Section 13.1, the temperature dependence of the magnetization $M(T)$ generally obeys the Bloch equation (13.4). The line through the data in Fig. 13.40b is a fit to this equation. The Curie temperature estimated from the fit where $M(T)$ becomes zero is 1000 K, which is probably an overestimate. Interestingly, the ferromagnetism can be experimentally observed up to 800 K, which means that the oligomers produced in the presence of oxygen are much more stable than those produced in a vacuum. It was seen earlier that the polymers produced in vacuo returned to the monomer state in the vicinity of 200°C. This Curie temperature is the highest temperature ever observed for an organic ferromagnetic material. An interesting feature of the result is that the ferromagnetic phase is not produced when the photolysis is done in a vacuum. This suggests that the photolysis in air involves a photodissociation of O_2 with a subsequent attachment of atomic oxygen to the C_{60} molecules of the oligomer. There are two questions about this result. Why is the linear polymer producing the ferromagnetic phase more stable than the phase produced in vacuo, which is unstable above 200°C, and where does the unpaired electron come from that is needed to form a ferromagnetic phase? A possible explanation may lie in the fact that the C_{60} anion is actually more stable than neutral C_{60}. The C_{60} anion is paramagnetic and can be detected by electron paramagnetic resonance (EPR), and every sample of pure C_{60} shows such an EPR, signal. It is possible that some of the C_{60} anions are incorporated into the oligomers providing the source of the unpaired electron spin. Molecular orbital calculations of the stability of C_{60} dimers show that negatively charged dimers are considerably more stable than are neutral dimers against dissociation into two C_{60} molecules, which could explain why the ferromagnetic phase is much more stable.

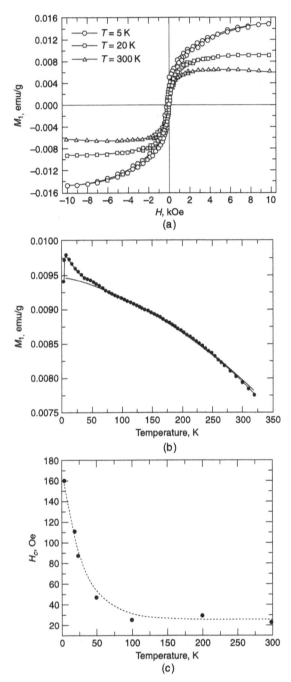

Figure 13.40. (a) Magnetization versus dc magnetic field at three temperatures in C_{60} photo-lyzed in air; (b) temperature dependence of magnetization in a 300-G field (the line through data is a fit to the Bloch equation); (c) temperature dependence of coercive field. [Adapted from F. J. Owens et al., *Phys. Rev.* **B69**, 033403 (2004).]

13.5. ANTIFERROMAGNETIC NANOPARTICLES

In the antiferromagnetic state the magnetic moments of the individual atoms of the material have an antiparallel arrangement as shown in Fig.13.1d for a two-dimensional system. In the antiferromagnetic state the material has no net magnetic moment and the susceptibility will be zero. Antiferromagnetic nanoparticles display unusual behavior compared to ferromagnetic nanoparticles. In 1961 Néel proposed that antiferromagnetic nanoparticles could display weak ferromagnetism and superparamagnetism. We saw earlier that as particle size decreases, the percentage of atoms on the surface of the particle increases. Néel proposed that for antiferromagnetic nanoparticles the antiparallel orientation of the moments on the surface would be lost and that the moments of the surface atoms would actually align ferromagnetically parallel to the axis of the antiferromagnetic alignment in the interior of the particle. This idea means that an antiferromagnetic nanoparticle can be viewed as having an inner antiferromagnetic core surrounded by an outer shell of uncompensated spins. Experimental measurements in nickel oxide (NiO) confirm this model. Bulk NiO has a rhombohedral crystal structure and becomes antiferromagnetic below 523 K. Above that temperature it is paramagnetic and the crystal structure is cubic. Figure 13.41 shows a measurement of the dc susceptibility of NiO nanoparticles versus temperature for particles of two different sizes that were zero-field-cooled (ZFC) and measured in a field of 100 G. The presence of a nonzero susceptibility means the antiferromagnetic particles are displaying ferromagnetic behavior. For particles greater than 100 nm, the susceptibility is zero. For the smaller particles, the figure shows that the susceptibility increases with increasing temperature to a maximum and then decreases. The dc magnetic field at which the maximum occurs shifts to higher temperature as the particle size gets smaller, and the magnitude of the

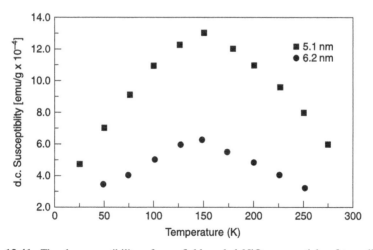

Figure 13.41. The dc susceptibility of zero-field-cooled NiO nanoparticle of two different sizes measured in a 100-G field as a function of increasing temperature. (Adapted from S. D. Tiwari and K. P. Rajeev, unpublished.)

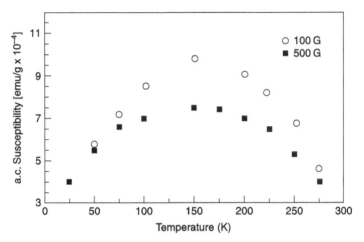

Figure 13.42. A measurement of temperature dependence of ac susceptibility for a NiO nanoparticle in two different magnetic fields. (Adapted from S. D. Tiwari and K. P. Rajeev, unpublished.)

susceptibility increases with reduced particle size. The data show clearly that the behavior is not a result of superparamagnetism because the temperature at which the system transitions to superparamagnetic behavior, the blocking temperature, increases with increasing particle size as shown in Fig. 13.9, contrary to the behavior shown in Fig. 13.41. The increase in magnitude of the susceptibility is a result of the increase in the number of Ni atoms on the surface having uncompensated spins. Figure 13.42 shows a measurement of the ac susceptibility for a particle of one size in two different dc magnetic fields. It is seen that the peak shifts slightly to lower temperature with increasing dc magnetic field. The relationship of the peak temperature has been shown to follow an equation of the form

$$B = c \left[\frac{1 - T}{T_s} \right]^{3/2} \tag{13.10}$$

This is referred to as the *de Almeida–Thouless line*, and it describes the relation of the field and temperature for a transition to a spin glass state, which is essentially a state where the magnetic moments of the surface atoms are disordered. Thus the peak temperature is the temperature at which the surface moments begin to lose their ferromagnetic order. A similar relationship has been shown to describe the melting of the vortex lattice in superconducting materials, which will be discussed in Section 15.4.

PROBLEMS

13.1. Calculate the total number of atoms in an iron nanoparticle versus the diameter of the particle assuming a FCC structure using Table 1.1. Plot the number of atoms versus particle diameter from 0.744 to 4.22 nm.

13.2. Calculate the number of atoms on the surface of iron nanoparticles from 0.74 to 4.22 nm in diameter. Plot the percentage of surface atoms versus the diameter of the particle.

13.3. Taking the magnetic moment of the iron atom as 2.22 Bohr magnetons, and assuming that all iron atom spins are ferromagnetically aligned except the iron atoms on the surface whose spins are randomly oriented, calculate the average magentization versus particle size at 0 K. Plot the magnetization versus particle size from 0.74 to 4.22 nm. Why does the average magnetization decrease at small particle sizes?

13.4. Consider a powder of 12-nm iron nanoparticles that is cooled from room temperature to 0 K in a 100-G dc magnetic field that aligns the magnetization of the ferromagnetic particles parallel to the dc magnetic field. As the temperature of the powder is raised, calculate the fraction of nanoparticles whose magnetization becomes randomly oriented as a function of the temperature.

13.5. Calculate the number of iron atoms in the bulk and on the surface of a spherical iron nanoparicle with a radius $r = 4.7$ nm.

13.6. Consider a powder of consisting of 10^4 of the particles from Problem 13.5 that is cooled in zero dc magnetic field. What is the total magnetic moment at 4.2 K? Now, at 4.2 K a 1000 G magnetic field is applied to the powder. Calculate the total magnetic moment of the powder as a function of increasing temperature.

13.7. The same powder of particles from Problem 13.6 is cooled to 4.2 K in a 1000-G dc magnetic field. What is the total magnetic moment of the powder at 4.2 K, and in which direction does it point? Calculate the total magnetic moment at a number of temperatures up to room temperature.

13.8. The same powder of the 4.7-nm particles is suspended in 20 mL of a viscous liquid that freezes at 260 K. When the liquid is cooled to 200 K, what is its total magnetization? A magnetic field is applied at 200 K to the frozen liquid. Now, what is the magnetization of the liquid? The same suspension is cooled from room temperature to 200 K in a 1000-G dc magnetic field, and the field is then removed. What is the magnetization of the liquid? In what direction does it point?

CHAPTER 14

Nanoelectronics, Spintronics, Molecular Electronics, and Photonics

14.1. NANOELECTRONICS

14.1.1. N and P Doping and PN Junctions

In Section 1.4.1 we discussed N and P doping of silicon. Let us briefly review it here. Silicon, the major component of most transistor devices, has four valence electrons, which are shared in the covalent bonds with four nearest-neighbor silicon atoms in the silicon lattice. If the lattice is doped with a phosphorous atom, which has five valence electrons, an extra electron is available to contribute to the conductivity of the lattice. The donor energy level of this extra electron lies in the bandgap just below the conduction band as shown in Fig. 1.24 such that a very small amount of thermal energy $k_B T$ will excite it to the conduction band. This kind of doped semiconductor is called an *N-type semiconductor*. If silicon atoms are replaced by aluminum, which has three valence electrons, one bond will be missing an electron and is referred to as a *hole*. The acceptor energy level of the hole, as shown in Fig. 1.24, is just above the valence band, and this hole can also contribute to the conductivity of the material. This kind of doping is referred to as *P doping*.

Now consider a piece of silicon in which donor impurities (N) are introduced on one side and acceptors (P) on the other side forming what is called a *P–N junction* as illustrated in Fig. 14.1a. When the junction is formed, there is a density gradient of holes and electrons at the interface between the N and P regions. Because of this, holes will diffuse to the N side and electrons to the P side. As a result of the displacement of charge, an electric field will be produced across the junction, and this field will increase until it becomes large enough to prevent further diffusion, establishing equilibrium. There will be a net positive charge on the N side and a net negative charge on the P side. As a result, charge cannot move in this region, which is referred to as the *depletion region*. Now let us apply a voltage across the junction such that the P side is connected to the negative terminal and the N side to the positive terminal as shown in Fig. 14.2. This is referred to as *reverse bias*. Under this situation the holes and electrons will move away from the P–N

The Physics and Chemistry of Nanosolids. By Frank J. Owens and Charles P. Poole, Jr.
Copyright © 2008 John Wiley & Sons, Inc.

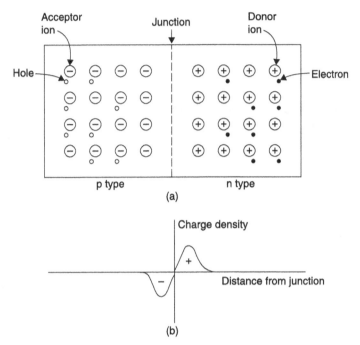

Figure 14.1. (a) Schematic illustration of a P–N junction and (b) the charge distribution at the interface due to the net flow of holes and electrons across the junction.

interface and no current can flow across the junction. If we reverse the polarity of the voltage shown in Fig. 14.2, current will flow.

Figure 14.3 shows a typical current–voltage relationship for a P–N junction. We see from the figure that the application of +0.15 V produces a current flow in the forward direction of 10 mA, which increases very rapidly with a further increase in the applied (bias) voltage. In contrast to this, application of the reverse bias of −0.15 V produces the totally negligible flow in the reverse direction of only −0.01 mA, which does not increase with further increases in negative applied

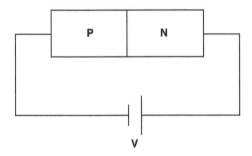

Figure 14.2. A P–N junction with a reverse bias voltage applied to it.

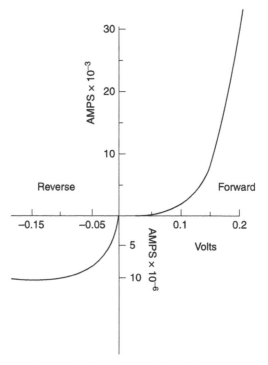

Figure 14.3. Current–voltage relationship for a typical P–N junction with an applied forward bias (right), and an applied reverse bias (left). Note the factor of 1000 difference between the ordinate scales for the forward and the reverse biases.

voltages. Thus we can say that, for all practical purposes, a P–N junction does not conduct when subjected to a reverse bias; it is the basis of a rectifying diode.

14.1.2. MOSFET

Now let us consider the basic switching element used on computer chips called a MOSFET, which stands for *metal oxide semiconductor field effect transistor*. Figure 14.4 is a simple illustration of a MOSFET, which consists of a P-type semiconductor with an N-type one, on each side. On the top of the P-type semiconductor is a thin layer of silicon oxide on top of which is a thin layer of a metal connected to an electrical lead to which a voltage can be applied. There is also a metal electrical contact on top of each N-doped semiconductor. If a voltage is applied across the two N types, points *a* and *b* in the figure, no current will flow because one N–P junction is always reverse-biased regardless of the polarity of the voltage across *a* and *b*. Now, if a positive voltage is applied to the metallic contact on the P-doped material, called the "gate," the holes in the gate will be pushed down to the bottom of the gate and there will be a narrow channel under the silicon oxide layers that will be N-doped, and thus current can flow across the junction through this layer when a voltage is

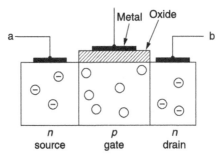

Figure 14.4. A schematic illustration of a metal oxide semiconductor field effect transistor, abbreviated MOSFET.

applied across points *a* and *b*. The device will be in an ON state in this case and an OFF state when no voltage is applied to the gate. Devices like this, which display ON and OFF states representing the 1 and 0 of computer logic, respectively, are the fundamental elements of computer chips. Most of the power consumption in a MOSFET occurs when it changes from ON to OFF. In order to reduce power consumption, MOSFETs are arranged in pairs called *complementary MOSFETs* (CMOSs). The pairs consist of PNP and NPN MOSFETs, where the outer layers are the source and the drain. If a negative voltage is applied to the gate of both devices, the PNP switches ON and the NPN device switches OFF. This results in little or no net power dissipation for the complementary pair on switching. Paired MOSFETs or CMOSs are widely used on computer chips. A *computer chip* is a large array of such switches electrically connected to one another. Now, you might ask what all this has to do with nanotechnology. The smaller one can make these switches, the MOSFETs, the more switches that can be put on a chip, the closer together they will be, and thus the faster the processing time of the chip. At the present time MOSFETs are approaching nanometer dimensions. The clock speed of chips has increased by about 29% a year since 1980 because methods have been developed to fabricate smaller and smaller switches on a chip. Figure 14.5 is a plot of the increase in the clock speed of computer chips from 1980 to 2005.

14.1.3. Scaling of MOSFETs

Figure 14.6 is a plot of the decrease in the nominal size of these devices, such as the length of the gate from 1980 to 2006, showing that at present the switches are <100 nm long. But how much smaller can they get? It is possible that there may be some small size at which the MOSFET stops working in the desired way. Let us examine some of the challenges in reducing the size of MOSFETs even further. If both the horizontal and vertical dimensions are reduced by a factor α, called the *scaling factor*, and the area of the chip remains the same, the number of switches on the chip will increase by α^2. If we keep the voltage of the smaller device the same as that of a larger device, we are interested in how the other parameters of the device scale

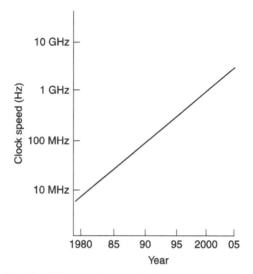

Figure 14.5. Plot of continual increase in speed of computer microprocessors versus years since 1980.

with α. Table 14.1 lists some of the changes in parameters with α. Notice that power dissipated by each device increases linearly with α. This means that the power density on the chip increases as α^3, which means that the heat generated on the chip increases by α^3. Of course, heating is not a good thing because it can cause thermal excitation of holes or electrons across the bandgap and produce conductivity in the OFF state. One way around this problem would be to reduce the gate voltage. However, the gate voltage cannot be reduced too much because then thermal fluctuations of the voltage may become a significant fraction of the voltage and the device may fluctuate

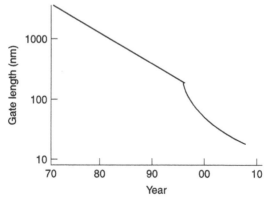

Figure 14.6. A plot of the reduction of the size of MOSFETs since 1980 showing that present devices have nanometer dimensions.

**TABLE 14.1. How the Various Parameters of a MOSFET
Scale with Size If the Voltage is Kept Constant**

Parameter	Scaling factor
Dimension	$1/\alpha$
Voltage	1
Doping concentration	α
E-field	α
Capacitance	$1/\alpha$
Current	α
Power dissipated	α
Power density	α^3
Current per gate width	α^2

between the ON and OFF states, an undesirable state of affairs. Another problem concerns the depletion region where there are no conducting electrons or holes. As the size of the device decreases, the width of the depletion region does not change, and hence the depletion region will become a larger fraction of the gate region. If the depletion region stretches across the gate region, the device will not work because there will be no holes that can be displaced by the gate voltage. Typically the depletion region in a P–N junction is in the order of 10^{-6} m, a micrometer. Most estimates of the minimum gate length at which a MOSFET will work are about one-fifth of a micrometer (0.2 μm), which will allow in excess of 10^9 switches on a chip, enabling a clock speed in the gigahertz region. So at present the fastest desktop computers are getting close to this limit since they have a clock speed close to 4 GHz. To go beyond this new technology may be needed such as molecule-based devices, which will be discussed later in this chapter.

There are, however, other factors that may limit significant enhancement of speed using MOSFET-like switches such as the *interconnects*, that is, the wires that connect all the ON/OFF switches in the array. The array of switches is actually an RC circuit because the interconnects have resistance R and the switches have capacitance C. In the depletion region at the interface of the P–N junction for a reverse bias, one side of the junction has a positive charge and the other side a negative charge as illustrated in Fig. 14.1b. So the MOSFET is effectively a capacitor. The current in an RC circuit has the form

$$I = I_0 \exp \frac{-t}{RC} \tag{14.1}$$

The larger RC, the more quickly the current decays, and it may not reach a neighboring switch. The capacitance due to the depletion region is given by

$$C_T = \frac{\varepsilon A}{W} \tag{14.2}$$

where A is the interfacial area between the N–P junction, ε is the dielectric constant of silicon, and W is the width of the depletion region, essentially the distance between the maximum and minimum of the charge density distribution in Fig. 14.1b. Since W does not change with reduced size, C_T decreases as the size is reduced because the area decreases as α^2. However, because of the reduction in thickness of the oxide layer, the overall reduction of C_T is α. So the reduction in the capacitance with the size of the MOSFET is a good thing in that it contributes to reducing the decay of the current with time.

However the scaling of the resistance with reduced size of the interconnects contributes to increasing RC. The resistance of a wire of length L and cross-sectional area A is given by

$$R = \frac{\rho L}{A} \tag{14.3}$$

where ρ is the resisitivity of the material of which the wire is made, and is an intrinsic property of the material. If the thickness, width, and length are scaled by α, the overall resistance will increase by α. Hence reduction in the sizes of the switches and interconnects does not increase the decay time of the current flowing in the interconnects. Some possible solutions to the interconnect problem are to use metal carbon nanotubes as the connecting wires because they have very low resistance and angstrom diameters. Of course, this will be possible only if methods are developed to separate the metal and insulating tubes from each other, and a method has been devised to attach the tubes to the MOSFETs. Another possibility is to use interconnects made of superconducting materials, which have no resistance. If the wires were made of the copper oxide (cuprate) superconductors, the chip would have to be cooled to at least liquid nitrogen temperature, which is not very practical for desktop computers. Another problem that occurs when MOSFETs are reduced to nanometer dimension is leakage. The thin oxide layer under the metal gate shown in Fig. 14.4 is SiO_2, which forms an insulating layer between the metal electrode and the conducting channel in the P-doped silicon of the gate. As chips are packed more densely and the size of the MOSFETs is reduced to nanometer dimensions, the thickness of the SiO_2 layer will be reduced. It may be necessary to make the thickness 2 nm or less. When the insulating layer gets this thin, current can tunnel through the layer from the metal to the P-doped region of the gate. This is referred to as *leakage current*. Below $\sim$4 nm thickness the leakage current increases exponentially for approximately every 0.3 nm reduction in the thickness of the insulating layer. This gate current can cause an increase in power consumption of the MOSFET and detrimentally affect its performance. Once leakage current occurs, it can produce defects in the SiO_2 such as microscopic electron traps. When the defect density becomes sufficiently high, the oxide no longer acts as a good insulator and the MOSFET does not function properly. Thus leakage current reduces the performance lifetime of a MOSFET. More recently there has been some progress in dealing with this problem. A hafnium compound having a high dielectric constant has been found that may replace SiO_2 and allow thinner insulating layers.

14.2. SPINTRONICS

14.2.1. Definition and Examples of Spintronic Devices

Spintronics is a term coined by the Defense Advanced Research Projects Agency (DARPA), a Defense Department funding agency that supports high-risk advanced research, for electronic devices that depend on the difference in conductivity of conduction electrons depending on their spin orientation with respect to some axis such as an applied magnetic field. It turns out that electrons with spinup and spindown with respect to the field direction have different transport properties in certain kinds of nanometer-thick layered materials. In metallic ferromagnetic materials, the spinup and spindown conduction electrons act as two independent families of charge carriers. The ferromagnetic exchange interaction causes a splitting of the spinup electron conduction band from the spindown electron conduction band. This results in different band structures and density of states at the Fermi level. *Spin polarization* (SP) is defined as the ratio of the difference of the spinup minus the spindown population to the total number of carriers at the Fermi level:

$$SP = \frac{N_{up} - N_{down}}{N_{up} + N_{down}} \tag{14.4}$$

If SP is not zero, there will be a different number of electrons in the two different spin channels. The scattering of the spinup and spindown electrons is different in these structures, which means that the conductivity is different. Many of the structures where this difference is evident are of nanometer dimensions, in particular, layered materials of nanometer thickness. In Chapter 13, on magnetism, we have already discussed a number of nanostructured materials where electron spin plays an important role in their properties. Magnetoresistance was discussed in materials that consisted of nanometer-thick layers of a magnetic metal and a nonmagnetic metal as illustrated in Fig. 13.26. It was shown that for a Fe–Cr layered system the magnetoresistance depended on the thickness of the layers with a maximum effect occurring at about 8 nm thickness. The magnetoresistance effect occurs because of the dependence of electron scattering (the cause of resistance) on the orientation of the electron spin with respect to the direction of magnetization in the ferromagnetic layer. A device based on this observation would be called a *Spintronic* device. Magnetotunneling junctions were also briefly discussed in Chapter 13, and consist of a thin layer of two magnetic materials separated by an insulating layer. A typical system might consist of a layered structure such as $Fe/Al_2O_3/Fe$. As we have seen, the tunneling current across the junction depends on the relative orientation of the magnetizations in the ferromagnetic layers. We will discuss these junctions in a little more detail later on.

14.2.2. Magnetic Storage and Spin Valves

Another place where spin is employed is in the hard drives of computers, which store information based on the orientation of the magnetization of rod-shaped particles of

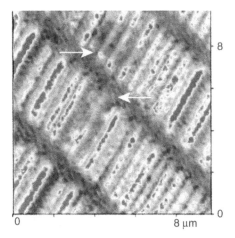

Figure 14.7. A magnetic force microscope image of a hard drive of a computer showing aligned magnetic particles. [Adapted from J. Sharma et al., *Appl. Phys. Lett.* **82**, 2236 (2003).]

nanometer thickness. A typical hard drive consists of an array 1-μm-long and 70-nm-wide particles. Figure 14.7 is a magnetic force microscope image of a hard drive. This particular hard drive was damaged by irradiation, and the arrows point to damaged regions. The device that reads the information on the hard drive has to be sensitive to the small magnetic fields generated by the particles. One device that has been developed to read hard drives is called a *spin valve*. It consists of nanometer-thick layers of materials, which collectively display giant magnetoresistance. The structure is illustrated in Fig. 14.8. It consists of an antiferromagnetic layer on a ferromagnetic layer followed by a conducting layer and another ferromagnetic layer all deposited on a substrate. The magnetization of the ferromagnetic layer adjacent to the

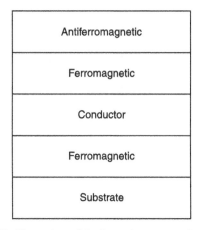

Figure 14.8. Illustration of the layered structure of a spin valve.

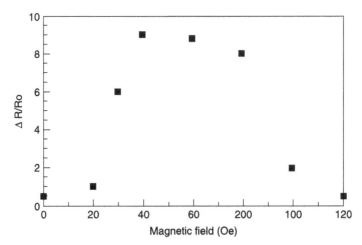

Figure 14.9. Change in the resistance versus dc magnetic field of a layered structure of Co/Cu/Co in which the direction of magnetization of one of the Co layers is pinned parallel to the interface between the layers. [Adapted from B. Dieny et al., *J. Appl. Phys.* **69**, 4774 (1991).]

antiferromagnetic layer is pinned parallel to the layer interface. In the other magnetic layer, referred to as the "free layer," the direction of magnetization can be changed by a relatively small dc magnetic field. As the magnetization in the two layers changes from parallel to antiparallel, the resistance of the spin valve changes by 5–10%. Now let us look at the behavior of some actual systems. Figure 14.9 shows the resistance versus dc magnetic field of a Co/Cu/Co structure in which the direction of magnetization of one of the ferromagnetic layers is pinned by exchange interaction with an adjacent antiferromagnetic material. The observed change in the resistance depends on the thickness of the layers. Figure 14.10 shows how the thickness of the Co layer in a spin valve structure having layers of composition Si/Co(x)/Cu(2.2)/ NiFe(4.7)/FeMn(7.8)/Cu(1.5) affects the magnetoresistance. The numbers in parentheses are the thicknesses of the layers in nanometers. Notice that magnetoresistance is optimum at a ∼10 nm thickness of the first cobalt layer. The thickness of the other layers also affects the magnetoresistance. Figure 14.11 shows how the magnetic moment of the free Co layers depends on the layer thickness. These spin valve devices, which are sensitive to small magnetic fields, are the basis of the reading of information on magnetic hard drives. It should be emphasized that these are truly nanometer-structured materials.

Magnetic tunneling junctions, which we have briefly discussed, can also be used to read information on magnetic storage disks. In this application the device is somewhat different from that shown in Fig. 13.35. One of the ferromagnetic layers will have an antiferromagnetic layer deposited on it to pin the direction of the magnetization in that layer. The tunneling current depends on the relative orientation of the

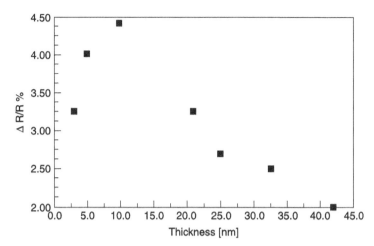

Figure 14.10. Magnitude of the relative change in resistance in a fixed applied magnetic field versus the thickness of the free magnetic layer for the spin valve structure discussed in the text. [Adapted from B. Dieny et al., *J. Appl. Phys.* **69**, 4774 (1991).]

magnetization of the two ferromagnetic metal layers on either side of the insulating layer. In Chapter 11 we discussed the tunneling that can occur between two nonferromagnetic metal layers on each side of an insulating layer. Tunneling can occur only when there are available states on one side of the junction for the electrons to go to. This occurs because the applied electric field lowers the Fermi level of the metal on one side of the junction. This must also occur in the magnetic tunneling junction,

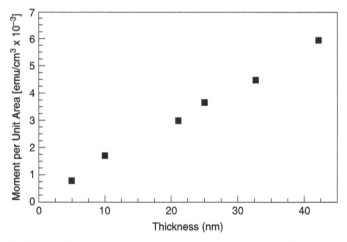

Figure 14.11. Plot of the magnetic moment of the same spin valve of Fig. 14.10 versus the thickness of the free magnetic layer. [Adapted from B. Dieny et al., *J. Appl. Phys.* **69**, 4774 (1991).]

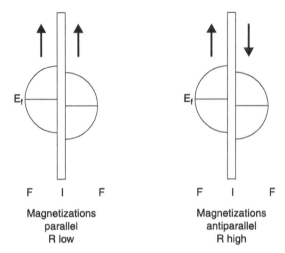

Figure 14.12. Illustration of the density of states and Fermi level for a magnetic tunnel junction in which the magnetizations are parallel (left) and antiparallel (right) to each other.

where the Fermi-level shift is a result of the relative orientation of the electron spins in the ferromagnetic layers, which depends on the relative orientation of the magnetization in the layers. Figure 14.12 illustrates the effect of the orientation of the magnetization in the layers on the energy level near the Fermi level. The magnetization of the layer on the left side of the junction is pinned parallel to the interface between the junctions. A small external field cannot change its orientation. The layer on the right side, called the *soft layer*, can undergo a change in orientation caused by an external dc magnetic field. When the directions of magnetization on both sides of the layer are parallel, current can flow from left to right and the junction will have a low resistance. This occurs because there are available spinup empty states on the right because of the lower Fermi level. When the magnetizations are antiparallel, there are no available spinup states on the right side, so current cannot flow and the junction has a high resistance. Figure 14.13 shows the tunneling resistance versus magnetic field at room temperature for a junction of $Co/Al_2O_3/NiFe$.

Magnetic tunnel junctions could also be used as information storage devices called magnetic random-access memory (MRAM). Because current can be turned ON and OFF by small magnetic fields, magnetic tunnel junctions can be used to represent the binary 1 and 0 of computer logic. Figure 14.14 illustrates a possible array of tunnel junctions that could be used as MRAM. It consists of layers of nanometer-thick metal strips overlaid on each other, as shown in the figure. Magnet tunnel junctions connect the horizontal and vertical strips where they cross over each other. Current flowing through the strips produces magnetic fields at the junctions. If the direction of current flow in the strips is such that magnetic fields on both sides of the junction are parallel, then current can flow between the horizontal and vertical strips. The relative orientation of the spins in the two ferromagnetic layers of the junction stores the information bit.

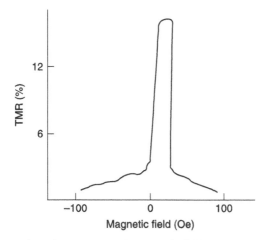

Figure 14.13. Change in resistance versus dc magnetic field at room temperature for a magnetic tunnel junction consisting of $Co/Al_2O_3/NiFe$ nanometer-thick layers. [Adapted from J. Nassar et al., *Appl. Phys. Lett.* **73**, 689 (1998).]

14.2.3. Dilute Magnetic Semiconductors

As we have discussed above, the switching elements in the logic circuits of computer processors are semiconducting materials involving an ON/OFF flow of electrons or holes. On the other hand, the storage of information employs hard disks, which depend on the alignment of magnetic nanoparticles. The possibility of combining both the switching and storage of information in one material has generated much interest in the development of dilute magnetic semiconductors (DMSs). These are semiconducting materials such as GaN, GaP, or ZnO doped with ions having unpaired electrons such as Mn^{2+} or Cu^{2+} at levels of 2–6% by weight. A driving

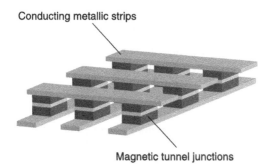

Figure 14.14. A proposed magnetic storage device based on magnetic tunnel junctions that exist at the overlapping regions where vertical and horizontal metallic strips cross over each other. The relative orientation of the spins in the two ferromagnetic layers stores the information bit.

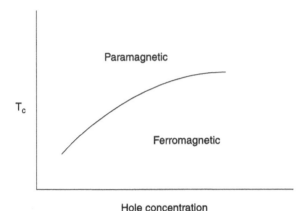

Hole concentration

Figure 14.15. A schematic plot of the dependence of the Curie temperature of a dilute magnetic semiconductor on hole concentration.

force behind the experimental efforts was a theoretical prediction that P-type semiconductors, when doped with magnetic ions, could be ferromagnetic; that the Curie temperature (T_c) depended on the hole concentration; and that at sufficiently high hole concentrations, such doped materials could be ferromagnetic above room temperature. A schematic illustration of the prediction of the dependence of the Curie temperature on hole concentration is shown in Fig. 14.15. An essential requirement for a ferromagnetic semiconductor to be useful is that the Curie temperature be well above room temperature.

Ferromagnetism having a T_c of 110 K was observed in GaAs doped with manganese. Later GaP doped with manganese by ion implantation was reported as having an estimated T_c of 385 K. In the ion implantation method GaP is subjected to a 250-keV beam of Mn ions. This method of synthesis is not inexpensive and not easily scaled up. It was subsequently shown that Mn^{2+} could be doped into ZnO and GaP by a relatively simple sintering process. This process is a solid-state reaction process where MnO and GaP are mixed in the ratio of typically 0.03 of the molecular weight of MnO to 1 molecular weight of GaP. The material is finely ground for many hours in a mortar and pestle and then pressed into a pellet. The pellet in an aluminum oxide boat was heated in an oven at 500°C for 4 h and then rapidly quenched to room temperature. Since Ga is +3 ion and Mn is +2 ion, doping with manganese should hole-dope the material. Evidence for hole doping can be obtained from Raman spectroscopy. Figure 14.16 shows the Raman spectrum of the transverse optical mode (lower-frequency mode) and the longitudinal optic mode (higher-frequency mode) in undoped GaP and Mn-doped GaP. The smaller Raman line is in the doped material. The frequency of the longitudinal optic mode has shifted down in the doped material. This is because the longitudinal optic mode is coupled to the plasma mode. As discussed in Chapter 7, the plasma mode is a collective vibration of the conduction electrons with its frequency proportional to the concentration of conduction electrons given by Eq. (7.45). So, if the electron concentration is

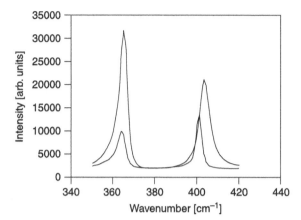

Figure 14.16. Raman spectra of the transverse optical (lower frequency) and longitudinal vibrational mode (higher frequency) of undoped GaP (larger spectrum) Mn^{2+}-doped GaP (smaller spectrum). [Adapted from F. J. Owens, *J. Phys. Chem. Solids* **66**, 793 (2005).]

reduced as a result of hole doping, the plasma frequency is reduced and the longitudinal optic mode frequency decreases. Figure 14.17 shows a measurement of the magnetization of the sample as a function of dc magnetic field at 300 K clearly showing that the material is ferromagnetic at room temperature. Ferromagnetic resonance measurements were used to study the ferromagnetism above room temperature, and the Curie temperature was determined to be 600 K.

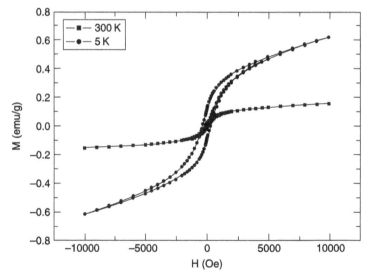

Figure 14.17. SQUID magnetometry measurement of the magnetization versus the dc magnetic field at 300 and 5 K in GaP doped with Mn^{2+}. (Courtsey of K. V. Rao, Royal Institute of Technology, Sweden.)

It is to be recalled that the dominant interaction that produces ferromagnetic ordering is the exchange interaction, which is a short-range nearest-neighbor interaction between the spins of the paramagnetic ions. Thus the question arises as to how the spins of the manganese ions are coupled to form ferromagnetic ions if they are dispersed in the lattice at low concentrations and widely separated. One possibility is that the Mn ions are not dispersed in the lattice but form clusters that have been observed to be ferromagnetic. However, the Mn clusters are observed to be ferromagnetic only at lower temperatures, and thus cannot account for the observation of ferromagnetism above room temperature. More recently copper-doped GaP has been observed to be ferromagnetic. Since there is no known evidence of ferromagnetism in copper clusters, this suggests that such clusters are not the origin of the ferromagnetism. The explanation of ferromagnetism in dilute magnetic semiconductors is still a matter of research. The present idea is that the conduction holes mediate the coupling between the spins of the paramagnetic ions. Unlike the paramagnetic dopant, the holes move through the lattice, acting as go-betweens coupling the spins of the spatially separated paramagnetic ions. The fact that the Curie temperature depends on hole concentration (Fig. 14.15) supports this idea. Further support for the idea comes from the observation that shining light on the materials, which causes excitations across the bandgap and thus changes hole concentrations, affects the magnitude of the magnetization.

Dilute magnetic semiconductors can be the basis of magnetic storage devices. One idea that has been proposed is to use them as electrically controlled magnetic switches where a small voltage can order or disorder spins of the paramagnetic ions. The device illustrated in Fig. 14.18 consists of a layer of metal, the gate, on a thin layer of an insulator, which is on top of a dilute magnetic semiconductor followed by a layer of the same material that does not contain paramagnetic ions. When a positive voltage is applied to the gate, holes are pushed into the undoped semiconducting layer and the doped layer becomes non-ferromagnetic. When a negative voltage is applied to the gate, the hole concentration increases under the insulating layer and the doped layer becomes ferromagnetic. In effect, a small voltage can control the existence of

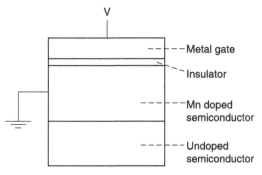

Figure 14.18. Proposed design of a device for using a dilute magnetic semiconductor as an electrically controlled magnetic switch. [Adapted from H. Ohono et al., *Nature* **408**, 944 (2000).]

ferromagnetic ordering in the device. Such a device could be the basis of a magnetic storage device. The fact that the magnetization can also be affected by light opens the possibility of an optical read–write storage device.

14.3. MOLECULAR SWITCHES AND ELECTRONICS

14.3.1. Molecular Switches

As we have discussed above, MOSFETs may not work when they have sizes below 10 nm. One approach to the development of smaller switching devices is to use molecules that can switch. This has motivated an effort to synthesize molecules that display switching behavior. This behavior might form the basis for information storage and logic circuitry in computers. A molecule that can exist in two different states, such as two different conformations, and can be converted reversibly between the two states by external stimuli, such as light or a voltage, can be used to store information. In order for the molecule to be used as a zero or one digital state, necessary for binary logic, the change between the states due to external stimuli must be fast and reversible. The two states must be thermally stable and be able to switch back and forth many times. Furthermore, the two states must be distinguishable by some probe, which is referred to as the *read mode*. Figure 14.19 is a schematic representation of the basic elements of a molecular switch, in which stimulus S_1 brings about a conversion from state 0 to state 1, and stimulus S_2 induces the reverse conversion. The two states are read by some probe, R_1 and R_2. There are a number of different kinds of molecular switch possibilities.

An example of a molecular switch is provided by the azobenzene molecule, which has the two isomeric forms sketched in Fig. 14.20a. Notice that the cis isomer is shorter than the trans isomer. Isomers are molecules having the same atoms and the same number of chemical bonds, but a different equilibrium geometry. The cis and trans forms of azobenzene can be distinguished by the difference in their optical absorption spectra. The molecule can be transformed from the trans isomer to the cis isomer by subjecting it to 313 nm ultraviolet light, and it returns to the original trans form when exposed to light of wavelength >380 nm. Unfortunately, the cis form of azobenzene is not thermally stable, and a slight warming causes it to return to the trans form, so optical methods of switching of this molecule are not of practical use for applications in computing.

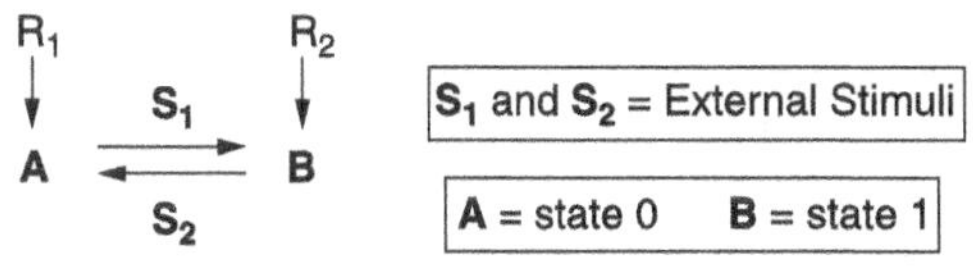

Figure 14.19. Schematic representation of the elements of a molecular switch. An external stimulus S_1 changes a molecule from state 0 to state 1, and S_2 returns the molecule to state 0. The symbols R_1 and R_2 indicate the light used to detect the state of the switch.

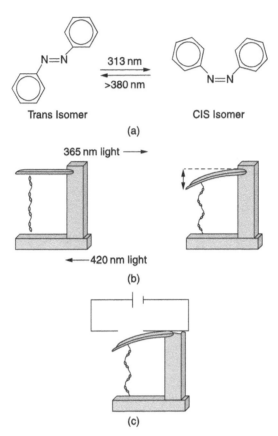

Figure 14.20. (a) Cis–trans UV-light-induced isomerization of azobenzene; (b) a molecular machine based on light-induced isomeric changes of the azobenzene polymer that contracts when it is converted to the cis form; (c) light-activated switch in an electric circuit based on the light-induced cis–trans transition in polyazobenzene.

The azobenzene polymer, formed from the monomers shown in Fig. 14.20a, contracts when it converts to the cis form when subjected to 365 nm light. When the cis form is subjected to 420 nm light, it converts to the trans form and expands. This light-induced expansion and contraction can be used to convert optical energy to mechanical energy in the device shown in Fig. 14.20b. The polymer is attached to a small flexible cantilever and to a permanent base. When it is subjected to 365 nm light, the polymer contracts and the beam bends. It is restored to the horizontal position by 420 nm light. If the beam were made of a conducting material and incorporated into a circuit as shown in Fig. 14.20c, it could be a device that uses light to turn current off and on with different wavelengths. However, this is only an idea and such a device has not yet been built.

A chiroptical molecular switch, such as the one sketched in Fig. 14.21, uses circularly polarized light (CPL) to bring about changes between isomers. The application

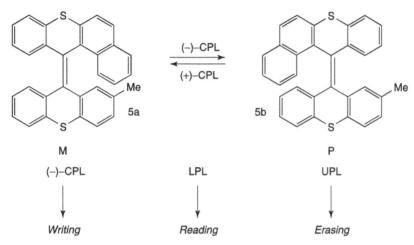

Figure 14.21. Sketch of a molecule that can be switched between two states denoted by *M* and *P* by using circularly polarized light (CPL). The molecular switch position is read using linearly polarized (LPL), and the information can be erased using unpolarized light (UPL). (With permission from M. Gomez-Lopez and F. J. Stoddart, in *Handbook of Nanostructured Materials and Nanotechnology*, H. S. Nalwa, ed., Academic Press, Boston, 2000, Vol. 5, Chapter 3, p. 231.)

of left circularly polarized light $(-)$-CPL to the molecular conformation *M* on the left side of the figure causes a rotation of the four-ring group on the top from a right-handed helical structure to a left-handed helical arrangement *P*, as shown. Right circularly polarized light $(+)$-CPL brings about the reverse transformation. Linearly polarized light (LPL) can be used to read the switch by monitoring the change in the axis of the light polarizer. The system can be erased using unpolarized light (UPL).

Conformational changes involving rearrangements of the bonding in a molecule can also be the basis of molecular switching. When the colorless spiropyran, shown on the left in Fig. 14.22, is subjected to UV light hv_1, the carbon–oxygen

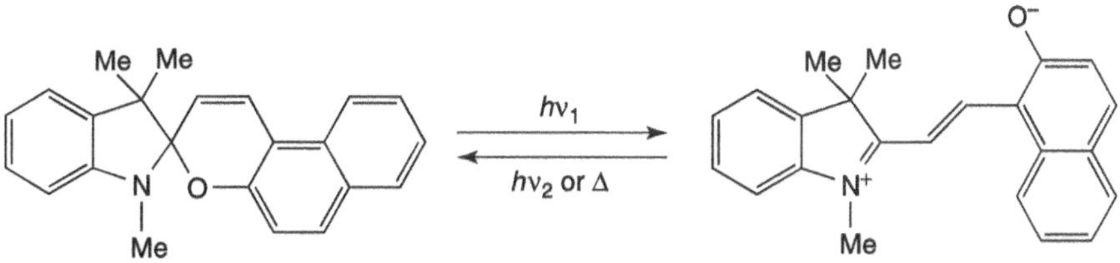

Figure 14.22. Photochemical switching of spiropyran (left) to merocyanine (right) by ultraviolet light hv_1 where red light (hv_2) or heat (Δ) induces the reverse direction conformational change in the molecule. [From M. Gomez-Lopez and F. J. Stoddart, in *Handbook of Nanostructured Materials and Nanotechnology*, H. S. Nalwa, ed., Academic Press, Boston, 2000, Vol. 5, Chapter. 3, p. 233.)

bond opens, forming merocyanine, shown on the right in Fig. 14.22. When the merocynanine is subjected to visible (red) light, hv_2, or heat (Δ) the spiropyran reforms.

A catenane molecule has been used to make a molecular switch that can be turned on and off with the application of a voltage. A *catenane* is a molecule with a molecular ring mechanically interlinked with another molecular ring, as shown by the example sketched in Fig. 14.23. Its two different switched states are shown in Figs. 14.23a and 14.23b. This molecule is 0.5 nm long and 1 nm wide, making it in effect a nanoswitch. For this application a monolayer of the catenane anchored

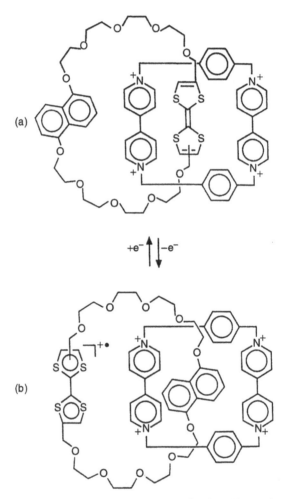

Figure 14.23. Illustration of a switchable catenane molecule, which undergoes a change in conformation when subjected to a voltage that induces oxidation ($-e^-$) from the upper isomeric state to the lower state, and reduction ($+e^-$) for the reverse transformation. [With permission from R. Dagani, *Chemi. Eng. News* 28 (Oct. 16, 2000).]

with amphiphilic phospholipid counterions is sandwiched between two electrodes. The structure in Fig. 14.23a is the open switch position because this configuration does not conduct electricity as well as does the structure shown in Fig. 14.23b. When the molecule is oxidized by applying a voltage, which removes an electron, the tetrathiafulvalene group, which contains the sulfurs, becomes positively ionized and is thus electrostatically repelled by the cyclophane group, the ring containing the nitrogen atoms. This causes the change in structure shown in Fig. 14.23, which essentially involves a rotation of the ring on the left side of the molecule to the right side.

14.3.2. Molecular Electronics

Molecular electronics becomes possible if molecules can conduct electricity and be switched on and off. The STM has been used to measure the conductivity of long chainlike molecules. A monolayer of octanethiol is formed on a gold surface by self-assembly. The sulfur group at the end of the molecule bonds to the surface in the manner illustrated in Figs. 4.16 and 4.17. Some of the molecules are then removed using a solvent technique and replaced with 1,8-octanedithiol, which has sulfur groups at both ends of the chain. A gold-coated STM tip is scanned over the top of the monolayer to find the 1,8-octanedithiol. The tip is then put in contact with the end of the molecule, forming an electric circuit between the tip and the flat gold surface. The octanethiol molecules, which are bound only to the bottom gold electrode, serve as molecular insulators, electrically isolating the octane-dithiol wires. The voltage between the tip and the bottom gold electrode is then increased and the current measured. The results yield five distinct families of curves, each an integral multiple of the fundamental curve, which is the dashed curve in the Fig. 14.24. In the figure we show only the top and bottom curves. The fundamental curve corresponds to electrical conduction through a single molecule; the other curve, to conduction through two or more such molecules. It should be noted that the current is quite low, and the resistance of the molecule is estimated to be 900 MΩ.

Having developed the capability to measure electrical conduction through a chain molecule, researchers began to address the question of whether a molecule could be designed to switch the conductivity on and off. They used the relatively simple molecule sketched in Fig. 14.25, which contains a thiol group (SH—) that can be attached to gold by losing a hydrogen atom. The molecule, 2-amino-4-ethynylphenyl-4-ethy-nylphenyl-5-nitro-1-benzenethiolate, consists of three benzene rings linked in a row by triple-bonded carbon atoms. The middle ring has an amino (NH$_2$) group, which is an electron donor, pushing electric charge toward the ring. On the other side is an electron acceptor nitro (NO$_2$) group, which withdraws electrons from the ring. The net result is that the center ring has a large electric dipole moment. Figure 14.26 shows the current–voltage characteristics of this molecule, which is attached to gold electrodes at each end. There is an onset of current at 1.6 V, then a pronounced increase, followed by a sudden drop at 2.1 V. The result was observed at 60 K but not at room temperature. The effect is called *negative differential resistance*. The

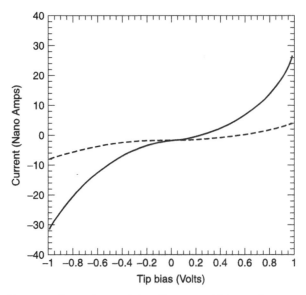

Figure 14.24. Current–voltage characteristics of an octanethiol monolayer on a gold substrate measured by STM using a gold-coated tip probe. Five curves are actually observed, but only two, the lowest (- - -) and the highest are shown here. The solid curve corresoponds to 4 times the current of the dashed curve. [Adapted from X. D. Cui et. al., *Science* **294**, 571 (2001).]

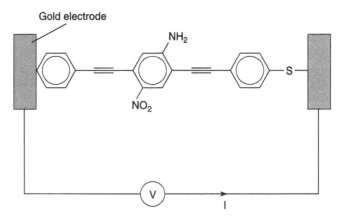

Figure 14.25. Illustration of an electronic switch made of a conducting molecule bonded at each end to gold electrodes. Initially it is nonconducting; however, when the voltage is sufficient to add an electron from the gold electrode to the molecule, it becomes conducting. A further voltage increase makes it nonconducitng again by adding a second electron. [Adapted from J. Chen, *Science* **286**, 1550 (1999).]

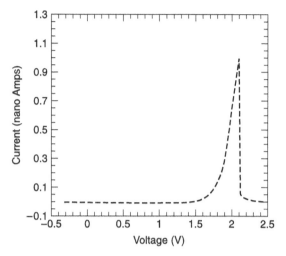

Figure 14.26. Current–voltage characteristics of the electronic switch shown in Fig. 14.25. [Adapted from J. Chen, *Science* **286**, 1550 (1999).]

proposed mechanism for the effect is that the molecule is initially nonconducting, and at the voltage where a current peak is observed the molecule gains an electron, forming a radical ion, and becomes conducting. As the voltage is increased further, a second electron is added, and the molecule forms a nonconducting dianion.

Of course, demonstrating that a molecule can conduct electricity, and that the conduction can be switched on and off, is not enough to develop a computer. The molecular switches have to be connected together to form logic gates. A rotaxane molecule, shown in Fig. 14.27, which can change conformation when it gains and loses an electron, has been used to make switching devices that can be connected together. A schematic cross section of an individual switch is shown in Fig. 14.28. The device consists of 6 μm diameter Al wires patterned on a silica substrate. The Al wires have a layer of Al_2O_3 on them. On top of this is a monolayer of rotaxane molecules shown in Fig. 14.27. On top of this is a 5-nm layer of Ti followed by a 0.1-μm layer of aluminum.

Figure 14.29 shows the current–voltage characteristics of this device. The application of −2 V, the read mode, caused a sharp increase in current. Applying 0.7 V could open the switch. The difference in the current between being open and closed was a factor of 60–80. A number of these switches were wired together in arrays to form logic gates, essential elements of a computer. Two switches (*A* and *B*) connected as shown in Fig. 14.30 can function as an AND gate. In an AND gate both switches have to be ON (i.e., in ON state) for an output voltage to exist. There should be little or no response when both switches are OFF, or only one switch is ON. Figure 14.30 shows the response of the gate for the different switch positions, *A* and *B*, (called *address levels*). There is no current flow for both switches OFF ($A = B = 0$), and very little current for only one switch ON ($A = 0$ and $B = 1$, or

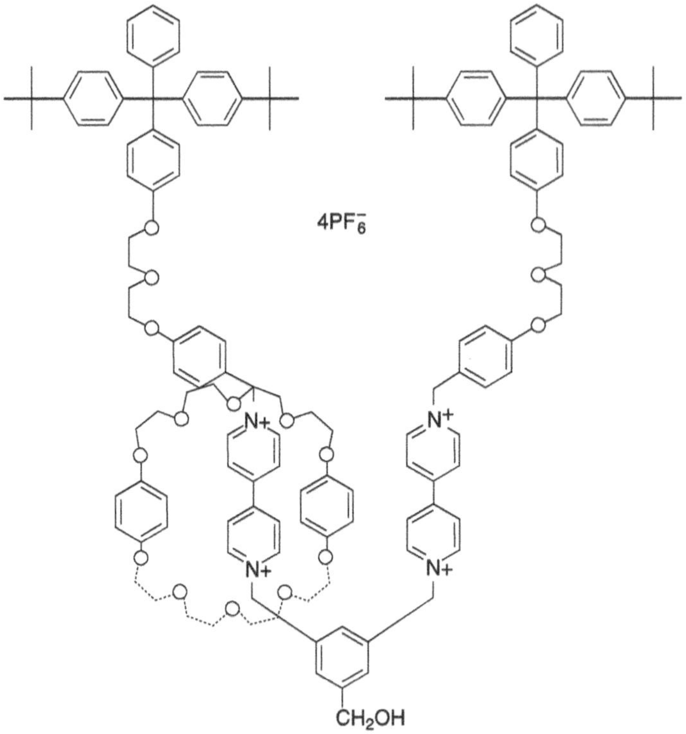

$4PF_6^-$

CH_2OH

Figure 14.27. Rotaxane molecule used in the switch illustrated in Fig. 14.28 to make molecular based logic gates. [With permission from C. P. Collier et al., *Science* **285**, 391 (1999).]

$A = 1$ and $B = 0$). Only the combination $A = 1$ and $B = 1$ for both switches ON provides an appreciable output current, showing that the device can function as an AND gate. These results demonstrate the potential of molecular switching devices for future computer technology.

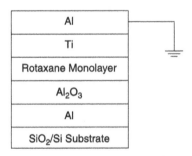

Al
Ti
Rotaxane Monolayer
Al_2O_3
Al
SiO_2/Si Substrate

Figure 14.28. Cross section of a molecular switch prepared by lithography, incorporating rotaxane molecules that have the structure sketched in Fig. 14.27 sandwiched between aluminum electrodes, and mounted on a substrate. The upper electrode has a layer of titanium, and the lower one is coated with aluminum oxide. [Adapted from C. P. Collier et al., *Science* **285**, 391 (1999).]

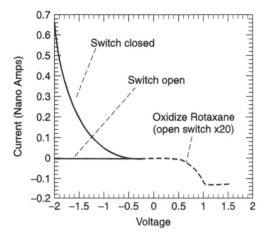

Figure 14.29. Current–voltage characteristics of the molecular switch shown in Fig. 14.28 [Adapted from C. P. Collier et al., *Science* **285**, 391 (1999).]

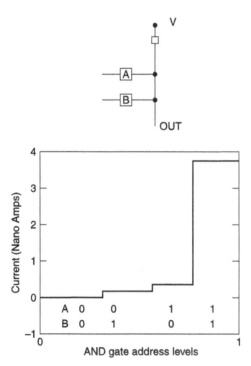

Figure 14.30. Measured truth table for an AND gate constructed from two molecular switches *A* and *B* wired as shown at the top. The lower plot shows that there is an output (high current) only when both switches are ON ($A = 1$, $B = 1$). The address levels listed toward the bottom refer to the possible ON (1) and OFF (0) arrangements of switches *A* and *B*. [With permission from C. P. Collier et al., *Science* **285**, 391 (1999).]

14.3.3. Mechanism of Conduction through a Molecule

The physics of the mechanism of conduction through a molecule bonded on each end to a metal electrode such as gold is challenging and a matter of current research. Here we provide a qualitative picture of how an electron may move from an electrode through a molecule to the other electrode. An isolated molecule, as discussed in Chapter 3, has a discrete set of energy levels. The highest occupied molecular orbitals (HOMOs) are separated from the lowest unoccupied molecular orbitals (LUMOs) by an energy gap, the HUMO–LUMO gap (HLG). When the molecule is bonded to the two metal electrodes, the discrete energy levels of the molecule are broadened but not as much as the energy levels in the metal electrode, and thus still retain their individual identity. The difference in the workfunction between the metal and the molecule causes charge transfer and band alignment between the energy levels of the metal and the molecule. The energy levels of the system essentially equilibrate such that the Fermi level of the metal typically lies inside the HLG of the molecule as shown

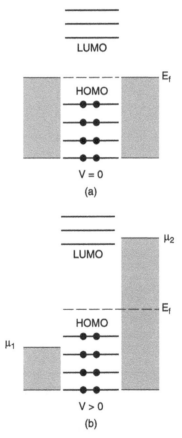

Figure 14.31. (a) Schematic of energy levels of a molecule bonded to two electrodes (shown darkened) and (b) effect of a positive voltage across the electrodes on the energy levels.

schematically in Fig. 14.31a. When a positive voltage is applied across the electrodes, electrons in the metal connected to the positive terminal gain energy and there is a new higher Fermi energy called the *chemical potential*, μ_2, as shown in Fig. 14.31b. If μ_2 is greater than the first LUMO of the molecule as shown, an electron can move to the molecule from the right electrode (reduced). The electron in the molecule can then move to the left electrode where the energy of μ_1 is lower than that of the LUMO and current begins to flow.

14.4. PHOTONIC CRYSTALS

A photonic crystal consists of a lattice of dielectric particles with separations on the order of the wavelength of visible light. Such crystals have interesting optical properties. Before discussing these properties, we will say a few words about the reflection of waves of electrons in ordinary metallic crystal lattices.

The wavefunction of an electron in a metal can be written in the free-electron approximation as

$$\Psi_{k[r]} = \left[\frac{1}{V}\right]^{1/2} e^{ik\cdot r} \tag{14.5}$$

where V is the volume of the solid, the momentum $p = \hbar k$, and the wavevector k is related to the wavelength λ by the expression $k = 2\pi/\lambda$. In the nearly free-electron model of metals the valence or conduction electrons are treated as noninteracting free electrons moving in a periodic potential arising from the positively charged ion cores. Figure 14.32 shows a plot of the energy versus the wavevector for a one-dimensional lattice of the same ions. The energy is proportional to the square of the wavevector, $E = \hbar^2 k^2/8\pi^2 m$, except near the band edge, where $k = \pm \pi/a$. The important result is that there is an energy gap of width E_g, meaning that there are certain wavelengths or wavevectors that will not propagate in the lattice. This is a result of Bragg reflections. Consider a series of parallel planes in a lattice

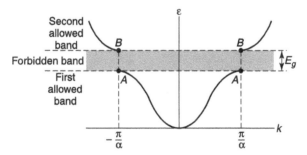

Figure 14.32. Dispersion curve of energy E plotted versus wavevector k for a one-dimensional line of atoms.

separated by a distance d containing the atoms of the lattice. The path difference between two waves reflected from adjacent planes is $2d \sin \Theta$, where Θ is the angle of incidence of the wavevector to the planes. If the path difference $2d \sin \Theta$ is a half-wavelength $\left(\frac{\lambda}{2}\right)$, the reflected waves will destructively interfere, and will not be able to propagate in the lattice, so there will be an energy gap. This is a result of the lattice periodicity and the wave nature of the electrons.

In 1987 Yablonovitch proposed the idea of building a lattice with separations such that light could undergo Bragg reflections in the lattice. For visible light, this requires a lattice dimension of about $\sim 0.5\,\mu$m or ~ 500 nm. This is 100 times larger than the spacing in atomic crystals but still 100 times smaller than the thickness of a human hair. Such crystals have to be artificially fabricated by methods such as electron-beam lithography or X-ray lithography. Essentially, a photonic crystal is a periodic array of dielectric particles having separations on the order of 500 nm. The materials are patterned to have symmetry and periodicity in their dielectric constants. The first three-dimensional photonic crystal was fabricated by Yablonovitch for microwave wavelengths. The fabrication consisted of covering a block of a dielectric material with a mask consisting of an ordered array of holes and drilling through these holes in the block on three perpendicular facets. A technique consisting in stacking micromachined wafers of silicon at consistent separations has been used to build the photonic structures. Another approach is to build the lattice out of isolated dielectric materials that are not in contact. Figure 14.33 illustrates a one-dimensional photonic crystal consisting of equal nanometer-thick layers of materials where the dielectric constant alternates from layer to layer between two values, one high and the other low. When light is incident in the z direction, it will be reflected from the interface between the high and low dielectric layers. If the thickness of the layers is a half-multiple of the wavelength of the incident light, the reflected light from the interface will destructively interfere with the incident light, and light of this wavelength will be unable to propagate through the material. In other words, there are certain wavelengths that cannot propagate through the crystal, and there will be frequency gaps in the transmission similar to the case of the electron waves in solids discussed above.

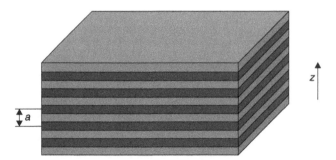

Figure 14.33. A one-dimensional photonic crystal consisting of alternating layers of materials having high and low dielectric constants. (Adapted from J. D. Joannopoulos, R. D. Meade, and J. N. Winn, *Photonic Crystals*, Princeton Univ. Press, 1995.)

Since the wavelength of light ranges from 200 nm in the UV to 900 nm in the near infrared, photonic crystals will have nanometer dimensions.

The description of the behavior of light in photonic crystals involves solving Maxwell's equations in a periodic dielectric structure. The associated Helmholtz equation obtained for the case of no external current sources is

$$\nabla^2 H(r) + \varepsilon \frac{\omega^2}{c^2} H(r) = 0 \tag{14.6}$$

where H is the magnetic field associated with the electromagnetic radiation, ε is the dielectric constant of the components constituting the photonic crystal, ω is the frequency of the light, and c is the velocity of light. This equation can be solved exactly for light in a photonic crystal primarily because there is little interaction between photons, and quite accurate predictions of the dispersion relationship are possible. The dispersion relationship is the dependence of the frequency or energy on the wavelength or k vector. Let us examine the nature of the solution of Eq. (14.6) for the photonic crystal shown in Fig. 14.33. We assume a solution of the form

$$H(z) = U(z)\exp[iKz] \tag{14.7}$$

where $U(z)$ is periodic in the layer separation, that is, $U(z) = U(z + Na)$, where a is the period of the layers as shown in Fig. 14.33. Substituting Eq. (14.7) into (14.6), we obtain the solution for the case where all layers have the same dielectric constant:

$$\omega(K) = \frac{CK}{[\varepsilon]^{1/2}} \tag{14.8}$$

The dispersion relationship is plotted in Fig 14.34a for this case and in Fig. 14.34b for the case where the dielectric constants of the alternating layers are 1 and 13. We see from Fig. 14.34 that when the dielectric constants alternate with the layers, there is a frequency region where light cannot propagate, a region referred to as the *photonic bandgap*. The existence of this photonic bandgap is the basis for many applications of photonic crystals. For example, a crystal with a bandgap could be a filter for certain wavelengths, passing only those frequencies not in the gap.

Figure 14.35 depicts a two-dimensional photonic crystal made of dielectric rods arranged in a square lattice. Figure 14.36 shows a plot of the dispersion relationship of alumina rods (Al_2O_3, $\varepsilon = 8.9$), having the structure shown in Fig. 14.35, with a radius of 0.37 mm, and a length of 100 mm for the transverse magnetic modes. This corresponds to the vibration of the magnetic H vector of the light. The separation between the centers of the rods is 1.87 mm. This lattice is designed for the microwave region, but the general properties would be similar at the smaller rod separations needed for visible light. The labels Γ and X refer to special symmetry points in k-space for the square lattice. The results show the existence of a photonic bandgap, which is essentially a range of frequencies where electromagnetic energy

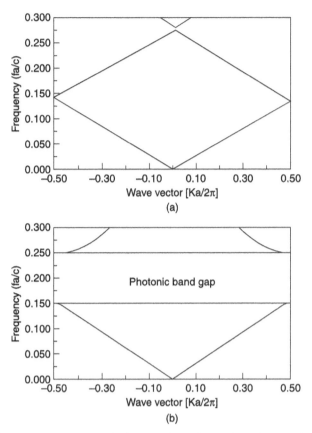

Figure 14.34. Dispersion relationship for one-dimensional photonic crystal where all the layers have the same dielectric constant (a) and where the alternate layers have dielectric constants of 1 and 13(b). The thickness of the layers is 0.5a. The ordinate scale is the frequency f multiplied by the lattice parameter a divided by the speed of light c.

cannot propagate in the lattice. The light power is intense below this bandgap, and in analogy with the terms *conduction band* and *valence band*, this region is called the *dielectric band*. Above the forbidden gap the light power is low, and the region there is referred to as the *air band*.

Now let us consider what would happen if a line defect were introduced into the lattice by removing a row of rods. The region where the rods have been removed would act like a waveguide, and there would now be an allowed frequency in the bandgap, as shown in Fig. 14.37. This is analogous to P and N doping of semiconductors, which puts an energy level in the energy gap. A waveguide is like a pipe that confines electromagnetic energy, enabling it to flow in one direction. An interesting feature of this waveguide is that the light in the photonic crystal guide can turn very sharp corners, unlike light traveling in a fiberoptic cable. Because the frequency of the light in the guide is in the forbidden gap, the light cannot escape into the crystal.

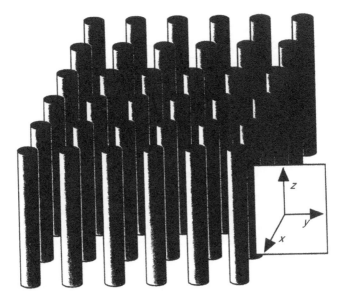

Figure 14.35. A two-dimensional photonic crystal made by arranging long cylinders of dielectric materials in a square lattice array.

It essentially has to turn the sharp corner. Fiberoptic cables rely on total internal reflection at the inner surface of the cable to move the light along. If the fiber is bent too much, the angle of incidence is too large for total internal reflection, and light escapes at the bend.

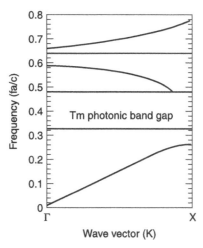

Figure 14.36. A part of the dispersion relationship of a photonic crystal mode (TM) of a photonic crystal made of a square lattice of alumina rods. The ordinate scale is the frequency f multiplied by the lattice parameter a divided by the speed of light c. [Adapted from J. D. Joannopoulos, *Nature* **386**, 143 (1997).]

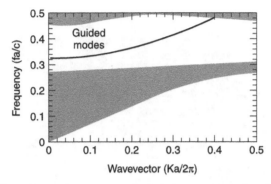

Figure 14.37. Effect of removing one row of rods from a square lattice of a photonic crystal that introduces a level (guided mode) in the forbidden gap. The ordinate scale is the frequency f multiplied by the lattice parameter a divided by the speed of light c. [Adapted from J. D. Joannopoulos, *Nature* **386**, 143 (1997).]

A resonant cavity can be created in a photonic crystal by removing one rod, or changing the radius of a rod. This also introduces an energy level into the gap. It turns out that the frequency of this level depends on the radius of the rod, as shown in Fig. 14.38. The air and dielectric bands discussed above are indicated on the figure. This provides a way to tune the frequency of the cavity. This ability to tune the light and concentrate it in small regions gives photonic crystals potential for use as filters and couplers in lasers. *Spontaneous emission* is the emission of light that occurs when an exited state decays to a lower-energy state. It is an essential part of the process of producing lasing. The ability to control spontaneous emission is necessary to produce lasing. The rate at which atoms decay depends on the coupling between the atom and the photon, and the density of the electromagnetic modes available for the emitted photon. Photonic crystals could be used to control each of these two factors independently.

Semiconductor technology is the basis of integrated electronic circuitry. The goal of putting more transistors on a chip requires further miniaturization. This unfortunately leads to higher resistances and more energy dissipation. One possible future direction would be to use light and photonic crystals for this technology. Light can travel much faster in a dielectric medium than an electron can in a wire, and it can carry a larger amount of information per second. The bandwidth of optical systems such as fiberoptic cable is in the terahertz range, in contrast to electron systems (current flow through wires), which is a few hundred kilohertz. Photonic crystals have the potential to become the basis of future optical integrated circuits.

PROBLEMS

14.1. In Fig. 14.3 it is shown that when a reverse bias is applied to a P–N junction, there is actually a small current flow. Explain the origin of this current.

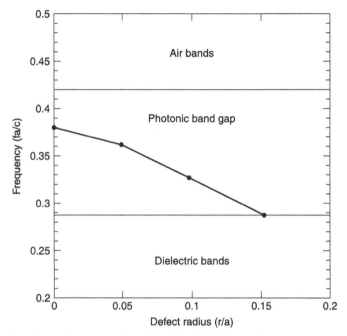

Figure 14.38. Dependence of frequency of localized states in the bandgap formed on the radius r of a single rod in the square lattice. The ordinate scale is the frequency f multiplied by the lattice parameter a divided by the speed of light c. [Adapted from J. D. Joannopoulos, *Nature* **386**.143 (1997).]

14.2. Given that a 60-cycle sinusoidal alternating voltage with a peak voltage of 0.15 V is applied to a PN junction having the current–voltage relationship shown in Fig. 14.3, make a plot of the time dependence of the current produced by the junction.

14.3. Calculate the resistance of a micrometer-long aluminum wire versus the diameter from 5 to 50 nm in 5-nm steps. Plot the resistance versus the diameter. Take the resistivity of wire to be 2.733×10^{-8} Ω·m.

14.4. For a current of 30 mA going through an aluminum wire, calculate the power dissipated and the heat generated in 2 min as a function of the diameter of the wire. Calculate the temperature of the wire versus the diameter. Take the heat capacity of aluminum as 0.904 J/g·K. At what diameter would the wire melt?

14.5. Design a system based on a diluate magnetic semiconductor in which light is used to store and erase information.

14.6. Consider a long rod in which the dielectric varies down the length of the rod as $\varepsilon = \varepsilon_0 \cos^2 (kz/2\pi)$, where the maxima of ε have a separation L. Plot the dispersion relationship in the first Brillion zone. At what value of k does the band gap appear?

REFERENCE

E. Yablonovitch, *Phys. Rev. Let.* **58**, 2059 (1987).

CHAPTER 15

Superconductivity in Nanomaterials

15.1. INTRODUCTION

In this chapter we present a brief overview of the properties of the superconducting state and show how many of these properties depend on lengths and structures of nanometer dimensions. More specifically, a superconductor is a normal conductor of electricity at high temperatures, and it acquires superconducting properties below a transition temperature T_c when its electrical resistance drops to zero, so it conducts electricity without any dissipation of energy. Superconductors may be considered as nanomaterials because their fundamental length parameters are in the nanorange, and because their basic magnetic and electrical properties arise from nanoscale entities. The former two are the coherence length ξ and the penetration depth λ, and the latter two are, respectively, quantized nanowires of magnetic flux Φ called *vortices*, and quantized electrical tunneling devices called *Josephson junctions*. These various factors will all be discussed in turn.

15.2. ZERO RESISTANCE

Let us consider normal electrical conductivity before we consider superconductivity. In a metal the valence electrons are relatively weakly bound to the atoms of the lattice, and the application of a sufficiently strong electric field E will cause these outer electrons, called *conduction electrons*, to move through the lattice. The electric field exerts a force $F = eE$ on the electrons, where e is the electronic charge, and in the absence of any resistance the velocity of each electron should continuously increase. However, this does not happen because the electrons collide with the vibrating atoms as well as with impurity atoms and defects in the lattice, and are scattered out of the path of flow. The result is that the electrons acquire a limiting velocity v, producing a current density $J = nev$, where n is the volume density of electrons. The metal acquires a resistance R as a result of these scattering processes. Because the atoms are vibrating about their equilibrium positions in the lattice, and at higher temperatures the vibrating atoms have larger amplitudes of vibration, the probability for

The Physics and Chemistry of Nanosolids. By Frank J. Owens and Charles P. Poole, Jr.
Copyright © 2008 John Wiley & Sons, Inc.

electrons to scatter from them increases, and hence the resistance increases with the temperature.

The resistance of a metal does not become zero as the temperature is lowered to absolute zero even though the lattice vibrations freeze out; rather, R approaches a limiting value. This constant resistance at low temperature, called the *residual resistance*, is a result of scattering from imperfections and defects in the lattice. In a superconductor, however, there is a transition temperature T_c at which the resistance to direct current flow as well as low-frequency ac current becomes zero. Figure 15.1 is a plot of the dc resistance normalized to the room-temperature value versus temperature for the superconductor $Hg_{0.8}Pb_{0.2}Ba_2Ca_2Cu_3O_{8+x}$, which reaches zero resistance at 130 K, the highest transition temperature of any superconductor at ambient pressure.

Figure 15.2 shows the crystal structure of the parent compound $HgBa_2Ca_2Cu_3O_{8+x}$. It is a member of a chemical family of superconductors called *cuprates*, in which superconductivity, was discovered in 1986. The characteristic of the cuprate superconductors that distinguishes them from the lower-temperature classical superconductors is the two-dimensional copper oxide layers in which most of the electrical conduction takes place. It is believed that this two-dimensional structure plays a crucial role in causing the higher transition temperature. Prior to 1986 superconductivity had been observed in many elemental metals and alloys. The highest transition temperature in these materials was 23.2 K in the niobium germanium alloy, Nb_3Ge.

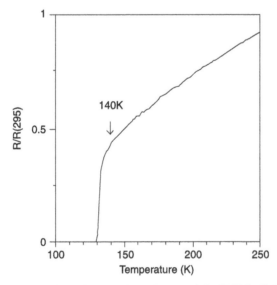

Figure 15.1. Temperature dependence of the resistance of the HgPbBaCaCuO superconductor normalized to its value at 295 K. [Adapted from Z. Iqbal et al., *Phys. Rev.* **B49**, 12331 (1994).]

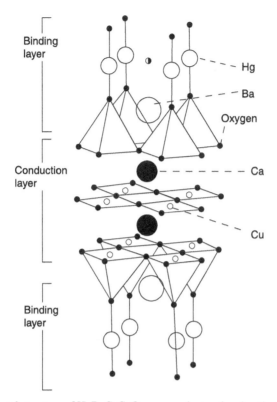

Figure 15.2. Crystal structure of HgBaCaCuO superconductor showing the two-dimensional copper oxide planes.

15.2.1. The Superconducting Gap

Prior to the development of the modern theory of superconductivity, experiments had shown that the carriers of current in the superconducting state have a charge $2e$, twice the electron charge. This means that below T_c the electrons at the Fermi surface of the metal, which carry the current, are bound together in pairs called *Cooper pairs*. The existence of these bound pairs alters the energy bandgap picture of the metal in the superconducting state. In a metal the top occupied, called the conduction band, is not full, and the Fermi level delineates the energy of the uppermost filled state in the band. In the superconducting state the presence of bound electron pairs implies that there must be an energy gap, $E_g = 2\Delta$, at the Fermi level. The magnitude of this superconducting gap corresponds to the binding energy of the electron pairs. It is the energy difference between the normal electrons and the bound electron pairs at the Fermi level.

For most low-temperature superconductors the Bardeen–Cooper–Schrieffer (BCS) theory[1] in the weak coupling limit predicts that the bandgap $E_g(0)$ at absolute zero is related to the transition temperature T_c by the expression

$$E_g(0) = 3.528 \, k_B T_c \tag{15.1}$$

As the temperature is raised above absolute zero, the superconducting gap decreases in magnitude in the manner illustrated in Fig. 15.3, which presents a plot of $E_g(T)/E_g(0)$ versus reduced temperature T/T_c for the element tantalum. The temperature dependence of the gap follows an approximate $(T_c - T)^{1/2}$ relation, which is the well-known mean field result for the order parameter of a second-order phase transition. This suggests that the superconducting transition is the second-order variety for which the latent heat is zero.

15.2.2. Cooper Pairs

One of the major problems in the development of an understanding of superconductivity is how to explain why two negatively charged electrons can be bound into pairs despite the repulsive electrostatic or Coulombic force between them. In 1956, a year before the formulation of the BCS theory, L. N. Cooper[2] showed how lattice phonons could bring about the binding of two electrons near the Fermi level. Measurements of the transition temperatures of several isotopes of the element mercury had demonstrated that the transition temperature shifts to lower values as the mass of the

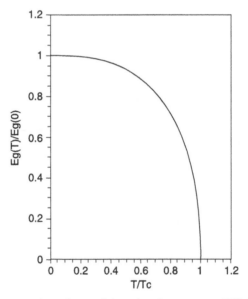

Figure 15.3. Temperature dependence of the reduced energy gap $E(T)/E(0)$ versus reduced temperature T/T_c for superconducting tantalum. [Adapted from P. Townsend and J. Sutton, *Phys. Rev.* **128**, 591 (1962).]

nucleus increases. More specifically, the experiment showed that the shift of T_c is proportional to $(m)^{-1/2}$, where m is the mass of the mercury isotope. We saw in Section 1.1.6 that the vibrational frequencies of atoms and ions in lattices are inversely proportional to the square root of the atom's mass. This isotope effect provided a critical piece of evidence in support of the role of lattice phonons in superconductivity. The BCS theory proposed that the Cooper pair binding arises from a phonon coupling mechanism. However, this mechanism may not account for superconductivity in cuprate superconductors, where the cause of the superconductivity is still under discussion.

A classical (i.e., non-quantum-mechanical) description can be used to provide some insight into how lattice phonons can cause the binding of the electrons into pairs. Because the valence electrons have detached themselves from the atoms and move freely through the lattice, the atoms of the metal have acquired a positive charge. When the conduction electrons move past these positively charged atoms, the atoms are attracted to the electrons, and there is a slight shift in the positions of the atoms toward the passing electrons. The situation is illustrated in Fig. 15.4. This distorted region is slightly more positively charged than the remainder of the lattice, and it follows the electron as it moves through the lattice. This more positive region may attract a distant electron and cause it to follow the distortion as it moves through the lattice, in effect forming a bound electron pair. The binding energy of the two electrons is in the order of 10^{-4} eV, and the separation of the electrons in the pair is about 100 nm in classical superconductors. This means that Copper pairs can be considered as nanostructured particles. Thus the quantum-mechanical wavelength of the Cooper pairs is much longer than the diameters and spacings of the atoms of the solid. As a result, the Cooper pairs do not "see" the atoms of the lattice and are not scattered by them. The spins of the electrons of the pair are oppositely aligned, so a bound Cooper pair has zero spin, and hence is a boson. This means that at absolute zero all Cooper pairs will be in the ground state and have the same

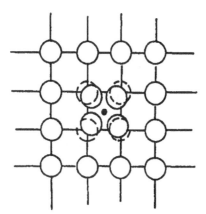

Figure 15.4. Illustration of how a conduction electron moving through a lattice distorts the lattice along its path.

energy and therefore the same wavelength. Thus the wavelength of the pairs not only is very large, but is also the same for every pair. Further, it turns out that the phase of the wave of every pair is the same as that of every other pair. The Copper pairs have a phase coherence analogous to the waves of light produced by a laser. In other words, the motion of the pairs in the lattice is correlated. It is this remarkable property of the quantum-mechanical wave describing the Copper pairs that accounts for their movement through the lattice without scattering, and the resulting zero resistance of the superconducting state. The BCS theory explained how these circumstances reduce the energy of the super electrons below the value of the gap energy.

15.3. THE MEISSNER EFFECT

15.3.1. Magnetic Field Exclusion

The second major characteristic of the superconducting state is called the *Meissner Effect*.[3] If a superconducting material is cooled below its transition temperature in an applied magnetic field

$$B_{app} = \mu_0 H_{app} \tag{15.2}$$

the magnetic flux density B_{in} within the bulk of the material will be expelled below the transition temperature T_c. This behavior is most commonly observed by measuring the temperature dependence of either the magnetization M or the dimensionless susceptibility

$$\chi = \frac{M}{H_{in}} \tag{15.3}$$

of the sample. These various quantities have the following relationships inside the superconductor

$$B_{in} = \mu_0(H_{in} + M) = \mu_0 H_{in}(1 + \chi) \tag{15.4}$$

where the mks system of units is used (for the cgs system, χ is replaced by $4\pi\chi$). For a perfect superconductor the internal field $B_{in} = 0$ and the dimensionless susceptibility $\chi = -1$. This means that for the magnetization or magnetic moment per unit volume, we have $M = -H_{in}$. The material, in effect, behaves like a perfect diamagnet. Figure 15.5 shows the results of a measurement of the temperature dependence of the magnetization for a single crystal of a YBaCuO cuprate superconductor that has a transition temperature of 90 K.

Although magnetic flux is excluded from the bulk, it penetrates the surface layers of the superconductor. Fritz London used the two-fluid model of superconductivity and Maxwell's equations to explain the Meissner effect and the flux penetration of surface layers. The two-fluid model envisions the superconducting state as having a mixture of normal electrons and superconducting electrons whose fraction increases

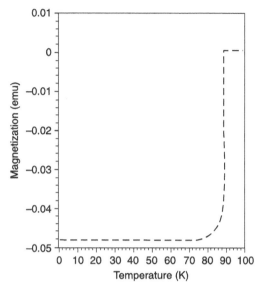

Figure 15.5. Temperature dependence of the magnetization M of a single crystal of the YBaCuO superconductor as the temperature is lowered below the transition temperature.

as the temperature is lowered below T_c. The temperature dependence for the penetration depth λ shown in Fig. 15.6 is given by

$$\lambda(T) = \frac{\lambda_0}{(1 - [T/T_c]^4)^{1/2}} \tag{15.5}$$

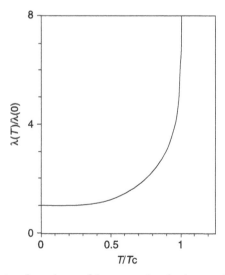

Figure 15.6. Temperature dependence of the penetration depth normalized to its value at 0 K.

TABLE 15.1. Coherence Length ξ, Penetration Depth λ, and Ginzburg–Landau Parameter κ of Various Superconductors

Material	Type	T_c, K	ξ, nm	λ, nm	$\kappa = \lambda/\xi$
Cd	I	0.56	760	110	0.14
Al	I	1.18	1520	40	0.03
Sn	I	3.72	180	42	0.23
Pb	I	7.20	82	39	0.48
Nb	II	9.25	39	52	1.28
Pb–Bi	II	8.3	20	200	10
Nb–Ti	II	9.5	4	300	75
$PbMo_6S_8$ (Chevrel)	II	15	2	200	100
V_3Si (A15)	II	16	3	60	20
Nb_3Ge (A15)	II	23.2	3	90	30
K_3C_{60}	II	19	2.6	240	92
Rb_3C_{60}	II	29.6	2.0	247	124
$(La_{0.925}Sr_{0.075})_2O_4$	II	37	2.0	200	100
$YBa_2Cu_3O_7$	II	89	1.8	170	95
$HgBa_2Ca_2Cu_3O_{8+\delta}$	II	131			100

Source: Data from C. P. Poole, Jr. et al., *Superconductivity*, Academic Press, San Diego, 1995, Table 9.1. Some data are averages of several determinations.

where λ_0 is the penetration depth at absolute zero. Table 15.1 shows that the penetration depths of various superconductors fall in the range from 39 nm to 300 nm. Besides λ, there is another important parameter called the *coherence length* ξ, which is a characteristic of the superconducting state. The boundary between a superconductor in contact with a normal metal is not sharp because the density of superconducting pairs increases in a gradual manner from the surface inward into the bulk of the superconductor. The coherence length ξ is a measure of the distance over which the density of Cooper pairs builds up to its final value. It is also the average distance between the two electrons of a Cooper pair. From Table 15.1 we see that coherence lengths are also in the range of nanometers. Thus we see that the fundamental parameters that determine many of the properties of superconductors are in the order of nanometers.

15.3.2. Type I and Type II Superconductors

We have been discussing "perfect" superconductors for which $\chi = -1$. More generally, there are two types of superconductors characterized by the Ginsburg–Landau parameter

$$\kappa = \frac{\lambda}{\xi} \tag{15.6}$$

with $\kappa < \left(\frac{1}{2}\right)^{1/2}$ for type I and $\kappa > \left(\frac{1}{2}\right)^{1/2}$ for type II superconductors. The two types are also differentiated by their behavior in an applied magnetic field. In type I, as the external magnetic flux density B_{appl} is increased, it does not penetrate the bulk of the material in the superconducting state until a field is reached called the *critical field* B_c, above which the superconducting state no longer exists. The situation is illustrated in Fig. 15.7. In a type II superconductor no magnetic field penetrates the bulk of the sample until a field B_{c1}, referred to as the *lower critical field*, is reached. At B_{c1} flux density begins to penetrate the sample but does not remove all of the superconducting state. As the field is further increased to a value B_{c2}, referred to as the *upper critical field*, flux density continues to penetrate the sample. At B_{c2} the superconducting state is totally removed. The copper oxide superconductors, which can be superconducting as high as 133 K, are type II superconductors. For a type II superconductor, B_{c2} can be quite large, as shown by the data listed in Table 15.2. In $YBa_2Cu_3O_{7+x}$, for example, at 0 K the value of B_{c2} is estimated to be 670 T. Elements are type I, and alloys and compounds are type II. The element niobium is an exception, being type II.

The upper critical magnetic field B_{c2} is temperature-dependent, usually having a temperature dependence that can be described by

$$B_{c2} = B_{c2}(0)(1 - [T/T_c]^2) \qquad (15.7)$$

where $B_{c2}(0)$ is the critical magnetic field at absolute zero. For magnetic fields below B_{c1} a type II superconductor totally excludes magnetic flux and is said to be in the Meissner state, while for fields between B_{c1} and B_{c2} there is partial flux penetration

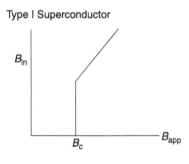

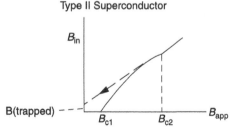

Figure 15.7. Plot of magnetic field B_{in} inside type I and type II superconductors as a function of external applied magnetic field B_{app}.

TABLE 15.2. Critical Fields of Some Selected Superconductors

Material	T_C, K	B_{C1}, mT	B_C, mT	B_{C2}, T
Nb wire, RRR = 750	9.3	181	370	2.0
$In_{0.95}Pb_{0.05}$ (alloy)	3.7	32	37.5	0.049
CTa (NaCl structure)	10	22	81	0.46
V_3Si (A15)	16	55	670	23
Nb_3Sn (A15)	18	35	440	23
$NbSe_2$	7.2	7	204	17.4
UBe_{13} (heavy fermion)	0.9			6.0
K_3C_{60} (buckyball)	19	13	380	32
Rb_3C_{60} (buckyball)	30	12	440	57
$HgBa_2CuO_{4+\delta}$	99		10^3	> 35

Source: Data from C. P. Poole, Jr. et al., *Superconductivity*, Academic Press, San Diego, 1995, Table 9.3. Some data are averages of several determinations.

and it is said to be in the mixed state. In the mixed state there are thin tubelike normal regions in which the magnetic flux is present. A somewhat over simplified picture of these thin threads of flux, called *vortices*, is illustrated in Fig. 15.8 for a slab in the superconducting state located in an external applied magnetic field. The magnetic field lines pass through the slab in restricted regions corresponding to vortices.

In order to understand why some superconductors are type I and others are type II, let us consider the free energy at the interface between a normal metal and a super-conductor. The situation illustrated in Fig. 15.9(a), shows how the density of Copper pairs varies with the distance into the surface of the superconductor, characterized by the coherence length ξ, and how the dc magnetic field decreases into the surface characterized by the penetration depth λ. In zero dc magnetic field the Gibbs free

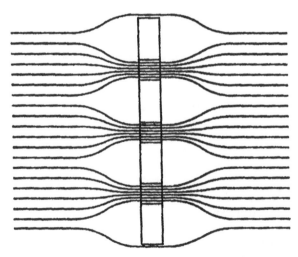

Figure 15.8. Illustration of the passage of magnetic field lines B through vortices in the mixed state of a type II superconducting thin film.

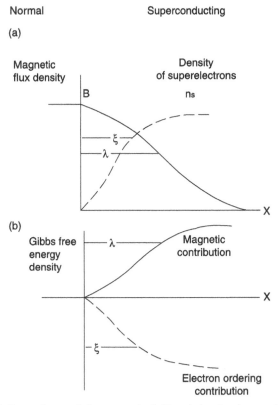

Figure 15.9. (a) Dependence of the magnetic field and the density n_s of superconducting electrons (COOPER PAIRS) on the distance x from the normal-superconducting interface for a type II superconductor and (b) the change in the magnetic and electron ordering contributions as a function of distance from the interface.

energy per unit volume of a material in the superconducting state G_s is lower than that of the normal state G_n because of the ordering of the electrons into Cooper pairs. When a magnetic field $B_{app} = \mu_0 H_{app}$ is applied, the free energy per unit volume is increased by an amount $\frac{1}{2}\mu_0 H_{app}^2$ because the magnetization of the superconductor is opposite to the direction of the applied dc magnetic field. However, at the interface between a superconductor and a normal metal the system must be in equilibrium, and the Gibbs free energies per unit area G^* must be equal on both sides of the interface: $G_n^* = G_s^*$. The surface energy per unit area due to the ordering of the electrons into pairs can be approximated as $\mu_0 H_c^2 \xi/2$ and the surface energy due to the applied field is approximately $\mu_0 H_c^2 \lambda/2$. The net free energy per unit area at the surface is thus

$$G_{surf} = \frac{1}{2}\mu_0 H_c^2(\xi - \lambda) \qquad (15.8)$$

The relative magnitudes of ξ and λ will determine whether the surface energy is positive or negative. In a type I superconductor ξ is greater than λ and the surface energy is positive. The formation of normal regions in the bulk of the superconductor, which increase the area of the normal–superconducting interface, would further increase the Gibbs free energy, and thus not be energetically favorable. When λ is greater than ξ, the surface energy is negative and the appearance of normal regions in the bulk of the superconductor would reduce the free energy, and it is thus energetically favorable to form the tubular normal regions or vortices through which magnetic flux can thread.

15.4. PROPERTIES OF FLUX

15.4.1. Quantization of Flux

In the mixed state the flux penetrating each tubelike normal region can only have the value Φ_0 given by $h/2e = 2.07 \times 10^{-15}$ Wb, where a weber is a tesla (meter)^2. In other words, the flux in the mixed state is quantized. As the applied magnetic field is increased, the flux density in the superconductor can increase only in discrete amounts Φ_0. Each normal tube through which flux penetrates can only enclose a total flux of Φ_0. Therefore increasing the dc magnetic field increases the number of normal tubular regions in the superconductor.

The quantization of flux is a direct result of the coherence of the wavefunctions describing the Cooper pairs. Consider a small superconducting ring of radius R cooled below the superconducting state and carrying a supercurrent of density J. Since the wavefunction of every Cooper pair is in phase with that of every other pair, an exact integral multiple N of the wavelength λ of the pair is required to fit around the circumference $2\pi R$ of the ring; otherwise the waves would not be coherent. In effect, the current in the ring is quantized with the quantization condition

$$N\lambda = 2\pi R \tag{15.9}$$

Since $\lambda = h/P$ where P is the momentum of a Cooper pair, we have in effect a Bohr-like momentum quantization condition for the current in the ring:

$$Nh = 2\pi RP \tag{15.10}$$

The energy E of a current I flowing in a loop can be written in terms of the current I and flux Φ through the loop:

$$E = \frac{I\Phi}{2} \tag{15.11}$$

Since the current I for n electrons moving around the loop is $nve/2\pi R$, Eq. (15.11) becomes

$$E = \frac{\Phi\, nev}{4\pi R} \tag{15.12}$$

The kinetic energy of n electrons moving around the ring can also be written

$$E = \frac{nmv^2}{2} = \frac{nPv}{4} \tag{15.13}$$

where $P = 2mv$ is the momentum of a Cooper pair that contains two electrons. A comparison of Eqs. (15.12) and (15.13) gives the momentum P:

$$P = \frac{\Phi e}{\pi R} \tag{15.14}$$

Substituting this into Eq. (15.10) and using the flux quantization condition $\Phi = N\,\Phi_0$ provides the expression for the unit quantum of flux:

$$\Phi_0 = \frac{h}{2e} \tag{15.15}$$

15.4.2. Vortex Configurations

A vortex has a core of normal material of radius ξ where the magnetic field is the strongest, and outside this core the magnetic field decreases with distance r from the core. This surrounding field is appreciable in magnitude for $r < \lambda$, and drops to a very small value for $r \gg \lambda$. Thus the penetration depth λ may be regarded as the effective radius of a vortex. More quantitatively, the magnetic field strength outside the core falls off with distance in accordance with the expression

$$B(r) = \frac{\Phi_0}{2\pi\lambda^2} K_0\left(\frac{r}{\lambda}\right) \tag{15.16}$$

where $K_0(r/\lambda)$ is the zero-order modified Bessel function. At large distances, $r \gg \lambda$, $K_0(x)$ has the asymptotic behavior $e^{-x}/(2x/\pi)^{1/2}$, and we can express the magnetic field around the vortex as follows:

$$B(r) = \frac{\Phi_0}{2\pi\lambda^2}\frac{\exp\left(-r/\lambda\right)}{(2r/\pi\lambda)^{1/2}} \tag{15.17}$$

As seen in Table 15.1, the value of λ ranges from 39 to 300 nm depending on the type of superconductor, so vortices can be considered as quantum wires, which are discussed in Chapter 9. Thus, while superconductors themselves are not nanomaterials, they have in the superconducting state substructures that are of nanometer dimensions. This distance dependence of $B(r)$ is sketched in Fig. 15.10. The factor

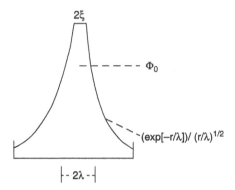

Figure 15.10. Dependence of the magnetic field on the distance from center around an individual vortex.

$\Phi_0/2\pi\lambda^2$ in Eq. (15.17) is related to the lower critical field through the expression

$$B_{c1} = \frac{\Phi_0 \ell n \kappa}{4\pi\lambda^2} \tag{15.18}$$

where $\kappa = \lambda/\xi$. The upper critical field has an expression similar to that for the coherence length replacing the penetration depth

$$B_{c2} = \frac{\Phi_0}{2\pi\xi^2} \tag{15.19}$$

At the lower critical field the vortices are separated by the approximate distance $d \cong \lambda$, and at the upper critical field the density of vortices is so high that the cores are almost touching.

15.4.3. Flux Creep and Flux Flow

In the mixed state below a certain value of temperature and magnetic field, the vortices are symmetrically arranged in a lattice called the *Abrikosov lattice*. The intersection points of the vortices with a plane perpendicular to the applied dc magnetic field forms a triangular lattice, illustrated in Fig. 15.11a. For a given magnetic field there will be some temperature at which this arrangement becomes disordered as illustrated in Fig. 15.11b. The vortex lattice is said to have melted, and the vortices can move around in a random fashion resembling the motion of the molecules of a liquid. This phase is called the *vortex fluid phase*. The temperature at which the vortex lattice melts depends on the magnetic field strength. It is found that the dc magnetic field $H = B/\mu_0$ and the temperature at which the melting occurs are related by

$$1 - \frac{T}{T_c} = AH^q \tag{15.20}$$

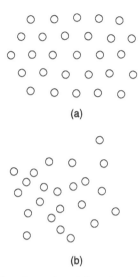

(a)

(b)

Figure 15.11. Arrangement of vortices viewed from top in a type II superconductor for the ordered vortex lattice (a) and the melted vortex lattice (b). The dc magnetic field is perpendicular to the plane of the page.

where A is a constant that depends on the particular superconductor. The exponent q often has the value $\frac{2}{3}$, but has varied from $\frac{1}{2}$ to $\frac{3}{4}$ in measurements made with a variety of cuprates. The $H-T$ line described by this equation is called the irreversibility line, because below it the magnetic properties such as the magnetization as a function of dc magnetic field display hysteresis. The magnetization is not reversible in the sense that when the dc field is decreasing at a certain point the value of the magnetization is different from what it is when the dc field is increasing at that same point. Above the $H-T$ irreversibility line the magnetic properties are reversible. Figure 15.12 shows a measurement of the irreversibility line in the $YBa_2Cu_3O_x$ superconductor.

If a current density $\mathbf{J}$ is sent through a superconductor at some angle to the applied dc magnetic field, there will be a force per unit length exerted on each vortex given by

$$\mathbf{F} = \mathbf{J} \times \mathbf{\Phi_0} \tag{15.21}$$

where $\mathbf{\Phi_0}$ is a vector that is aligned in the vortex direction, and represents the quantized flux enclosed by it. If this force is strong enough, the vortices may move. This movement of vortices will introduce an effective resistance to the current flow in the superconducting state. The force in Eq. (15.21) does not change when the applied field increases, but the density of vortices does, which means that the total force arising from all the vortices increases. As a result, the resistance to current flow becomes greater with increasing strength of the applied dc magnetic field. The increase in resistance will also depend on whether the vortices are in the lattice

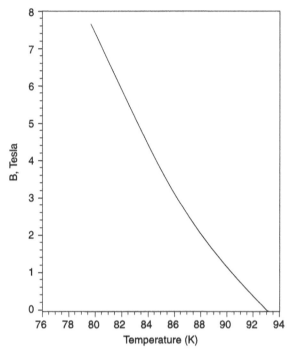

Figure 15.12. Plot of the irreversibility line for a single crystal of YBa$_2$Cu$_3$O$_{7-x}$ [Adapted from T. K. Worthington et al., *Phys. Rev. Lett.* **59**, 1160 (1987).]

phase or the liquid phase. Vortex motion occurs much more easily in the liquid, so the resistance is higher for this phase.

The melted vortex phase can be modeled as a two-dimensional liquid with the vortices having a viscous drag coefficient η. In equilibrium the viscous force per unit length will equal the driving force given by Eq. (15.21), and for **J** perpendicular to Φ we have the scalar expression

$$J\Phi = \eta v_v \tag{15.22}$$

where v_v is the velocity of the vortices. The moving vortices produce an electric field **E** given by he vector expression

$$\mathbf{E} = \mathbf{v}_v \times \mathbf{B} \tag{15.23}$$

Since the resisitivity ρ is E/J, the velocity can be eliminated between Eqs. (15.22) and (15.23), giving an expression for the resistivity induced by motion of the vortices in the fluid phase under the condition that B and J are perpendicular to each other:

$$\rho = \frac{B\Phi}{\eta} \tag{15.24}$$

This simple model predicts that at a fixed temperature below T_c the resistivity will increase linearly with the strength of the applied dc magnetic field in the fluid vortex phase.

In the vortex lattice state the vortices are more strongly bound at their lattice sites and can not move as easily as in the fluid state. The vortices are considered to be trapped in potential wells, and their movement can occur only by a thermally activated hopping of bundles of vortices . Thus their velocity can be described by

$$v_v = v_0 \exp \frac{F_0}{k_B T} \tag{15.25}$$

where F_0 is the height of the energy barrier that has to be overcome for flux bundles to hop. In the vortex lattice phase a larger force is required to induce changes in the resistance of the superconductor. The change in resistivity will be

$$\rho \sim \frac{B v_0}{J} \exp \frac{F_0}{k_B T} \tag{15.26}$$

which is also linear in the applied field. Figure 15.13 shows a measurement of the resistance versus temperature in both zero magnetic field, and in a field of 10 T for the HgBaCaCuO superconductor.

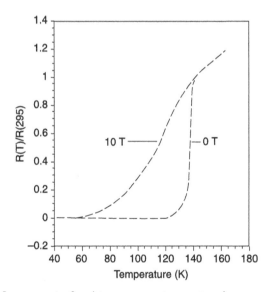

Figure 15.13. Measurement of resistance versus temperature in a zero applied field and in a magnetic field of 10 T for the $n = 3$ Hg–Ba–Ca–Cu–O superconductor. (T. Datta, unpublished.)

15.4.4. Vortex Pinning

Vortices often have their cores held in place at so-called pinning centers, so they are unable to move about the lattice. There are various kinds of pinning centers such as columnar defects, grain boundaries, inclusions, oxygen vacancies, point defects, screw dislocations, and twin boundaries. Most of these centers have dimensions of nanometers. Doping and irradiating a superconductor can increase the number of pining centers. Vortices become bound to these centers by a force called a *pinning force* whose nature depends on the type of center. The Lorentz force of Eq. (15.21) from the presence of a transport current can depin or release a vortex from its pinning center, and raising the temperature induces thermal vibrations of vortices, which can have the same effect. Pinning is beneficial because it hinders or prevents the vortex motion that is responsible for the resistive effects mentioned above.

Nanoparticles of $BaSnO_3$ dispersed in the superconductor $YBa_2Cu_3O_{7-\delta}$ act as pinning centers that bring about improved critical current densities.[4] A similar improvement has been obtained in the current-carrying capacity of MgB_2 below T_c by doping it with short lengths of carbon nanotubes.[5] The critical current in MgB_2 is also enhanced by reducing the grain size into the nanometer region.[6]

15.5. DEPENDENCE OF SUPERCONDUCTING PROPERTIES ON SIZE EFFECTS

We have been discussing superconductivity in bulk materials, that is, materials in which all three dimensions are large, at least in the millimeter or centimeter range. The question arises as to how the properties of a superconductor, such as zero resistance and perfect diamagnetism, are effected by reducing one or more of its dimensions to the micrometer or the nanometer range of size. The obvious answer is that size effects become important when they approach coherence length ξ or penetration depth λ values. We know from the data in Table 15.1 that these two fundamental lengths both have values in nanometers for all known superconductors, and the coherence length in particular is in the very low nanometer range ($1.8\,\text{nm} \leq \xi \leq 4\,\text{nm}$) for most type II materials. We are interested in the size dependence of the superconducting properties of materials in isolation, as well as of materials in close contact with one another.

15.6. RESISTIVITY AND SHEET RESISTANCE

One aspect of the size effect dependence of superconductivity is how the electrical resistivity of a material depends on the material's dimensions. The resistance R of a material such as a wire or a plate of length L and cross-sectional area A is related to the resistivity ρ through the expression

$$R = \frac{\rho L}{A} \tag{15.27}$$

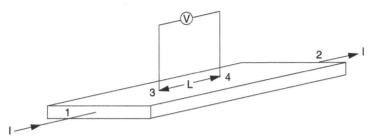

Figure 15.14. Experimental arrangement for a four-probe resistivity determination. (From C. P. Poole, Jr. et al., *Superconductivity*, Academic Press, San Diego, 1995, p. 30.)

A common way to measure this resistivity is the four-probe method sketched in Fig. 15.14. The resistance R between points 3 and 4 is the ratio V/I of the voltage V between these two points, and the current I that enters the material at point 1 and leaves at point 2. The voltage is measured by a voltmeter, the current is measured by an ammeter, their ratio is the resistance, and the resistivity ρ is calculated from Eq. (15.27). When uniform, steady-state current flow is ensured, this four-probe measurement technique is superior to the more common two-probe method, in which the voltage determination is carried out at the entrance and exit points 1 and 2 of the current. This is because the arrangement eliminates the effect of resistance at the points where wires are attached to the sample, an effect referred to as *contact resistance*.

An axially symmetric superconductor has a longitudinal resistivity ρ_c for current flow along its axial or c direction, and a transverse resistivity ρ_{ab} for current flow perpendicular to this direction. Some cuprate high-temperature superconductors have tetragonal crystallographic structures and hence are axially symmetric, while others are orthorhombic but close to tetragonal. For example, the orthorhombic compound $YBa_2Cu_3O_{7-\delta}$ with $\delta = 0.1$ has the lattice constants $a = 0.383$ nm, $b = 0.388$ nm, and $c = 1.168$ nm. The CuO_2 planes lie in the ab plane, perpendicular to the crystallographic c direction. Current flows much more readily along these CuO_2 planes than it does perpendicularly to them, and for the compound $YBa_2Cu_3O_{6.9}$ the resistivity ratio has the approximate value $\rho_c/\rho_{ab} \approx 100$.

When a current flows along a film of thickness d through a square region of surface that has the dimensions $a \times a$, as shown in Fig. 15.15, it encounters the resistance R_S, which, from Eq. (15.27), has the value

$$R_S = \frac{\rho a}{ad} = \frac{\rho}{d} \qquad (15.28)$$

The quantity $R_S = \rho/d$ is called the *sheet resistance*, or the *resistance per square*, because it applies to a square section of film, and is independent of the length of the side a. It is analogous to the surface resistance $R_S = \rho/\delta$ of a metallic surface interacting with an incident high-frequency electromagnetic wave, where δ is the skin depth of the metal at the frequency of the wave.

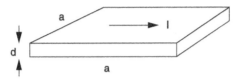

Figure 15.15. Geometric arrangement and current flow direction for a sheet resistance determination. (From C. P. Poole, Jr. et al., *Superconductivity*, Academic Press, San Diego, 1995, p. 32.)

There is a quantum of resistance R_Q, which is given by

$$R_Q = \frac{h}{4e^2} = 6.4532 \text{ k}\Omega \tag{15.29}$$

where the quantity $2e$ is the charge of a Cooper pair, the charge carrier of a supercurrent. For a film to become superconducting, it must be thick enough so that its sheet resistance just above the superconducting transition temperature T_c is less than the quantum value (15.29). When the film becomes so thin that $R_S > R_Q$, then it will no longer superconduct. If the sheet resistance of the film is only slightly less than the quantum value, then its superconducting transition temperature T_c will be very low, much below the bulk value.

The element bismuth superconducts only under pressure and as a film. The sheet resistance R_S of a film of Bi attains the quantum value R_Q when its thickness d has the

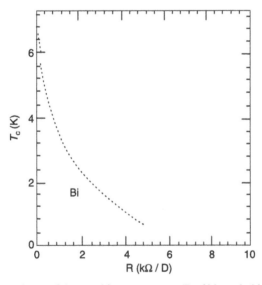

Figure 15.16. Dependence of the transition temperature T_c of bismuth thin films on the sheet resistance. [From D. B. Haviland et al., *Phys. Rev. Lett.* **62**, 2180 (1989).]

critical dimension $d_c = 0.673$ nm, and much thicker films ($\rho/d \ll h/4e^2$) have a transition temperature $T_c = 6.1$ K. Figure 15.16 indicates how the transition temperature of a Bi film depends on the sheet resistance. The higher the value of R_S, the lower the value of T_c. The lower part of Fig. 15.17 presents the transition from the normal to the superconducting state for films of bismuth varying in thickness from 7.427 nm down to the critical value $d_c = 0.673$ nm where $R_S = R_Q$. The transition to the superconducting state is fairly sharp over this entire temperature range. The upper part of

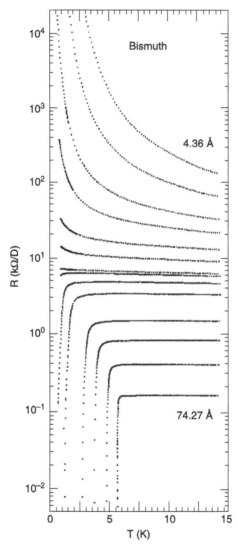

Figure 15.17. Temperature dependence of the sheet resistance of films of Bi deposited on Ge as a function of the film thickness in the range from 0.436 to 7.427 nm. [From D. B. Haviland et al., *Phys. Rev. Lett.* **62**, 2180 (1989).]

the figure presents data for thinner films with thickness $d < d_c$ for which the sheet resistance is always greater than the quantum value R_Q, so they never become superconducting. Instead, their sheet resistance increases as the temperature is lowered, as shown.

Thin films of bismuth may be considered as quantum wells that have superconducting transition temperatures that are lower than the values expected if the decrease in thickness were not playing a role. The existence of a quantum of resistance plays a critical role in bringing about this quantum well behavior.

15.7. PROXIMITY EFFECT

When films of two different superconductors are in direct contact with each other, the wavefunction of one penetrates into the other, and there is some interchange of Cooper pairs. This can influence the transition temperatures of the two superconductors, and in some cases cause them to exhibit the same intermediate value of T_c. When a normal metal film is in contact with a superconducting film, a transfer of electrons and electron pairs across the interface can reduce T_c of the superconductor. These findings are called *proximity effects*, and they are especially pronounced for films with thicknesses in the nanorange.

To elucidate the lowering of T_c by the presence of a normal metal, Werthamer[7] studied a series of composite films, each consisting of a normal metal copper film of thickness d_n in contact with a superconducting lead film of thickness d_s. For bulk lead, we find from Table 15.1 that the transition temperature $T_c = 7.2$ K, the coherence length $\xi = 82$ nm, and the penetration depth $\lambda = 39$ nm. Measurements were made of the transition temperature T_c' for various values of d_s and d_n, and the results presented in Fig. 15.18 show that thinner superconducting layers have lower transition temperatures T_c'. The presence of the normal layer of copper reduces T_c' for each thickness of lead, an effect that becomes more pronounced for increasing thicknesses of copper until the copper attains a thickness of about the penetration depth of lead, namely, $d_n = \lambda = 39$ nm. Greater thicknesses of the copper layer have no additional effect. The pronounced drop in T_c' for the ultrathin lead layer with $d_s = 7$ nm causes it to go normal for $d_n = 12$ nm. Figure 15.19 plots the reduction of the limiting values of T_c' as a function of d_s for $d_n = 80$ nm.

Similar experiments have been carried out with layers of the high-temperature superconductor $YBa_2Cu_3O_{7-\delta}$ and its isomorphous nonsuperconducting counterpart $PrBa_2Cu_3O_{7-\delta}$. Thicknesses were recorded in terms of the number of CuO_2 layers N_Y in the yttrium compound and the number N_{Pr} in the praesodymium compound. Since each unit cell contains two CuO_2 layers, the corresponding numbers of unit cells are $\frac{1}{2}N_Y$ and $\frac{1}{2}N_{Pr}$, respectively. Figure 15.20a plots the reduction in transition temperature T_c'/T_c versus N_{Pr} for $N_Y = 2,4,6,8$. These four layer numbers correspond to $YBa_2Cu_3O_{7-\delta}$ film thicknesses of $d = 1.17$ nm, 2.34 nm,. 3.51 nm, and 4.68 nm, respectively, since the unit cell has the length $c = 1.17$ nm. In addition, $N_{Pr} = 32$ corresponds to the thickness 18.7 nm. The four curves in

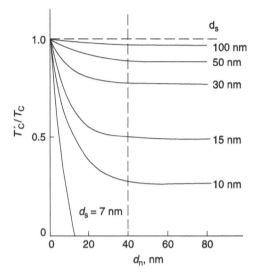

Figure 15.18. Proximity effect for a PbCu composite illustrating how the critical temperature T_c of superconducting Pb in a copper–lead composite relative to T_c of bulk lead depends on the copper film thickness d_n for several Pb thicknesses d_s from 7 to 100 nm. The vertical dashed line indicates the characteristic thickness L_0. [Adapted from N. R. Werthamer, *Phys. Rev.* **132**, 2440 (1963).]

Fig. 15.20a for the high-temperature superconductor case are very similar in shape to the five upper curves in Fig. 15.18 for the lead/copper composite cases. The $N_{Pr} = 32$ case plotted in Fig. 15.20b is the analog of the Pb/Cu composite limiting value plot of T_c'/T_c versus d_s presented in Fig. 15.19.

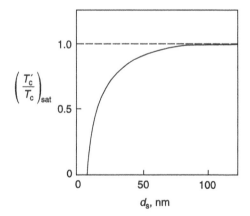

Figure 15.19. Dependence of the limiting value of the transition temperature $(T_c'/T_c)_{sat}$ of the PbCu composite of Fig. 15.18 on thickness d_s of the superconducting component Pb. (Data from Fig. 15.18, figure from C. P. Poole, Jr., et al., *Superconductivity*, Academic Press, San Diego, 1995, p. 425.)

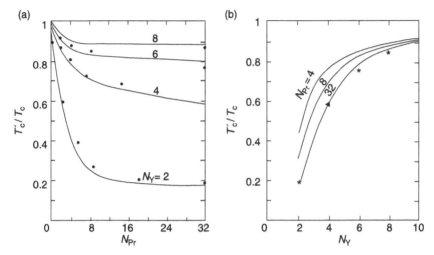

Figure 15.20. Normalized transition temperature T'_c/T_c of $YBa_2Cu_3O_{7-\delta}/PrBa_2Cu_3O_{7-\delta}$ layers plotted versus the number of Cu–O planes in (a) $PrBa_2Cu_3O_{7-\delta}$ and (b) $YBa_2Cu_3O_{7-\delta}$. The calculated curves are drawn to fit the data points. [From J. F. Wu et al., *Phys. Rev.* **B44**, 411 (1991).]

15.8. SUPERCONDUCTORS AS NANOMATERIALS

The proximity effect data presented in the previous section provide us with some insight into the properties of superconductors considered as quantum wells, and earlier discussions of vortices in this chapter focused on some quantum wire aspects of superconductors. These materials are defined by their properties of perfect electrical conductivity (zero resistance) and perfect diamagnetism (magnetic field exclusion). The former electrical property has associated with it a quantum of resistance $R_Q = h/4e^2$ given by Eq. (15.29), and the latter magnetic property has associated with it a quantum of magnetic flux $\Phi_0 = h/2e$ given by Eq. (15.15). For a material to be superconducting, its sheet resistance $R_S = \rho/d$ must be less than the quantum value R_Q. When the sheet resistance of a superconducting film starts to approach this quantum value, the film becomes a quantum well, and its transition temperature T_c decreases significantly from its bulk or thick-film value. When a type II superconductor is located in an applied magnetic field B_{app} whose magnitude lies between the lower critical field $B_{c1} = \Phi_0 \, Ln \, \kappa/4\pi\lambda^2$ and the upper critical field $B_{c2} = \Phi_0/2\pi\xi^2$, it is no longer a perfect diamagnet. Instead the magnetic field penetrates in the form of discrete quantum wires called *vortices*, each of which encloses one quantum of flux Φ_0. Thus, from the electrical perspective, superconductors can function as quantum wells, and from a magnetic perspective their properties are dominated by the presence of quantum wires.

When the dimensions of a superconductor become smaller in size than the penetration depth λ, then an applied magnetic field will penetrate its bulk if it is type I, and will penetrate its bulk below the lower critical field B_{c2} if it is type II.

15.9. TUNNELING AND JOSEPHSON JUNCTIONS

15.9.1. Tunneling

Tunneling or barrier penetration, which is discussed in Sections 11.4.1 and 13.2.2, is a process whereby an electron confined to a region by an energy barrier is able to penetrate the barrier through a quantum-mechanical process, and emerge on the other side. The example shown in Fig. 15.21 involves electrons with kinetic energy $E_{KE} = \frac{1}{2}mv^2$ confined to a region on the left side of the barrier by the potential V_b, where $\frac{1}{2}mv^2 < eV_b$. We show in the figure an electon tunneling through the barrier to the right side, where it ends up with the same kinetic energy. Such a phenomenon can take place because there is a quantum-mechanical probability that the electron will penetrate the barrier and escape, as is explained in standard quantum-mechanical texts.

Tunneling can occur through an insulating layer between two normal metals, called N–I–N tunneling, in accordance with the arrangement shown in Fig 15.22. Figure 15.23 shows one electron reflected by the barrier, and one electron tunneling through it. The reflection process is more probable than the penetration one. Since this barrier is biased, with the potential $V = 0$ on the right and the potential $+V$ on the left, the current flow J_{12} to the right exceeds the current flow J_{21} to the left in Fig. 15.22 to give the net current flow J_{12} to the right indicated at the top of Fig. 15.23. Because by convention current flow is considered to be the movement of positive charges, and the actual carriers are negative electrons, the electron motion is shown opposite in direction to the current flow in Fig. 15.23. Tunneling can also occur between a normal metal and a superconductor (N–I–S tunneling), and between two superconductors (S–I–S tunneling).

15.9.2. Weak Links

When an S–I–S barrier layer of the type shown in Fig. 15.22 is too thick, then the two superconductors lose contact with each other, and no tunneling takes place. If the barrier is only a monolayer of foreign atoms, there is strong contact, and current flows easily across the interface. Of special interest is the intermediate case, called a *weak link* or *microbridge*, in which the coupling across the barrier is weak, and tunneling

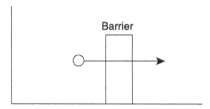

Figure 15.21. Tunneling of electrons through a barrier when the kinetic energy is less than the barrier energy eV_b. (From C. P. Poole, Jr. et al., *Superconductivity*, Academic Press, San Diego, 1995, p. 402.)

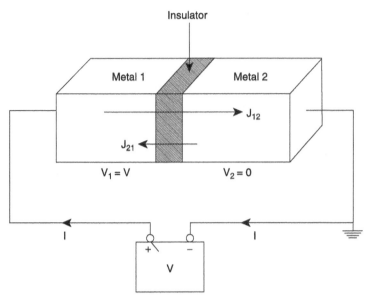

Figure 15.22. Metal–metal tunnel junction in the form of a thin uniform insulating layer between two metals M_1 and M_2. Tunneling current densities J_{12} and J_{21} in both directions are shown, with $J_{12} > J_{21}$ for the bias indicated. (From C. P. Poole, Jr. et al., *Superconductivity*, Academic Press, San Diego, 1995, p. 411.)

occurs. A typical microbridge thickness is a coherence length ξ, and hence high-temperature cuprate superconductors require much thinner barriers than do classical materials. Recall from Table 15.1 that $\xi \approx 2\,\text{nm}$ for cuprates, and $\xi = 82\,\text{nm}$ for lead, so weak links are indeed nanostructures. Other types of barriers also exist.

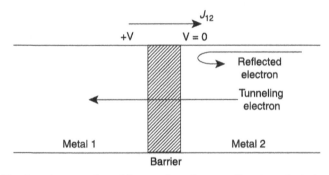

Figure 15.23. One electron reflected from, and another tunneling through, the insulating layer of a tunnel junction for the bias of Fig. 15.22. The tunneling of the electron is in the direction from the negative to the positive side of the junction, but the corresponding current flow J_{12} is from + to − because it is based on the convention of positive charge carriers. (From C. P. Poole, Jr. et al., *Superconductivity*, Academic Press, San Diego, 1995, p. 412.)

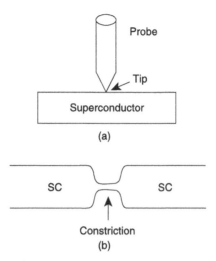

Figure 15.24. Tunnel junction formed from (a) a probe tip in contact with the superconducting surface and (b) a constricted region of a superconductor. (From C. P. Poole, Jr. et al., *Superconductivity*, Academic Press, San Diego, 1995, p. 421.)

Figure 15.24a shows a barrier consisting of a tip of a probe on a superconductor surface. Such a probe might be a normal metal monitoring the condition of the superconductor surface, as employed by a scanning tunneling microscope. Figure 15.24b illustrates a constriction present in a superconductor or between two superconductors.

15.9.3. Josephson Effect

The Josephson effect is the flow of a super current I_s given by

$$I_s = I_c \sin \Delta\theta \qquad (15.30)$$

across a weak link in the absence of an applied potential difference, where I_s is the maximum supercurrent that the junction can support and $\Delta\theta$ is the phase difference between the wavefunctions of the Cooper pairs on each side of the junction. If the voltage difference V is present across the junction, then the difference in phase $\Delta\theta$ will evolve in time in accordance with the equation

$$\frac{d(\Delta\theta)}{dt} = \frac{2eV}{h} \qquad (15.31)$$

These two expressions, called *Josephson relations*, are the basic equations that govern the tunneling of Cooper pairs. The steady current flow $I_s = I_c \sin \Delta\theta$ across the weak link in the absence of a voltage difference is referred to as the *dc Josephson effect*, and the presence of an oscillatory current

$$I_s(t) = I_c\sin(\omega_J t - \Delta\theta_0) \tag{15.32}$$

across the junction in the presence of a voltage difference V is called the *ac Josephson effect*. The corresponding characteristic frequency $\nu_J = \omega_J/2\pi$

$$\nu_J = \frac{2eV}{h} = \frac{V}{\Phi_0} \tag{15.33}$$

is known as the *Josephson frequency*, where Φ_0 from Eq. (15.15) is the quantum of magnetic flux. This expression (15.33) provides the following practical formula

$$\frac{\nu_J}{V} = 483.6\text{MHz}/\mu\text{V} \tag{15.34}$$

for the frequency corresponding to a particular applied bias voltage.

An inverse ac Josephson effect occurs when a voltage V is induced across an unbiased junction by radiation incident on the junction. There are also macroscopic quantum interference effects involving a tunneling current density J with an oscillatory dependence $\sin(\pi\Phi/\Phi_0)$ on the applied magnetic flux Φ.

15.9.4. Josephson Junctions

We have been using the term *Josephson junction* to denote a weak link, such as the probe tip or constriction shown in Fig. 15.24, across which Josephson tunneling takes place. The same term, *Josephson junction*, is also sometimes employed to designate a device based on the Josephson effect. Such a device might function, for example, as a field effect transistor (FET). In this role a Josephson junction is the superconducting analog of a P–N semiconductor junction. In some practical applications stacks of superconducting thin films separated by barrier layers constitute the electrodes.[8]

15.9.5. Ultrasmall Josephson Junctions

When a Josephson junction becomes small enough, the tunneling of individual electrons can be observed. Consider a microbridge of the type sketched in Fig. 15.22 with a cross-sectional area of $(100\,\text{nm})^2$ and a thickness of $d = 0.1\,\text{nm}$, much less than a coherence length ξ. It has a capacitance $C = \varepsilon_0 A/d$ of $\sim 10^{-15}$ F. For a single-electron transfer, the change in voltage corresponds to $\Delta V = e/C = 0.16\,\text{mV}$, which is an appreciable fraction of a typical junction voltage. This change can be sufficient to impede the tunneling of the next electron, thereby blocking the smooth flow of current. This phenomenon is referred to as a *Coulomb blockade*. The plot of the current I versus the bias voltage V across the ultrasmall Josephson junction shown in Fig. 15.25 exhibits steps or fluctuations as a result of this blockade. These fluctuations are enhanced in the differential plot dI/dV versus V, which is the first derivative of the I–V plot, as shown. The step feature on the I–V plot is called a *Coulomb staircase*.

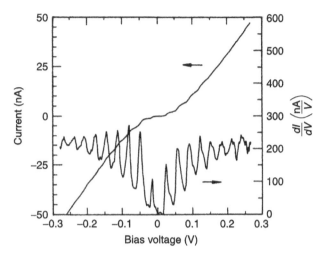

Figure 15.25. Coulomb staircase structure in plots of I versus V and dI/dT versus V for electrons tunneling between a granular lead film and the tip of a scanning tunneling microscope. [From K.A. McGreer et al., *Phys. Rev.* **B39**, 12260 (1989).]

15.10. SUPERCONDUCTING QUANTUM INTERFERENCE DEVICE (SQUID)

It was mentioned above that the current I through a Josephson junction has an oscillatory dependence on the magnetic flux Φ through the junction. This effect permits

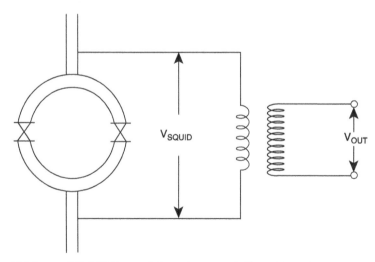

Figure 15.26. An RF SQUID consisting of two weak links. The RF output voltage is a measure of the change in the loading of the circuit produced by a change in magnetic flux through the loop. (From C. P. Poole, Jr. et al., *Superconductivity*, Academic Press, San Diego, 1995, p. 452.)

magnetic fields to be measured to a very high degree of accuracy, to better than 10^{-11} T. The measurement device can be a superconducting loop containing one or two Josephson junctions, The device is called a SQUID, which is an acronym for superconducting quantum interference device. The radiofrequency SQUID shown in Fig. 15.26 measures the change in magnetic flux through the superconducting loop by detecting the change in loading of the circuit when the weak links sense this change in flux. Nanoscale SQUIDS have been made using nanobridge Josephson junctions.

15.11. BUCKMINISTERFULLERENES

15.11.1. The Structure of C_{60} and Its Crystal

The C_{60} molecule has been named *fullerene* after the architect and inventor R. Buckminister Fuller, who designed the geodesic dome that resembles the structure of C_{60}. Originally the molecule was called *buckminsterfullerene*, but this name is a bit unwieldy, so it has been shortened to *fullerene*. It also has the nickname "buckyball." The sketch of the molecule presented in Fig. 10.5 shows that it has 12 pentagonal (five-sided) and 20 hexagonal (six-sided) faces symmetrically arrayed to form a molecular ball. In fact, a soccer ball has the same geometric configuration as does fullerene. These ball-like molecules bind with each other in the solid state to form a crystal lattice having the face-centered cubic (FCC) structure with the unit cell sketched in Fig. 10.9. In the lattice each C_{60} molecule is separated from its nearest neighbor by 1 nm (the distance between their centers is 1 nm), and they are held together by the weak van der Waals force that was discussed earlier. Because C_{60} is soluble in benzene and toluene, single crystals of it can be grown by slow evaporation from these solutions.

15.11.2. Alkali-Doped C_{60}

In the fcc fullerene lattice structure 26% of the volume of the unit cell is empty, so much smaller atoms such as Na or K can easily fit into the empty spaces between the molecular balls of the material. When C_{60} crystals and potassium metal are placed in evacuated tubes and heated to 400°C, potassium vapor diffuses into these empty spaces to form the compound K_3C_{60}. The C_{60} crystal is an insulator, but when doped with alkali atoms, it becomes electrically conducting. These alkali atoms occupy two vacant tetrahedral sites and one larger octahedral site per C_{60} molecule, as explained in Section 1.1.2. In the smaller tetrahedral site the alkali atom has four surrounding C_{60} balls, and in the larger octahedral site there are six surrounding C_{60} molecules. When C_{60} is doped in this manner, the potassium atoms ionize to form K^+, and their electrons become associated with the C_{60} molecules, which become the triply negative ions C_{60}^{3-}. Thus each C_{60} has three extra electrons that are loosely bonded to it and can move freely through the C_{60}

lattice, making it electrically conducting. In this case the C_{60} lattice is said to be electron-doped.

15.11.3. Superconductivity in C_{60}

In 1991 when A. F. Hebard and his coworkers[9] at Bell Telephone Laboratories doped C_{60} crystals with potassium by the methods described above and tested them for superconductivity, the most surprising of all evidence was found for a superconducting transition at 18 K. Figure 15.27 shows the drop in magnetization indicative of the onset of superconductivity. A new class of superconducting materials had been found having a simple cubic structure and containing only two elements. Not long after the initial report, it was found that many alkali atoms could be doped into the lattice, and the transition temperature increased to as high as 33 K in Cs_2RbC_{60}. As the radius of the dopant alkali atom increases, the cubic C_{60} lattice expands, and the superconducting transition temperature goes up. Figure 15.28 is a plot of the transition temperature versus the lattice parameter.

It was mentioned above that graphite consists of parallel planar graphitic sheets of carbon atoms. It is possible to put other atoms between the planes of these sheets, a procedure called *intercalation*. When intercalated with potassium atoms, crystalline graphite becomes superconducting at the extremely low temperature of a few tenths of a Kelvin.

Larger fullerenes such as C_{70}, C_{76}, C_{80}, and C_{84} have also been found, and a C_{20} dodecahedral carbon molecule has been synthesized by gas-phase dissociation of $C_{20}HBr_{13}$. $C_{36}H_4$ has also been made by pulsed laser ablation of graphite. A solid phase of C_{22} has been identified in which the lattice consists of C_{20} molecules bonded together by an intermediate carbon atom. One interesting aspect of the existence of these smaller fullerenes is the prediction that they might possibly be superconductors with high transition temperatures when appropriately doped.

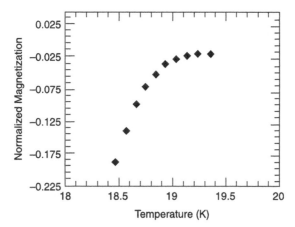

Figure 15.27. Magnetization versus temperature plot for K_3C_{60} showing the transition to the superconducting state. [Adapted from A. F. Hubbard, *Phys. Today* **29** (Nov. 1992).]

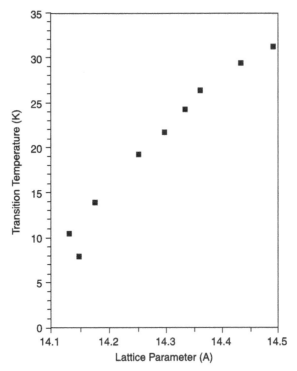

Figure 15.28. Plot of the transition temperature of alkali metal (A)-doped A_3C_{60} versus the lattice parameter a, where $10 = 1$ nm. [Adapted from A. F. Hebard, *Phys. Today* **29** (Nov 1992).]

PROBLEMS

15.1. A superconducting film has dimensions $5 \text{ cm} \times 5 \text{ cm} \times 10 \text{ mm}$. It has an internal magnetic field B with the value of 6 T. How may vortices constitute this field for the following two conditions: (a) the field lies in the plane of the film and (b) the field direction is perpendicular to the plane?

15.2. A type II superconductor has a coherence length $\xi = 5$ nm and a penetration depth $\lambda = 85$ nm. What are the values of the lower critical field, the thermodynamic critical field, and the upper critical field?

15.3. A current of 6 A flows in the plane of a superconducting film of dimensions $7 \text{ cm} \times 7 \text{ cm} \times 5 \text{ mm}$. The film is placed in an applied magnetic field that forms an angle of $45°$ with respect to the direction perpendicular to the film. How much total force does it exert on each vortex?

15.4. A superconductor has a coherence length $\xi = 4$ nm, a Ginzburg–Landau parameter $\kappa = 50$, and a transition temperature $T_c = 20$ K. Find the penetration depth and the lower critical field at the temperature $T = 10$ K.

15.5. A microbridge has an area $A = 50\,\text{nm}^2$ and a separation $d = 0.2\,\text{nm}$. Find the change in voltage for the transfer of one electron across it.

REFERENCES

1. J. Bardeen, L. N. Cooper, and J. R. Schrieffer, *Phys. Rev.* **108**, 1175 (1957).

2. L. N. Cooper, *Phys. Rev.* **104**, 1189 (1956).

3. W. Meissner and R. Ochsenfeld, *Naturwissenschaften* **21**, 787 (1933).

4. C. V. Varanasi, P. N. Barnes, J, Burke, L. Brunke, I. Maartense, T. J. Haugan, E. A. Stinzianni, K. A. Dunn, and P. Haldar, *Supercond. Sci. Technol.* **19**, L37 (2006).

5. W. K. Yeoh, J. H. Kim, J. Horvat, S. X. Dou, and P. Munroe, *Supercond. Sci. Technol.* **19**, L5 (2006).

6. B. Lorenz, O. Perner, J. Eckert, and C. W. Chu, *Supercond. Sci. Technol.* **19**, 912 (2006).

7. N. R. Werthamer, *Phys. Rev.* **132**, 2440 (1963).

8. M. G. Blamire, *Supercond. Sci. Technol.* **19**, S132–S137 (2006).

9. A. F. Hebard et al., *Nature* **350**, 600 (1991).

Formulas for Dimensionality

A.1. INTRODUCTION

We saw in Chapter 9, on quantum wells, wires, and dots, that the dimensionality of the system has a pronounced effect on its properties. This is one of the dramatic new factors that makes nanoscience so interesting. In this appendix we gather together some of the important formulas that summarize the role played by dimensionality. These equations are relegated to the appendix because many of the readers of this work may not have the quantum-mechanical background needed to follow the development of the relevant mathematics. We will assume here that the reader has a familiarity with the reciprocal lattice or k-space, and we will also assume that the temperature is absolute zero, that is, T = 0 K.

A.2. DELOCALIZATION

The regions occupied by a system of conduction electrons delocalized in one-, two-, and three-dimensional coordinate space are the length, area, and volume given in the first column of Table A.1. Column 2 of the table gives the size of the unit cell in reciprocal or k-space, and column 3 gives the size of the Fermi region that is occupied by the delocalized electrons, where the Fermi energy E_F has the value $E_F = h^2 k_F^2 / 2m$, and in this region E < E_F. Finally. column 4 gives expressions for k^2 in the three systems. The numbers of electrons N in the occupied regions of column 3 at the temperature of absolute zero, as well as the density of states $D(E)$ defined by the expression $D(E) = dN(E)/dE$, are given in Table A.2. We see from this table that the density of states decreases with energy for one dimension, it is constant for two-dimensions, and it increases with increasing energy for three dimensions. Thus the number of electrons and the density of states as functions of energy have quite different behaviors for the three cases, as indicated by the plots of Figs. 9.12, 9.13, and 9.18.

The Physics and Chemistry of Nanosolids. By Frank J. Owens and Charles P. Poole, Jr.
Copyright © 2008 John Wiley & Sons, Inc.

TABLE A.1. Properties of Coordinate and k-Space in One, Two and Three Dimensions

Coordinate Region	k-Space Unit Cell	Fermi Region	Value of k^2	Dimensions
Length L	$2\pi/L$	$2k_F$	k_X^2	1
Area $A = L^2$	$(2\pi/L)^2$	πk_F^2	$k_X^2 + k_Y^2$	2
Volume $V = L^3$	$(2\pi/L)^3$	$4\pi k_F^3/3$	$k_X^2 + k_Y^2 + k_Z^2$	3

TABLE A.2. Number of electrons $N(E)$ and Density of States $D(E) = dN(E)/dE$ as Function of Energy E for Electrons Delocalized in One, Two, and Three Spatial Dimensions, Where $A = L^2$ and $V = L^3$

Number of Electrons $N(E)$	Density of States $D(E)$	Delocalization Dimensions
$N(E) = \dfrac{4k_F}{2\pi/L} = \dfrac{2L}{\pi}\left[\dfrac{2m}{h^2}\right]^{1/2} E^{1/2}$	$D(E) = \dfrac{L}{\pi}\left[\dfrac{2m}{h^2}\right]^{1/2} E^{-1/2}$	1
$N(E) = \dfrac{2\pi k_F^2}{(2\pi/L)^2} = \dfrac{A}{2\pi}\left[\dfrac{2m}{h^2}\right] E$	$D(E) = \dfrac{A}{2\pi}\left[\dfrac{2m}{h^2}\right]$	2
$N(E) = \dfrac{2(4\pi k_F^3/3)}{(2\pi/L)^3} = \dfrac{V}{3\pi}\left[\dfrac{2m}{h^2}\right]^{3/2} E^{3/2}$	$D(E) = \dfrac{V}{2\pi^2}\left[\dfrac{2m}{h^2}\right]^{3/2} E^{1/2}$	3

A.3. SQUARE AND PARABOLIC WELLS

It will be appropriate to review some of the properties of square wells in one, two and three dimensions. Standard quantum-mechanical texts show that for an infinitely deep square potential well of width a in one dimension, the coordinate x has the range of values $-(1/2a) \le x \le (1/2)a$ inside the well, and the energies are given by

$$E_n = E_0 n^2 \tag{A.1}$$

as shown in Fig. 9.14, where $E_0 = \pi^2 h^2/2ma^2$ is the ground-state energy, and the quantum number $n = 1,2,3,\ldots$. There are even and odd wavefunctions ψ_n for the square well, and for the infinite square well we have

$$\psi_n = \cos\frac{n\pi x}{a} \qquad n = 1,3,5,\ldots \qquad \text{even parity} \tag{A.2}$$

$$\psi_n = \sin\frac{n\pi x}{a} \qquad n = 2,4,6,\ldots \qquad \text{odd parity} \tag{A.3}$$

These wavefunctions are sketched in Fig. 9.14 for the infinite well. Two- and three-dimensional infinite square wells have wavefunctions that are combinations of these, such as the following three-dimensional one with mixed parity

$$\psi_{nnn} = \cos\frac{n_x\pi x}{a}\,\sin\frac{n_y\pi y}{a}\,\sin\frac{n_z\pi z}{a} \tag{A.4}$$

where the three quantum numbers take on the values $n_i = 1, 2, 3, \ldots$

Square well expressions can also be written in two-dimensional polar coordinates and in three-dimensional spherical coordinates, and these are related to their Cartesian counterparts through the respective expressions:

$$\rho = (x^2 + y^2)^{1/2} \qquad \tan\phi = \frac{y}{x} \qquad\qquad \text{well radius } \rho = a \quad (A.5)$$

$$r = (x^2 + y^2 + z^2)^{1/2} \qquad \cos\theta = \frac{z}{r}, \quad \tan\phi = \frac{y}{x} \qquad \text{well radius } r = a \quad (A.6)$$

The wavefunctions for infinite square wells with circular geometries in two and three dimensions involve Bessel functions, $J_p(k_{pn}\rho)$, spherical Bessel functions $j_p(k_{pn}r)$ and spherical harmonics $Y_{nm}(\theta,\phi)$. Another commonly used potential well is the parabolic one, which is given by $V(x) = (1/2)kx^2$, $V(\rho) = (1/2)k\rho^2$, and $V(r) = (1/2)kr^2$ in one, two, and three dimensions, respectively. This is the well-known harmonic oscillator potential. The wavefunctions for a Cartesian geometry involve Hermite polynomials $H_N(u_i)$, where $\gamma = m\omega/h$, $u_x = \gamma^{1/2}x$, and similarly for u_y and u_z. If this problem is solved in spherical coordinates, then the Hermite polynomials group into Tesseral harmoincs, which are the real versions of spherical harmonics. The energies that are degenerate group into S, P, D, and other states, for $L = 0, 1, 2, \ldots$, with no dependence of the energy on the orbital angular momentum quantum number L.

A.4. PARTIAL CONFINEMENT

The conduction electrons in nanostructures can be partially confined and partially delocalized, depending on the shape and the dimensions of the structure. One limiting case is a quantum dot, in which they are totally confined, and the other limiting case is bulk material, in which they are all delocalized. The intermediate cases are a quantum wire that is long in one dimension and very small in its transverse directions, and a quantum well that is a flat plate nanosized in thickness and much larger in length and width. The quantum wire exhibits electron confinement in two dimensions and delocalization in one dimension, and the quantum well reverses these characteristics. Figures 9.12 and 9.13, respectively, provide plots of how they numbers of electrons $N(E)$ and the densities of states $D(E)$ for these cases depend on the energy. The degeneracies d_i in Table A.3 refer to potential well energy levels. (For further details, see L. Jacak, P. Hawrylak, and A. Wojs, *Quantum Dots*, Springer, Berlin, 1998, Section 3.1.)

TABLE A.3. Number of Electrons $N(E)$ and Density of States $D(E) = dN(E)/dE$ as Function of Energy E for Electrons Delocalized/Confined in Quantum Dots, Quantum Wires, Quantum Wells, and Bulk Material[a]

Type	Number of Electrons $N(E)$	Density of States $D(E)$	Dimensions Delocalized	Dimensions Confined
Dot	$N(E) = 2\sum d_i \Theta(E - E_{iW})$	$D(E) = 2\sum d_i \delta(E - E_{iW})^2$	0	3
Wire	$N(E) = \dfrac{2L}{\pi}\left[\dfrac{2m}{h}\right]^{1/2} \sum d_i(E - E_{iW})^{1/2}$	$D(E) = \dfrac{L}{\pi}\left[\dfrac{2m}{h^2}\right]^{1/2} \sum d_i(E - E_{iW})^{-1/2}$	1	2
Well	$N(E) = \dfrac{A}{2\pi}\left[\dfrac{2m}{h^2}\right] \sum d_i(E - E_{iW})$	$D(E) = \dfrac{A}{2\pi}\left[\dfrac{2m}{h^2}\right] \sum d_i$	2	1
Bulk	$N(E) = \dfrac{V}{3\pi^2}\left[\dfrac{2m}{h^2}\right]^{3/2} (E)^{3/2}$	$D(E) = \dfrac{V}{2\pi^2}\left[\dfrac{2m}{h^2}\right]^{3/2} (E)^{1/2}$	3	0

[a]The degeneracies of the confined (square or parabolic well) energy levels are given by d_i. The Heaviside step function $\Theta(x)$ is zero for $x < 0$ and one for $x > 0$ the delta function $\delta(x)$ is zero for $x \neq 0$ and infinity for $x = 0$ and integrates to a unit area.

The formulas presented in this appendix are for idealized cases of isotropic systems with circular Fermi limits in two dimensions, and spherical Fermi surfaces in three dimensions. The bulk case in the last row of Table A.3 assumes the presence of only one conduction band. In practical cases the bands are more numerous and more complex, but these expressions do serve to clarify the roles played by the effects of delocalization and electron confinement in nanostructures.

Tabulations of Semiconducting Material Properties

In this book we have discussed various types of nanostructures, many of which are semiconductors composed of the group IV elements Si or Ge, types III–V compounds such as GaAs or types II–VI compounds such as CdS. The properties of these materials often become modified when they are incorporated into nanoscale structures or devices. It is helpful to bring together in this appendix a set of tables surveying and comparing their various properties in the bulk state so that they can be referred to when needed throughout the chapters of the book.

The first five tables, Tables B.1–B.5, contain crystallographic and related data, and electronic information such as bandgaps, effective masses, mobilities; donor/acceptor ionization energies, and dielectric constants are presented in Tables B.6–B.11. These are the tables most often referred to throughout the book. For the convenience of the reader we gather together in Tables B.12–B.21 data on some additional properties that may be of interest, presented in the same format as the earlier tabulated data. These include indices of refraction (Table B.12), melting points and heats of formation (Tables B.13 and B.14), and some thermal properties such as Debye temperatures Θ_D, specific heats, and thermal conductivities (Tables B.15–B.17). The mechanical properties tabulated are linear expansion, volume compressibility, and microhardness (Tables B.18–B.20). Finally we list in Table B.21 the diamagnetic susceptibilities of several III–V compounds. Some tables include more limited data entries than others because complete data sets are not readily available for every property.

TABLE B.1. Lattice Constant a (in nm) for Types III–V (Left) and II–VI (Right) Semiconductors[a]

	P	As	Sb		S	Se	Te
Al	0.545	0.562	0.613	Zn	0.541	0.567	0.609
Ga	0.545	0.565	0.612	Cd	0.582	—	0.648
In	0.587	0.604	0.648	Hg	0.585	0.608	0.643

[a]The values for Si and Ge are 0.543 and 0.566 nm, respectively.
Source: Data from R. W. G. Wyckoff, *Crystal Structures*, Wiley, New York, 1963, Vol. 1, p. 110.

The Physics and Chemistry of Nanosolids. By Frank J. Owens and Charles P. Poole, Jr.
Copyright © 2008 John Wiley & Sons, Inc.

TABLE B.2. Atomic Radii from Monatomic Crystals and Ionic Radii of Several Elements Found in Semiconductors

Group	Atomic Number[a]	Atom	Radius	Ion	Radius
II	30	Zn	0.133	Zn^{2+}	0.074
II	48	Cd	0.149	Cd^{2+}	0.097
II	80	Hg	0.151	Hg^{2+}	0.110
III	13	Al	0.143	Al^{3+}	0.051
III	31	Ga	0.122	Ga^{3+}	0.062
III	49	In	0.163	In^{3+}	0.081
IV	14	Si	0.118		
IV	32	Ge	0.123		
V	15	P	0.110	P^{3-}	0.212
V	33	As	0.124	As^{3-}	0.222
V	51	Sb	0.145	Sb^{3-}	0.245
VI	16	S	0.101	S^{2-}	0.184
VI	34	Se	0.113	Se^{2-}	0.191
VI	52	Te	0.143	Te^{2-}	0.211

[a]Number of atoms.
Source: Data from *Handbook of Chemistry and Physics*, CRC Press, Boca Raton, FL, 2002, pp. F189, F-191.

TABLE B.3. Comparison of Nearest-Neighbor Distance $\left(\frac{1}{4}\right)3^{1/2}a$ of Several Semiconductor III–V and II–VI Binary Compounds AC^a

Compound	$\left(\frac{1}{4}\right)3^{1/2}a$ Distance	$A^{n-} + C^{n+}$ Ionic Sum	$A + C$ Monatomic Sum	Compound Type
ZnS	0.217	0.258	0.233	Semiconductor
AlP	0.236	0.263	0.253	Semiconductor
GaAs	0.245	0.284	0.246	Semiconductor
CdS	0.252	0.281	0.247	Semiconductor
HgTe	0.279	0.321	0.294	Semiconductor
InSb	0.281	0.326	0.308	Semiconductor
NaCl	0.282	0.278	0.350	Alkali halide
KBr	0.353	0.353	0.404	Alkali halide
RbI	0.367	0.367	0.419	Alkali halide
MgO	0.211	0.198	0.232	Alkaline-earth chalcogenide
CaS	0.285	0.283	0.299	Alkaline-earth chalcogenide
SrSe	0.301	0.303	0.419	Alkaline-earth chalcogenide

[a]These compounds are listed with the sums of the corresponding ionic radii, and with the sums of the corresponding monatomic lattice radii. For comparison purposes, data are shown for some alkali halides and alkaline-earth chalcogenides. The distances are in nanometers.

TABLE B.4. Molecular Mass of Group III–V and Group II–VI Compounds[a]

	P	As	Sb		S	Se	Te
Al	57.95	101.90	148.73	Zn	97.43	144.34	192.99
Ga	100.69	144.64	191.47	Cd	144.46	191.36	240.00
In	145.79	189.74	256.57	Hg	232.65	279.55	328.19

[a]The atomic masses of Si and Ge, respectively, are $\frac{1}{2}(56.172)$ and $\frac{1}{2}(145.18)$.
Source: Data from *Handbook of Chemistry and Physics*, CRC Press, Boca Raton, FL, 2002.

TABLE B.5. Density (in g/cm^3) of Group III–V and Group II–VI Compounds[a]

	P	As	Sb		S	Se	Te
Al	2.42	3.81	4.22	Zn	4.08	5.42	6.34
Ga	4.13	5.32	5.62	Cd	4.82	5.81	5.86
In	4.79	5.66	5.78	Hg	7.73	8.25	8.17

[a]The densities of Si and Ge, respectively, are 2.3283 and 5.3234.
Source: Data from *Handbook of Chemistry and Physics*, CRC Press, Boca Raton, FL, 2002.

TABLE B.6. Bandgaps E_g Expressed (in eV) for group III–V Semiconductors (Left) and Group II–VI Materials (Right)[a]

	P	As	Sb		S	Se	Te
Al	(2.45)/3.62	(2.15)/3.14	(1.63)/2.22	Zn	3.68	2.7	2.26
Ga	(2.27)/2.78	1.43	0.70	Cd	2.49	1.75	1.43
In	1.35	0.36	0.18	Hg	—	−0.061	−0.30 (4.4 K)

[a]Indirect bandgaps are given in parentheses. The values for Si and Ge are $(1.11)/3.48$ and $(0.66)/0.81$ eV, respectively.
Source: Data from P. Y. Yu and M. Cardona, *Fundamentals of Semiconductors*, Springer, Berlin, 2001, table on inside front cover.

TABLE B.7. Temperature Dependence of bandgap dE_g/dT (in meV/°C) (a) and Pressure Dependence of Bandgaps dE_g/dP (in the units meV/GPa) (b)[a]

	P	As	Sb		S	Se	Te
	a. Temperature Dependence dE_g/dT							
Al	—	(−0.4)/−0.51	(−3.5)	Zn	−0.47	−0.45	−0.52
Ga	(−0.52)/−0.65	−0.395	−0.37	Cd	−0.41	−0.36	−0.54
In	−0.29	−0.35	−0.29				
		Si(−0.28)	Ge (−0.37)/−0.4				
	b. Pressure Dependence dE_g/dP							
Al	—	(−5.1)/102	(−15)	Zn	57	70	83
Ga	(−14)/105	115	140	Cd	45	50	80
In	108	98	157				
		Si (−14)	Ge (50)/121				

[a]Corresponding values for Si and Ge are listed below the III–V values.
Source: Data from P. Y. Yu and M. Cardona, *Fundamentals of Semiconductors*, Springer, Berlin, 2001, table on inside front cover; see also *Handbook of Chemistry and Physics*, CRC Press, Boca Raton, FL, 2002.

TABLE B.8. Effective Masses m^* Relative to Free-Electron Mass m_e of Conduction Band Electrons and Three Types of Valence Band Holes[a]

	Electron Effective Mass			Heavy-Hole Effective Mass			Light-Hole Effective Mass			Split-off Hole Effective Mass			Spin–Orbit Splitting Δ_{SO} (eV)		
	P	As	Sb	P	As	Sb	P	As	Sb	P	As	Sb	P	As	Sb
Ga	—	0.067	0.047	0.57	0.53	0.8	0.18	0.05	0.05	0.25	0.15	0.12	0.08	0.34	0.75
In	0.073	0.026	0.014	0.58	0.4	0.42	0.12	0.026	0.016	0.12	0.14	0.43	0.11	0.38	0.81
	Ge 0.41			Si 0.54, Ge 0.34			Si 0.15, Ge 0.043			Si 0.23, Ge 0.09			Si 0.044, Ge 0.295		

[a]The spin–orbit splitting parameter Δ_{SO} is also listed. The split-off hole effective mass arises from spin–orbit coupling, which was not taken into account in this book. Corresponding values for Si and Ge are listed below the III–V values.

Source: Data from P. Y. Yu and M. Cardona, *Fundamentals of Semiconductors*, Springer, Berlin, 2001, pp. 71, 75, supplemented by G. Burns, *Solid State Physics*, Academic Press, New York, 1985, p. 312.

TABLE B.9. Mobilities at Room Temperature [in $cm^2/(V \cdot A \cdot s)$] for Types III–V and II–VI Semiconductors[a]

	P	As	Sb		S	Se	Te
Electron Mobilities
Al	80	1,200	200–400	Zn	180	540	340
Ga	300	8,800	4,000	Cd			1,200
In	4,600	33,000	78,000	Hg	250	20,000	25,000
Hole Mobilities
Al	—	420	550	Zn	5 (400°C)	28	100
Ga	150	400	1,400	Cd	—	—	50
In	150	460	750	Hg	—	≈1.5	350

[a]The electron mobilities are given first, and the hole mobilities are presented below them. The mobilities for Si and Ge for electrons are 1900 and 3800, respectively, and corresponding values for holes are 500 and 1820.
Source: Data from *Handbook of Chemistry and Physics*, CRC Press, Boca Raton, FL, 2002, pp. 12–101.

TABLE B.10. Ionization Energies (in meV) of Group III Acceptors and Group V Donors in Si and Ge

Ion	Group	Type	Ionization Energy (meV) Si	Ge
B	III	Acceptor	45	10
Al	III	Acceptor	67	11
Ga	III	Acceptor	71	11
In	III	Acceptor	155	12
P	V	Donor	45	13
As	V	Donor	54	14
Sb	V	Donor	43	10
Bi	V	Donor	71	13

Source: Data from G. Burns, *Solid State Physics*, Academic Press, New York, 1985, p. 172.

TABLE B.11. Static Relative Dielectric Constants $\varepsilon/\varepsilon_0$ for Types III–V and II–VI Semiconductors[a]

	P	As	Sb		S	Se	Te
Al	—	10.9	11	Zn	8.9	9.2	10.4
Ga	11.1	13.2	15.7	Cd	—	—	7.2
In	12.4	14.6	17.7				

[a]The values for Si and Ge are 11.8 and 16, respectively.
Source: Data from *Handbook of Chemistry and Physics*, CRC Press, Boca Raton, FL, 2002, pp. 12–101.

TABLE B.12. Index of Refraction n^a for Types III–V and II–VI Semiconductors[b]

	P	As	Sb		S	Se	Te
Al	—	—	3.2	Zn	2.36	2.89	3.56
Ga	3.2	3.30	3.8	Cd	—	—	2.50
In	3.1	3.5	3.96	Hg	2.85	—	—

[a]Optical region, $n = (\varepsilon/\varepsilon_0)^{1/2}$.
[b]The values for Si and Ge are 3.49 and 3.99, respectively.
Source: Data from *Handbook of Chemistry and Physics*, CRC Press, Boca Raton, FL, 2002.

TABLE B.13. Melting Points (in K) for Types III–V and II–VI Semiconductors[a]

	P	As	Sb		S	Se	Te
Al	2100	2013	1330	Zn	2100	1790	1568
Ga	1750	1510	980	Cd	1750	1512	1365
In	1330	1215	798	Hg	1820	1070	943

[a]The values for Si and Ge are 1685 and 1231 K, respectively.
Source: Data from *Handbook of Chemistry and Physics*, CRC Press, Boca Raton, FL, 2002.

TABLE B.14. Heats of Formation (in kJ/mol) at 300 K for Types III–V and II–VI Semiconductors[a]

	P	As	Sb		S	Se	Te
Al	—	627	585	Zn	477	422	376
Ga	635	535	493	Cd	—	—	339
In	560	477	447	Hg	—	247	242

[a]The values for Si and Ge are 324 and 291 kJ/mol, respectively.
Source: Data from *Handbook of Chemistry and Physics*, CRC Press, Boca Raton, FL, 2002.

TABLE B.15. Debye Temperature Θ_D (in K) for Types III–V and II–VI Semiconductors[a]

	P	As	Sb		S	Se	Te
Al	588	417	292	Zn	530	400	223
Ga	446	344	265	Cd	219	181	200
In	321	249	202	Hg	—	151	242

[a]The values for Si and Ge are 645 and 374 K, respectively.
Source: Data from *Handbook of Chemistry and Physics*, CRC Press, Boca Raton, FL, 2002.

TABLE B.16. Specific Heat C_P in the units J/(kg·A·K) at 300 K for Type III–V Semiconductors[a]

	P	A	Sb		S	Se	Te
Al	—	—	—	Zn	472	339	264
Ga	—	—	320	Cd	330	255	205
In	—	268	144	Hg	210	178	164

[a]The values for Si and Ge are 702 and 322 J/(kg·A·K), respectively.
Source: Data from *Handbook of Chemistry and Physics*, CRC Press, Boca Raton, FL, 2002.

TABLE B.17. Thermal Conductivity (in the units mW/(cm·A·K) at 300 K for Types III–V and II–VI Semiconductors[a]

	P	As	Sb		S	Se	Te
Al	920	840	600	Zn	251	140	108
Ga	752	560	270	Cd	200	90	59
In	800	290	160	Hg		10	20

[a]The values for Si and Ge are 1240 and 640 mW/(cm·A·K), respectively.
Source: Data from *Handbook of Chemistry and Physics*, CRC Press, Boca Raton, FL, 2002.

TABLE B.18. Linear Thermal Expansion Coefficient (in 10^{-6} K^{-1}) at 300 K for Type III–V and II–VI Semiconductors[a]

	P	As	Sb		S	Se	Te
Al	—	3.5	4.2	Zn	6.4	7.2	8.2
Ga	5.3	5.4	6.1	Cd	4.7	3.8	4.9
In	4.6	4.7	4.7	Hg	—	5.5	4.6

[a]The values for Si and Ge are 2.49×10^{-6} K^{-1} and 6.1×10^{-3} K^{-1}, respectively, and that for ZnTe is 6.6×10^{-3} K^{-1}.
Source: Data from *Handbook of Chemistry and Physics*, CRC Press, Boca Raton, FL, 2002.

TABLE B.19. Volume Compressibility (in 10^{-10} m^2/N) for Type III–V Semiconductors[a]

	P	As	Sb
Al	—	—	0.571
Ga	0.110	0.771	0.457
In	0.735	0.549	0.442

[a]The values for Si and Ge are 0.306 and 0.768×10^{-10} m^2/ N, respectively.
Source: Data from *Handbook of Chemistry and Physics*, CRC Press, Boca Raton, FL, 2002.

TABLE B.20. Microhardness (in the units N/mm²) of Types III–V and II–VI Semiconductors[a]

	P	As	Sb		S	Se	Te
Al	5.5 (M)	5000	4000	Zn	1780	1350	900
Ga	9450	7500	4480	Cd	1250	1300	600
In	4100	3300	2200	Hg	3 (M)	2.5 (M)	300

[a]The values for Si and Ge are 11,270 and 7,644 N/mm², respectively; M indicates microhardness values on the Mohs scale.
Source: Data from *Handbook of Chemistry and Physics*, CRC Press, Boca Raton, FL, 2002.

TABLE B.21. Atomic Magnetic Susceptibility (in 10^{-6} cgs) for Types III–V Semiconductors[a]

	P	As	Sb
Ga	−13.8	−16.2	−14.2
In	−22.8	−27.7	−32.9

[a]The values for Si and Ge are -3.9 and -0.12×10^{-6} cgs (cm · g · s), respectively, and the value for ZnS is -9.9×10^{-6} cgs.
Source: Data from *Handbook of Chemistry and Physics*, CRC Press, Boca Raton, FL, 2002.

Face-Centered Cubic and Hexagonal Close-Packed Nanoparticles

C.1. INTRODUCTION

In Chapter 1 we discussed the structures of face-centered nanoparticles. This appendix will fill in some additional details associated with these fcc particles, and we will examine the structures of their counterparts, namely, hexagonal close-packed (hcp) nanoparticles.

C.2. FACE-CENTERED CUBIC NANOPARTICLES

In Section 1.1.3 we showed how to construct a fcc 13-atom nanoparticle, and Figs. 1.5–1.8 present views of it from different perspectives. To construct the next-larger 55-atom fcc nanoparticle, a layer of 30 atoms is added around the three layers shown in Fig. 1.7, and 6-atom triangular arrays are added to the top and bottom, as depicted in Fig. C.1. The process is repeated to form the 147-atom fcc nanoparticle, and its middle, top, and bottom layers are sketched in Fig. C.2. A perspective sketch of the atoms on the surfaces of this nanoparticle viewed along the [111] axis is illustrated in Fig. 1.9, with the crystallographic planes labeled by their indices, as in Fig. 1.8. There are six square faces with square planar arrays of 16 atoms each, and eight equilateral triangle faces with planar hcp arrays of 10 atoms each. The atoms on all of these triangular and square faces have the nearest-neighbor separation $a/2^{1/2}$, where a is the lattice constant. Larger fcc nanoparticles with magic numbers 309, 561, ... of atoms have analogous arrays of atoms on their faces, with the numbers of surface atoms involved in each case listed in Table C.1. The numbers of atoms in the various planes of these nanoparticles are tabulated in Table C.2. The nanoparticles pictured in Figure 3.18 clearly display both the planar square and the planar hexagonal arrays of atoms corresponding to those on the faces of Fig. 1.9.

The Physics and Chemistry of Nanosolids. By Frank J. Owens and Charles P. Poole, Jr.
Copyright © 2008 John Wiley & Sons, Inc.

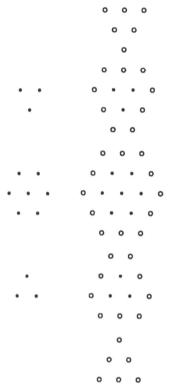

Figure C.1. Arrays of atoms in various layers of 13-atom (left) and 55-atom (right) fcc nanoparticles.

A way to visualize how the 42 surface atoms are added to the small 13-atom nanoparticle to form the 55-atom counterpart is in terms of the additions of atoms along the perimeters of the individual planes, as well as to the top and bottom, in the manner illustrated in Fig. C.1. Figure C.1 (left side) shows the three planes of the 13-atom nanoparticle, and Fig. C.1 (right side) sketches the five planes of the 55-atom nanoparticle. In (b) the center plane remains a regular hexagon, the two intermediate planes are irregular hexagons that alternate between having two and three atoms to a side, and the top and bottom planes are triangular in shape, as shown.

The next-larger fcc nanoparticle, with seven layers of atoms, 92 atoms on the surface covering the 55-atom core, corresponding to a total of 147 atoms, is constructed in an analogous manner. The centrally located hexagonal plane containing 37 atoms is a regular hexagon with four atoms along each side, and the top and bottom planes are equilateral triangles with 10 atoms each, as shown in Fig. C.2. The remaining four intermediate layers are irregular hexagons with four atoms along three of their sides, and either two or three atoms along their other three

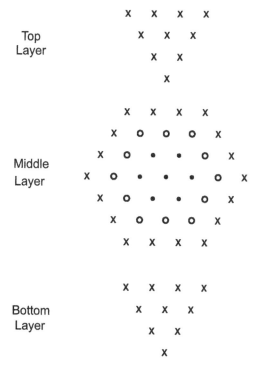

Figure C.2. Arrays of atoms in the top, middle, and bottom layers of a 147-atom fcc nanoparticle.

sides. If this seven-layer nanoparticle is viewed from the top, that is, along the [111] threefold axis, it has the appearance sketched in Fig. 1.9. Three of the sides exhibit a simple square planar atomic configuration, and the remaining three display a planar close-packed triangular lattice. These surfaces are shown in perspective in Fig. 1.9.

TABLE C.1. Number of Atoms on Faces of Face-Centered Cubic Nanoparticles

Total	Triangular Face	Square Face
13	3	4
55	6	9
147	10	16
309	15	25
561	21	36

TABLE C.2. Number of Atoms in Various Planes of Face-Centered Cubic Nanoparticles

Number of Atoms	1	13	55	147	309
Plane +4	—	—	—	—	15
Plane +3	—	—	—	10	25
Plane +2	—	—	6	18	36
Plane +1	—	3	12	27	48
Plane 0	1	7	19	37	61
Plane −1	—	3	12	27	48
Plane −2	—	—	6	18	36
Plane −3	—	—	—	10	25
Plane −4	—	—	—	—	15

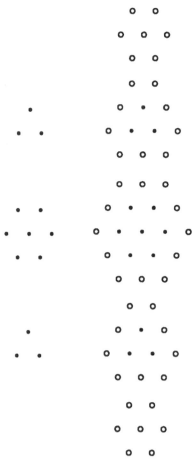

Figure C.3. Arrays of atoms in various layers of 13-atom (a) and 57-atom (b) hexagonal close-packed (hcp) nanoparticles.

C.3. HEXAGONAL CLOSE-PACKED NANOPARTICLES

We see from Fig. C.1a that the top and bottom triangular arrangement of atoms in the 13-atom fcc nanoparticle are rotated by 60° relative to each other. This occurs because the layers in a fcc structure are stacked in a threefold sequence $A-B-C-A-\ldots$, as was explained in Section 1.1.2. In the hcp structure the stacking sequence is twofold, namely, $A-B-A-B-\ldots$, so the layers in the three- and four-layer hcp nano-particles have the atom arrangements depicted in Fig. C.3. Note that in this four-layer hcp case there are 57 atoms, compared to 55 in the fcc case. More generally, when this layering procedure is used to construct nanoparticles with the hcp structure, then the structural magic numbers 1,13,57,153,321,581,... are usually somewhat larger than the fcc ones. In this case the central plane is a plane of mirror symmetry, and the axis of the nanoparticle is a threefold symmetry axis. The atom arrangements in the top and bottom planes are identical. When the number of planes is $n = 5,9,13,\ldots$, the perimeters of the top and bottom planes are always hexagonal in shape. In the fcc case they are always triangular. When viewed from the top, the 153-atom four-layer hcp nanoparticle has the appearance shown in Fig. C.4. We see from this figure that the sides of these hcp nanoparticles are not planar as in the fcc case, but rather are corrugated.

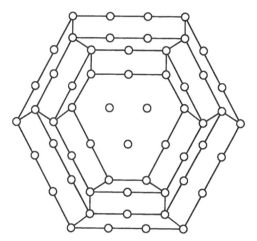

Figure C.4. Arrays of atoms on various faces of a 153-atom hcp nanoparticle.

The Physics and Chemistry of Nanosolids, By Frank J. Owens and Charles P. Poole, Jr.
Copyright © 2008 John Wiley & Sons, Inc.